GREAT DEBATES IN AMERICAN ENVIRONMENTAL HISTORY

GREAT DEBATES IN AMERICAN ENVIRONMENTAL HISTORY

Volume 1

Brian Black and Donna L. Lybecker

GREENWOOD PRESS
WESTPORT, CONNECTICUT • LONDON

Library of Congress Cataloging-in-Publication Data

Black, Brian.
　Great debates in American environmental history / Brian Black and Donna L. Lybecker.
　　p. cm.
　Includes bibliographical references and index.
　ISBN 978-0-313-33930-1 ((set) : alk. paper) — ISBN 978-0-313-33931-8 ((vol. 1) : alk. paper) —
ISBN 978-0-313-33932-5 ((vol. 2) : alk. paper)
　1. Human ecology—United States—History. 2. Nature—Effect of human beings on—United States. 3.
United States—Environmental conditions—History. I. Lybecker, Donna L. II. Title.
　GF503.B527 2008
　363.700973—dc22　　　2008002106

British Library Cataloguing in Publication Data is available.

Library of Congress Catalog Card Number: 2008002106
ISBN: 978-0-313-33930-1 (set)
　　　978-0-313-33931-8 (vol. 1)
　　　978-0-313-33932-5 (vol. 2)

First published in 2008

Greenwood Press, 88 Post Road West, Westport, CT 06881
An imprint of Greenwood Publishing Group, Inc.
www.greenwood.com

Printed in the United States of America

The paper used in this book complies with the
Permanent Paper Standard issued by the National
Information Standards Organization (Z39.48–1984).

10　9　8　7　6　5　4　3　2　1

CONTENTS

LIST OF ENTRIES

GUIDE TO RELATED TOPICS

Environmental Conservation

Pinchot Argues Conservation as a Development Strategy
Muir Argues for the Soul of Wilderness
Pinchot, Muir, and the Conservation Movement Meet at Hetch Hetchy
Public Learns Gardening and Rationing to Support the Cause
George Perkins Marsh Spurs Consideration of Industrialization
Discovering Alaska
Grazing Rights in the West and on Public Lands
Franklin D. Roosevelt Implements Conservation Policies
Debt for Nature and Development Swaps
Boone and Crockett Club Uses Virility to Attract Environmental Support
The Boy Scouts of America Involve Young Men in Outdoors
Progressives Demand Federal Action on Conservation
Yellowstone to Yukon Conservation Initiative

Influence and Effects of Aridity

Long-Term Implications of Sodbusting and Conversion of the American West for
    Agriculture
Using Chicago to Make the Great Lakes the Nation's Fifth Coast
Jefferson Argues for the Louisiana Purchase and American Expansion
American Leaders Create a Culture and System of Expansion
Mormons Create a Model for Interpreting the Aridity of the West
Settlement Systematizes and Simplifies the Ecology of the West
Powell and Efforts to Explore the Contours of the West

Urban Issues

City Beautiful, Urban Renewal, and the Effort to Reform the Modern City
Can the City of Boston Be Artificially Grown?
Alexander Hamilton Envisions an Industrial America
Modeling Public Works in Philadelphia
Bringing America's Nature Aesthetic to Life in Parks
The American Suburb, Sprawl Nation, and the Emergence
  of New Urbanism
Boston Harbor as Political Football
Can You Dig the Big Dig?
The Western United States, Urban Growth, and the New West
Olmsted and Vaux Design a Central Park for New York City
Olmsted Helps to Define the American Movement for Parks
Social Reformers Set Sights on Urban Problems
Russell Sage Studies Urban Problems in Pittsburgh

Exploration/Expansion to the West

Explorers Search the Resources of the American West
Jefferson Argues for the Louisiana Purchase and American Expansion
Lewis and Clark Seek to Know the Unknown Continent
The Land Ordinance of 1785-1787 Constructs the American Grid for
  Land "Disposal"
American Leaders Create a Culture and System of Expansion
Hitching the Nation's Future to the Railroad
Settlement Systematizes and Simplifies the Ecology of the West
Border Disputes and the Settlement of Texas
Gold Opens Up the West
Managing Resources in the Civil War
Powell and Efforts to Explore the Contours of the West
Frontier Thesis and American Meaning

Management of Resources

Pinchot Argues Conservation as a Development Strategy
Muir Argues for the Soul of Wilderness
Pinchot, Muir, and the Conservation Movement Meet at Hetch Hetchy
"Broad Arrow" Policy Conserves New England Forests and
  Foments Revolution
What Is the Importance of America's Natural History?
Internal Improvements in the Early Republic
Federalizing Forest Conservation
Turning to Fossil Fuels and an Electric Life
Foraging for Food in a Society of Mass Consumption
Energy Development on the Public Lands

Management of Water (Coasts, Canals, and Rivers)

Colonial and Indian Treatment of the Land and Rights

Agriculture

Populism and the Grassroots of the West
Grazing Rights in the West and on Public Lands

Politics and Social Reform

Constructing the Canal Age: Who Should Pay?
Working in a Coal Mine
Conceiving of Human Evolution
Franklin D. Roosevelt Implements Conservation Policies
Natural Backlash Fuels the National Environmental Policy Act and Environmental Policy of the 1970s
Involving the Federal Government in Public Health
Bald Eagle and the Importance of the Endangered Species Act
Love Canal and the Controversial Mandate for Superfund
Immigration along the United States–Mexico Border
The Green Party in the United States
Lyndon and Lady Bird Johnson Take Federal Action against Blight
Who Is Responsible for Fixing Polychlorinated Biphenyls in the Hudson River?
Debt for Nature and Development Swaps
The Convention on International Trade in Endangered Species of Wild Fauna and Flora
Dominican Republic–Central American Free Trade Agreement
Roadless Area Conservation Rule ("The Roadless Rule")
Wolves in Yellowstone National Park
Managing Wetlands and the Swampland Acts of the Mid-1800s
Social Reformers Set Sights on Urban Problems
Who Pays for Reclamation as a Federal Policy for the West?
Working for Workers' Rights
Populism and the Grassroots of the West
The Panama Canal Opens Routes of Trade and Ecological Change
Muckrakers Set the Tone for National Reform
Alice Hamilton Connects Social Reform with Human Health
Federal Efforts to Regulate Human Health
The Lacey Act Creates the First Federal Law for Wildlife Conservation
Progressives Demand Federal Action on Conservation
North American Free Trade Agreement, Human Rights, and Environmental Consequences
United States–Mexico Border

Ethics of Environmental Policies and Theories

William Penn and Roger Williams Establish a Unique American Model
George Perkins Marsh Spurs Consideration of Industrialization
Industrial Ethics and the Lessons of the Disaster in Johnstown
Romantics Claim That Nature's Value Derives from Beauty
Judging the Human Species: From Social Darwinism to the Movement for Civil Rights

Animals

Energy

Civil War

Miscellaneous

PREFACE

The perspective of these volumes is that of environmental history. It requires that we attach blinders to our view of history. As we selected events and issues for inclusion, we accentuated the changing idea of nature in American life. American definitions of terms such as nature or environment have changed a great deal over time. This book seeks to catalog such changes in definition, including romantics and extreme environmentalists who believed that God could be found in nature to wise users who saw no rationale for humans relenting on their use of natural resources. The issues covered in this volume seek to run the gamut of American ideas of nature and environment, including major policies that have been fashioned—and refashioned—to contend with social and environmental problems.

The essential common elements in each entry, then, are human users and the natural environment. In some of the topically organized entries, readers will be able to chart alterations in human approaches to the issue over time. In other entries, we have explored one event in depth and allow you to make the broader connections. Very often, government regulation or law making proves to be the outcome to changes in ideas about land use. We hope that these volumes will provide a valuable reference for policy makers as well as those involved in environmental law. Readers will see that some of these efforts have proven successful over time, whereas others have cleared the way for future adjustments. Throughout the volume, however, we have offered general themes with each entry, such as "Land Use," "Water Management," or "Planning." These, combined with the thematic table of contents, will assist readers who want to explore different dimensions of one issue in a variety of instances.

A clear theme in the entire volume is the need for an aware, educated public to speak out and take action when it deems it to be necessary, whether as a result of exploitation or inaction by others. We hope that these reference volumes will assist readers in approaching new environmental issues and challenges with a solid foundation of background knowledge.

Although the bibliography for this book will provide readers with a fine resource for further reading in environmental history, readers may also want to contact the American Society of Environmental History (http://www.h-net.org/~environ/ASEH/welcome_NN6.html).

We would like to thank the environmental studies students of Penn State Altoona for their unceasing interest and constant questioning of environmental topics. In addition, numerous colleagues and students at Gettysburg and Skidmore Colleges, Colorado State University, Penn State Altoona, and Idaho State University have taught, cajoled, and discussed many of these issues with us. We owe them an immense debt of gratitude for their contributions and inspiration. Finally, Chelsea Burket and Carole Bookhammer provided superior assistance in the preparation of this manuscript.

INTRODUCTION: WHAT LAY BELOW THE ARCTIC ICE CAP

> That man is, in fact, only a member of a biotic team is shown by an ecological interpretation of history. Many historical events, hitherto explained solely in terms of human enterprise, were actually biotic interactions between people and land. The characteristics of the land determined the facts quite as potently as the characteristics of the men who lived on it.
>
> —Aldo Leopold

Similar to other specialized approaches to studying the past, environmental history requires that we attach blinders to our view of history. In our survey of environmental historians, we can't be interested in every historical event; in fact, there are many essential events of the past that we will not discuss at all. Great characters of history will need to be entirely overlooked so that our focus might properly consider the specific factors that relate to environmental inquiry.

As the words of the 1930s biologist and naturalist Aldo Leopold declare, the perspective of environmental history accentuates the interaction between humans and nature over time—to explore events possibly hitherto considered to be of significance only to human actors. To do so, environmental historians overcome the limits of the human world by using scientific and biological concepts. Such concepts form the foundation of the historical stories that are told in the pages that follow.

At many junctures, these events elicited strong debate between alternative perspectives. The articles that follow are organized around these altercations, sometimes organized topically and at other times by specific events or time periods. Viewed chronologically, as they are organized in the two volumes, each of the debates dramatizes changing ideas about Americans' changing ideas of and relationship to the natural environment. Although they do not form a certain trajectory, these entries demonstrate changing ideas about free enterprise and about how active a role the federal government should take in American life. Following a clear high point of government involvement in resource management in the 1930s and again in the 1970s, the start of the twenty-first century brings new priorities and new possibilities. It is our contention that knowledge of past great debates of environmental history and policy will help us to better manage those that are certain to be around the next bend.

For instance, August 2007 may have witnessed the start of the next great resource debate. Eyeing future wealth (particularly energy resources), Russia sent two submersible ships two miles below the Arctic ice cap to perform a fairly simple task: planting a Russian flag. Similar to the United States planting its flag in the moon's soil a half century before, an enterprising nation has attempted to identify the next frontier of development in hopes of staking its claim ahead of any competitors. One reason for opportunity in the Arctic is global climate change, which is leading ice to recede at a record pace. Another reason is the region's appeal throughout human history: remoteness.

Five nations have cordoned off land in the Arctic (Denmark, Canada, the United States, Norway, and Russia). Frozen or not, crucial resources can be extracted there. Most appealing of all, there are few nongovernmental organizations or native residents to complain. Although activists, scientists, and health officials have taught us important lessons and helped to make the American living environment safer over the past few centuries, acquiring necessary resources and products has become somewhat more complicated. The remote, largely unspoken-for Arctic could be the next great frontier for development.

Of course, we can be certain that such a change will elicit debate and discussion, as will many other issues of importance to American land use. This book strives to construct a basic chronology of many debates in Americans' relationship with the natural environment. Some of the essays are organized very specifically on a moment at a specific time; in other essays, though, broader topics demonstrate the significance of specific events or policies. Although it is necessarily selective, these pages reveal a basic chronology in Americans' ability to consider some of the complications related to land use and development.

Brian Black, Hollidaysburg, PA
Donna Lybecker, Pocatello, ID

WAS THE LOSS OF THE ROANOKE COLONY CAUSED BY ENVIRONMENTAL FACTORS?

Time Period: 1600s
In This Corner: Settlers, British leaders
In the Other Corner: Native peoples
Other Interested Parties: Other nations interested in colonial settlement
General Environmental Issue(s): Climate, agriculture, environmental history

For decades, historians have debated the actual events of one of the first episodes in the history of Europeans' experience with the North American continent. Most historians must admit that, because of the existence of few written records, specific details about the first colony in American history are most likely permanently uncertain; however, in recent years, new ways of reading historical evidence have provided insight on one colonial episode in particular. These new interpretations suggest that the failure of the first colonial settlement in American history derived most clearly from a lack of understanding of the region's climate and residents.

Europeans' efforts to create colonies in the "New World" were a product of an economic model known as mercantilism. Stretching from the 1400s to the 1700s, this era relied on a basic technology: sailing ships. Although the ability to harness wind reached back to antiquity, additional innovation enabled sailors by the 1500s to move over much vaster areas. This capability, then, opened up a world of new opportunities.

With control of the wind and the transportation that it made possible, Asian and European civilizations reached out to create trade networks unlike any of the previous eras. This expansion made it possible for European leaders to gain wealth and resources from all over the world. Although shipping initially focused on moving goods or raw materials back and forth between points on the globe, it required little imagination to see the economic opportunities of establishing long-term economic relationships within a specific sphere of influence. What began as trading posts slowly matured into colonies. By definition, these colonies were permanent satellite settlements established abroad by nations to supply them with trade from another part of the world. It was most appealing, of course, to locate such colonies in areas and climates distinct from the supporting nation. For instance, colonies in tropical or temperate climates supplied European nations with crops and raw materials that were unavailable to them otherwise.

Atlantic System and Trade Network

In the logic of economic expansion that drove European powers abroad, New World resources became sources of military and political power at home. Efforts to collect and develop these resources allowed many leaders and merchants between 1400 and 1800 to suspend consideration of the rights of native peoples or of the human rights of laborers. By far, however, the most glaring example of this was the traders' ability to see humans, specifically Africans, as another commodity of value to be moved, sold, and collected.

Africans had been traded as slaves for centuries, reaching Europe via the Islamic-run, trans-Saharan trade routes. Between 1450 and the end of the nineteenth century, slaves were obtained from along the west coast of Africa with the full and active cooperation of African kings and merchants. In return, the African kings and merchants received various trade goods, including beads, cowrie shells (a type of money), textiles, brandy, horses, and guns.

By the mid-1400s, Portugal had a monopoly on the export of slaves from Africa. It is estimated that, during the four and a half centuries of the trans-Atlantic slave trade, Portugal was responsible for transporting more than 4.5 million Africans (approximately 40 percent of the total). However, during the eighteenth century when more than six million Africans were traded, Britain was the worst transgressor, estimated to have been responsible for trading almost 2.5 million Africans.

The system of trade that took shape in the Atlantic Ocean is often referred to as the "triangle trade." This term derives from a basic movement that currents and goods often followed. In this self-contained system, the eastward wind pattern, which blows on the southern part, came to be known as the "trade winds" because it enabled ships to cross the Atlantic. The westward wind pattern, blowing on the northern part, came to be known as the "westerlies."

Sailing ships were highly constrained by dominant wind patterns, and, therefore, the trading system followed this pattern. Manufactured commodities were exported from Europe to go in two directions: toward the African colonial centers and toward the American colonies. Slaves then left Africa bound primarily for Central and South American colonies (Brazil, West Indies). Tropical commodities (sugar, molasses) flowed from these colonies to the American colonies or to Europe. Within this Atlantic system, North America also exported tobacco, furs, indigo, and lumber to Europe.

Constructing the First Colony in North America

Amidst the Age of Sail's experience with colonization, settlers arrived in 1584 on an island south of Cape Hatteras, North Carolina, now known as Ocracoke. After exploring the area and burning some Indian villages to intimidate the residents, the English settlers discovered a natural break in the barrier reef due east of the southern tip of Roanoke Island. Prioritizing access to the sea, the settlers established a colony on the north end of Roanoke Island and appointed Ralph Lane as governor.

In the following spring, Sir Walter Raleigh, the primary funder of the early expedition, sent an additional 108 persons to Roanoke Island. Raleigh's cousin, Sir Richard Grenville, commanded the expedition of seven ships, which sailed from Plymouth, England, on April 9, 1585. After their successful arrival, Lane built a fort called "The New Fort in Virginia" near the shore on the east side of Roanoke Island between the "North Point" of the north end of the island and a "creek." Although it was referred to as a creek, the mouth of this waterway was deep enough to serve as an effective anchorage for small boats. It appeared a fine start to a successful port colony.

The settlers lived primitively. In fact, the fort was too small to hold most dwelling houses. Therefore, the houses of the early colonists were near but not inside the fort. Descriptions of the homes demonstrate an interest in permanence, with homes including thatched roofs and chimneys, possibly made of brick.

One of those included in the settlement, the naturalist Thomas Hariot, remarked that, although stone was not found on the island, there was good clay for making bricks, and lime could be made from nearby deposits of oyster shells.

From the intellectual point of conception, colonies such as Roanoke were strategically planned and managed. In 1584, Richard Hakluyt wrote his *Discourse of Western Planting* as a type of strategic primer for England's colonial aspirations. In listing the "things to be prepared for the voyadge" to Roanoke, Hakluyt stipulated that any colonial expedition should include "men experte in the arte of fortification." In addition, however, Hakluyt urged planners to keep their eye on the permanent stability of the colony by additionally including tradespeople, including, "makers of spades and shovells," "shipwrights," "millwrights, to make milles for spedy and cheape sawing of timber and boardes for trade, and first traficque of suertie," "millwrights, for corne milles," "Sawyers for common use," "Carpinters, for buildinges," "Brick makers," "Tile makers," "Thatchers," "Rough Masons," and "Lathmakers." In short, the colonial undertaking needed to be prepared to launch itself in complete self-sufficiency (National Park Service, *Fort Raleigh*).

Historians have assumed, therefore, that the Roanoke colony was a fairly diverse community composed of typical English thatched cottages and houses, such as were found in rural Elizabethan England. Existing records demonstrate that Roanoke was initially quite successful. Although there were some problems with their relations with certain native groups in the area, the colonists also worked with native peoples to plant crops and to make fish traps. One of these crops, of course, was tobacco, which colonists also learned how to smoke using Indian pipes or their own modeled after those of the Indians.

However, there was serious mistreatment of native groups by British settlers and many episodes of violence. Therefore, for generations, historians have assumed that, although the settlers' relations with natives had largely made settlement possible, it was also the most likely cause of the mysterious disappearance of the Roanoke settlers. The details of this disappearance are scarce and, therefore, do little to clarify any role of native peoples.

The murky details of 1586 begin with the settlers' overall relations with the natives in surrounding villages growing increasingly untenable. In an environment of concern and uncertainty, Grenville worsened the situation by burning the village of Aquascogok. In retaliation, surrounding natives stopped providing the settlers with basic rations and supplies on which the British had grown most dependent. The conflict worsened when fish traps in the area were robbed or destroyed. In a definitive sign of desperation, Lane divided up the people of the settlement and sent some of them to coastal areas where they could live on oysters and other shell fish. One group went to Hatoraske Island, another went to Croatoan Island, south of Cape Hatteras, and, finally, another party was sent to the mainland. Thus, the original settlement was already decomposing.

By June 1586, the colonists at the original settlement were engaged in open war with the natives. Grenville, after helping to escalate the conflict, traveled to England for supplies and was delayed in returning. The next written record comes from Captain Stafford who, on June 9, 1586, brought news that Sir Francis Drake was off the coast with a fleet of ships laden with booty from attacks on the Spanish West Indies and Florida.

Drake's fleet anchored near Roanoke Island and made Lane and his company a generous offer: he would give them one or two larger ships, a number of smaller boats, and sufficient ship masters, sailors, and supplies to afford another month's stay at Roanoke and a return

Sir Walter Raleigh, who organized the Roanoke expedition in 1587. Although the expedition was not successful, it led to future efforts that claimed the New World for European settlement. Library of Congress.

voyage to England, or he would give them all immediate return passage to England with his fleet. Not wanting to give up the Roanoke Island project, Lane accepted the first offer. However, as he took the supply-laden ship back to the colony, a storm blew it out to sea. Although Drake offered Lane another ship, it was too large for passage into the colony's port. With few options, Lane requested that Drake take him and the colonists back to England.

During this lapse in communication, Grenville at last arrived from England with three ships. He searched in vain for the colonists. Grenville found the places of colonial settlement desolate but, being "unwilling to lose the possession of the country which Englishmen had so long held," he left fifteen men on Roanoke Island, fully provisioned for two years, to hold the country for the queen while he returned to England. It is this small group of men and the settlement with which they were entrusted that have confounded generations of historians.

Still in England in 1587, Sir Walter Raleigh organized another colonial expedition of approximately 150 persons. A much different endeavor than the 1585 expedition, this group included women and children (at least seventeen women and nine children in the group that arrived safely in Virginia). In addition, the male members of the party were called "planters" and were offered approximately 500 acres of land. Finally, a lesser role was given to the military and more to business-oriented bureaucrats, a governor, and twelve assistants. In general terms, these changes suggested that colonization was becoming more of a corporate or business enterprise.

On July 22, 1587, a small portion of the new expedition arrived at Roanoke Island to confer with the men left behind by Grenville in 1586. They found the bones of one of them and no sign of the others. The party traveled to the site of Lane's fort and found that it had been razed "but all the houses standing unhurt, saving that the neather rooms of them, and also of the forte, were overgrown with Melons of divers sortes, and Deere within them, feeding on those Melons."

Despite these foreboding findings, the new party of settlers elected to fix the houses and resettle at Roanoke. This settlement proved to be largely successful and anticipated the financial and political arrangement of later colonies such as Jamestown. Therefore, the primary legacy of the early experiences at Roanoke was future settlement. The success of these later settlements was at least partly a result of the lessons learned at Roanoke.

Environmental History of the Roanoke Settlement

Historians in the twenty-first century are learning that one of these important lessons was the uniqueness of the climate in the American South. Trees in the southeastern Tidewater region were used to unravel at least a portion of the history of Roanoke. Growth rings of ancient trees have been used by archaeologists from the College of William and Mary to show that the late 1580s fell within the driest seven-year episode of the 800 years that they studied.

Climatologists from the University of Arkansas analyzed core samples taken from ancient bald cypress trees that covered the years 1185–1984 to ascertain information about rainfall and temperatures. This was then cross-referenced with the historical research done by archaeologists from William and Mary's Center for Archaeological Research. The findings demonstrated that great periods of drought confronted English settlers in both the Roanoke and Jamestown colonies.

The wedding of Pocahontas to John Rolfe represents the first between a European settler and a Native American. Library of Congress.

Although the failure of Roanoke has been attributed to many factors, including disagreements with native neighbors and poor planning, these recent findings demonstrate, at the very least, that the settlement would have had grave difficulties regardless of these other factors. The tree-ring reconstruction indicates that even the best-planned and supported colony would have been supremely challenged by the climatic conditions of 1587–1589.

Sources and Further Reading: National Park Service, *Fort Raleigh,* http://www.nps.gov/history/ history/online_books/hh/16/hh16d2.htm; State Library of North Carolina, *First English Settlement in the New World,* http://statelibrary.dcr.state.nc.us/nc/ncsites/english1.htm.

CASH CROPS AND NEAR DISASTER AT JAMESTOWN

Time Period: Early 1600s
In This Corner: English Crown and Virginia Company management
In the Other Corner: Entrepreneurial Virginia settlers
Other Interested Parties: Native peoples
General Environmental Issue(s): Land use, trade, colonial settlement

Particularly in the southern colonies, settlers to the New World in the 1600s sought to exploit the tropical climate's distinctiveness from Europe. If they viewed their movement to the New World as a business venture in which they risked their capital and their lives, why should they not grow the crops that promised the most return? Other farmers or even native traders, they hoped, would supply the settlements' food needs. For the common good, leaders of the settlement at Jamestown and elsewhere needed to step in and require southern agriculturalists to be self-sufficient first and profitable second.

The settlement at Jamestown began in June 1606 when English King James I granted a charter to a group of London entrepreneurs. With this royal support, the Virginia Company, as the business consortium was known, set out to establish a satellite English settlement in the Chesapeake region of North America. When 108 settlers left London in December 1606, they sought to settle Virginia primarily to clear a water route to India and Asia and to locate any gold supplies in the region.

The Virginia effort represented a vital expression of the evolving trade system now called mercantilism. Described as economic nationalism for the purpose of building a wealthy and powerful state, these ideas would eventually take written form in *Wealth of Nations* (1776), in which Adam Smith coined the term "mercantile system" to describe the system of political economy that sought to enrich the country by restraining imports and encouraging exports. This system dominated western European economic thought and policies from the sixteenth century to the late eighteenth century. The goal of these policies was to achieve a "favorable" balance of trade that would bring gold and silver into the country. In contrast to the previous agricultural system, the mercantile system served the interests of merchants and producers such as the British East India Company, whose activities were protected or encouraged by the state. Such companies, often referred to as joint stock companies, combined investors and capital with royal and military support. For instance, King James I granted the Virginia Company the power to appoint the Council of Virginia, the governor, and other officials and the responsibility to provide settlers, supplies, and ships for the venture.

The Jamestown settlers were primarily of an upper-middling economic standing. Historians argue that these gentry planned on doing minimal labor themselves to tame and settle wilderness. Ongoing archaeological research at the site of Jamestown suggests that at least some of the gentlemen and certainly many of the artisans, craftsmen, and laborers that accompanied them worked hard to make the colony succeed. Simply, the enterprise of settling a colony faced a host of challenges from many hostile forces (see above essay on Roanoke). In the case of Jamestown, these challenges combined with the mixed motivations of many settlers to give the colony a sputtering start.

The spot that the settlers selected on May 14, 1607, lay on the banks of the James River, approximately sixty miles from the mouth of the Chesapeake Bay. The appeal of this specific location was a harbor that was deep enough to hold their ships that would also keep them fairly concealed from the view of Spanish ships. In fact, the Spanish Armada was not the enemy who would prove most immediately problematic to settlers.

Although the Virginia settlers faced a long list of difficulties, their most immediate foe was Native Americans, particularly the Algonquin. Other groups, including the Powhatan, formed a vibrant trade with the European settlers, trading food for metal materials brought from Europe, particularly copper and iron. The constant danger, however, forced settlers to almost immediately place a priority on building wooden forts that they hoped would protect them from native invasion.

Over the course of the first year of settlement, colonists were decimated not by native attack but by disease and famine. Referred to as the "starving time," the winter of 1609 followed Smith's departure and nearly destroyed the settlement, with only sixty of the original 214 settlers surviving. In the spring, the survivors were kept from abandoning the colony only by the arrival of its new governor, Lord De La Ware. His supplies and stability helped the settlement effort considerably.

Relations with the Algonquin improved temporarily with the wedding of Chief Powhatan's daughter Pocahontas to tobacco entrepreneur John Rolfe. However, by 1622, Algonquins once again attacked the outer plantations, killing more than 300 of the settlers. Although Jamestown was spared this attack, the king determined that the Virginia Company was being mismanaged. He revoked the company's charter, and, in 1624, Virginia became a crown colony. Regardless of its administration, however, Jamestown had been founded to fill an economic role for the British empire.

Profit or Self-Sufficiency?

Behind the scenes of some of Jamestown's early difficulties lay an inner controversy about the basic purpose of the settlement. Cash crops, particularly the tobacco to which natives had introduced Jamestown's farmers, provided landholders with great economic opportunity. So great, in fact, that it proved a temptation that took them away from necessities. By 1618, Virginia exported 50,000 pounds of tobacco to England each year. The European market grew quickly, with many users hailing the "noxious weed" as a cure for assorted ailments and illnesses.

The desire of Virginia farmers to grow additional tobacco led to a period known as the "starving time," which taught the Virginia settlers important lessons about their new adopted climate. Although they would make adjustments to agriculture, southern farmers did not lose their passion for cash crops. In fact, this is partly responsible for the region's commitment to plantation agriculture and the cultural and social imperatives that it brought. In the face of the farmers' inability to use common sense and to first grow necessities, administrators of Jamestown implemented requirements to grow food crops.

Eagerness drove the tobacco growers. Raphe Hamor, a Virginia settler, in 1614 wrote the following:

> The valuable commodity tobacco, so much prized in England, which every man may plant and tend with a small part of his labor, will earn him both clothing and other necessities…. My own experience and trial of its goodness persuades me that no country under the sun can or does produce more pleasant, sweet, and strong tobacco than I have tasted there from my own planting. (Merchant 2005, 101)

In 1616, however, John Rolfe reacted to the colony's great difficulty when he demanded changes in farming. He wrote the following:

> To prevent the people—who are generally inclined to covet profit, especially after they have tasted the sweet results of the labors—from spending too much of their time and labor planting tobacco, which they know can be sold easily in England, and so neglecting their cultivation of grain and not having enough to eat, it is provided by the foresight and care of Sir Thomas Dale that no tenant or other person who must support himself shall plant any tobacco unless he cultivates, plants, and maintains every year for himself and every man servant ten acres of land in grain. (Merchant 2005, 101)

By following these guidelines, Rolfe argued, "They will be supplied with more than enough for their families and can harvest enough tobacco to buy clothing and other necessities."

To many settlers, the new rules seemed preposterous. Now that settlers had sailed halfway across the world and used their own capital to develop new farms, the Virginia Company wanted to force them to grow specific crops. Crops that were not profitable! From the planters' standpoint, the law may have seemed cruel. In fact, however, historians credit this restriction on agriculture with the very survival of the Virginia settlement.

Conclusion: Climate and the "Need" for Slaves

Southern agriculture remained unique through the mid-1800s. Since the beginning of colonial settlement, agriculture in the southern United States was defined by an imbalance between population and land. Because of climate considerations and also the preferences of some of the initial European settlements, southern planters focused on crops such as tobacco, rice, cotton, and sugar. Each of these crops required large tracts of land as well as many laborers. For these reasons, many planters organized their land into plantations. These vast agricultural colonies presented a very different model of economic development and the use of nature than the United State followed in other regions.

By definition, plantations are large agricultural estates cultivated by bonded or slave labor under central direction. After being used on the islands of the Caribbean, slavery on plantations was introduced into North America in the British colonies of Virginia, the Carolinas, and Georgia.

Jamestown hosted the first recorded arrival of Africans to North America. Records date this first trade in slaves to 1619, when a Dutch slave trader exchanged his cargo of Africans for food. The Africans became indentured servants, similar in legal position to many poor Englishmen who traded several years labor in exchange for passage to America. The large-scale use of slaves and the ensuing culture to maintain strict control over the black population did not fully emerge until the 1680s.

Although most farmers in the antebellum South did not own slaves, those who did dominated agricultural production. Planters who owned slaves also possessed power, not just to dominate other human beings and profit from their labor but also over the difficult environment. Slavery and exploitation of the environment went hand in hand.

Regional distinctions between the northern and southern portions of the United States became apparent during the 1800s. Primarily, whereas the northeastern and midwestern portions moved more deeply toward industrialization, the southern United States resisted infrastructural development and expanded the use of slaves to expand the growth of cotton through the South and to the Mississippi River. Historian Ted Steinberg wrote the following: "There is nothing the least bit natural about slave labor, but in the antebellum South, at least, it owed its rise to a climate that favored the growth of short-staple cotton. The development of the Cotton Belt rested on a set of climatic conditions; without them it is hard to imagine slavery taking on the role that it did in southern political culture" (Steinberg 1991, 87).

Even at Jamestown, settlers considered what style of agriculture to pursue. Their fateful choices eventually set the tone for the entire region.

Sources and Further Reading: Cowdrey, *This Land, This South*; Diamond, *Guns, Germs and Steel: The Fates of Human Societies*; Opie, *Nature's Nation*; Steinberg, *Down to Earth*.

EXPLORERS SEARCH THE RESOURCES OF THE AMERICAN WEST

Time Period: 1600–1800s
In This Corner: Explorer parties and their organizers
In the Other Corner: Native peoples, European monarchies
Other Interested Parties: Federal government, European monarchies
General Environmental Issue(s): Exploration, natural history

Explorers of the American West needed to have a certain audaciousness. Struggling against the odds of climate and the danger of moving outside their existing society, these men "boldly went where few had gone before them." Very often they were soldiers, but sometimes they were simply restless souls. Although there was no single debate about whether or not such exploration should proceed, each explorer strained against similar difficulties. Spanning different eras, these explorers' treks were initiated by a variety of rationales. Their accomplishment, however, was to expand settlement and, eventually, American society into the western United States.

Global Exploration and the New World

The first wave of international exploration came from Asia. In an era of outward expansion, national politics and culture played a significant role in how quickly each nation sought new opportunities. Although Portuguese and Spanish sailors are credited with being the first explorers of this era, China had expanded its worldview well in advance of Europe. In the early 1400s, Chinese Emperor Zhu Di sent out the greatest navy the world had ever seen. More than fifty years before the first Portuguese explorers, fleets of hundreds of Chinese junks explored India, Arabia, and East Africa. Seven epic Chinese naval expeditions from 1405 to 1433 explored and brought under the Chinese tributary system the vast periphery of the Indian Ocean.

To lead these expeditions, Emperor Zhu Di chose Admiral Zheng He. In 1403, the emperor issued orders to begin construction of an imperial fleet of warships and support ships to visit ports in the China seas and the Indian Ocean. Bearing vast amounts of gold and other treasures, and with a force of more than 37,000 officers and men under their command, the Chinese explorers built great ships and set sail from Suchow.

By mid-century, however, the Chinese view of this effort had changed wholesale. As a result, by 1474, the fleet was down to one-third of its earlier size. By 1503, the navy was down to one-tenth of this size. The anti-maritime party grew more powerful and made its power known through imperial edicts. In 1500, it was made a capital offense for a Chinese to go to sea in a ship with more than two masts without special permission. China's experience is the most dramatic example of the power of national authority over new efforts in trading.

The motives of the Western explorers and the Eastern fleets were very different. The Chinese were essentially on a tour of the civilized world. They wanted to conquer no one. Eventually, they wanted to collect rich gifts of tribute from other nations. However, their efforts were more symbolic than economic.

European explorers, conversely, were engaged in a bitter war with Islam and an eternal search for gold and profit. Ultimately, China came to see overseas activities as an unnecessary economic drain. For them, economic considerations were reserved for inland activities. The Europeans' exploration expanded as nations competed against each other; the Chinese, however, acknowledged no competitors.

Historian Robert Marks credits these different cultural approaches to the outside world for China's abandonment of the sea. He argues that the politics of China simply turned inward, after years of debate and infighting (Marks 2002). Zheng He called the ships back, and China, instead, concentrated on internal development within its vast nation. Europe, conversely, followed an entirely different model of economic development

Although navigation was still a relatively imprecise science, sailors were able to go farther and with more regularity than at any other time in human history. As the economy of Renaissance Europe developed, so too did the demand for imported goods and for new places to which to export local products.

Originally, the sailors of the Renaissance era took to the sea to supply Europeans with the many Asian spices, precious gems, and fine silk that were being purchased from foreign traders. The sea offered new networks of trade and new economic opportunities without the involvement of foreign traders. Essentially, European sailors could go directly to the source of the desired products.

Additionally, however, some mariners were drawn to the sea out of a curiosity to discover more about the world. These traders typically traveled on vessels carrying some of the world's first portable cannons. This allowed them to initiate a phase of "armed trade" in which coercion was freely used to open or enhance trade networks. These explorations led to trade for gold and ivory and, eventually, for slaves. Later, Portuguese sailors discovered the route around the southern tip of Africa that would take them to India entirely by sea.

By the end of the 1400s, the boundaries of the known world now seemed to expand exponentially with each passing year. In 1492, a trip to the East, made by sailing westward around the world, brought Christopher Columbus to the New World. Although the area that he unlocked became known as the Americas, Columbus had originally set out to find an all-water route to the East Indies. On spotting the Americas, Columbus believed he had reached his intended destination. His oversight would not be learned for a decade. His naming of the native peoples as "Indians" remains to this day.

Spanish explorers exuberantly followed Columbus to the Americas throughout the 1500s. The explorers sought to spread Christianity and to find resources of value, specifically gold and silver, to be traded and used as currency for trading. Spaniard Hernando Cortez discovered an abundance of gold among the Aztecs in what is now known as Mexico. Stories of more gold to be found led him and other Spanish explorers to conquer most of Mexico and Latin America. The discovery of silver led to the beginning of silver mining in Mexico and South America. Most of this silver was traded with China, which used it as its primary currency. Explorers in the New World also found new products to trade and grow, including corn, tomatoes, tobacco, and chocolate. In some areas, these and other crops were grown on farms called plantations that were run and owned by Europeans but worked by natives or imported labor. Such plantations took advantage of the non-European climates to grow exotic products to be sold in European markets. The push to explore and to eventually colonize the Americas was fueled by accounts written by many of the earliest explorers. This tradition, of course, begins with Columbus.

Mercantilism and the System of Global Trade

The buying and selling of slaves marked the low point of the basic rationalization of this new European trading system. Just as Europeans reinterpreted resources anywhere as commodities at home, slavery made sense because it was profitable.

From the markets and fairs of Europe, the interest in traded goods took form in the 1300s and fueled philosophical shifts in Europe that included an interest in individual profit.

Eventually, Adam Smith, John Locke, and others would shape the liberation of capital from royalty into an economic system today called capitalism. The priority of this system was the individual's opportunity to derive economic benefit and profit.

Combining this cultural and social desire with advancements in ocean navigation made the European worldview one shaped by profits and commodities. Eventually, this worldview became known as mercantilism, which is described as economic nationalism for the purpose of building a wealthy and powerful state. In *Wealth of Nations*, Adam Smith (1776) coined the term "mercantile system" to describe the system of political economy that sought to enrich the country by restraining imports and encouraging exports.

This system dominated western European economic thought and policies from the sixteenth to the late eighteenth centuries. The goal of these policies was to achieve a "favorable" balance of trade that would bring gold and silver into the country. In contrast to the previous agricultural system, the mercantile system served the interests of merchants and producers such as the British East India Company, whose activities were protected or encouraged by the state.

During the mercantilist period, military conflict between nation states was both more frequent and more extensive than at any time in history. The armies and navies of the main protagonists were no longer temporary forces raised to address a specific threat or objective but were full-time professional forces. Their job was to assist and protect economic interests. Therefore, each government's primary economic objective was to command a sufficient quantity of hard currency to support a military that would deter attacks by other countries and aid its own territorial expansion.

The evolution of new economic ideas had blossomed with the new interactions with other nations and resources. The mercantilist system knitted together participating nations in a way never before seen.

The first nation to seek out and develop the trade opportunities of the New World was Spain. The first Spaniard to see any part of the Southwest was shipwreck survivor Alvar Núñez Cabeza de Vaca and his three companions, including a North African named Estavan, whom historians speculate was very likely the first person from the Old World (Morocco) to reach the Colorado Plateau. Their arrival there came in 1536. Although he spent time imprisoned and enslaved, the recollections of de Vaca helped to attract subsequent explorers.

The first European to follow was Hernando de Soto, who explored the southeast region of North America for Spain, searching for gold, a suitable site for a colony, and an overland route from Mexico to the Atlantic. From 1539 to 1543, starting in Florida with more than 600 men, 200 horses, 300 pigs, and a pack of attack dogs, the expedition meandered for thousands of miles through the interior. At every point, the Spanish attacked Indian villages, pillaging, murdering, and commandeering food, supplies, and captives. They "discovered" the Mississippi River—a major challenge to cross—and continued west to Texas (without de Soto, who died from fever on the banks of the river). Finally, the surviving 300 men reached Mexico with no gold and no colony, having amassed only the hardened antagonism of the Indians. In these selections from the account by a Portuguese member of the expedition, known only as the "Fidalgo (gentleman) of Elvas," we read brief excerpts from the chapters recounting the mainland expedition from Florida to Mexico.

The stories sent back by these travelers and others included exaggerated stories of silver and gold that triggered the next wave of explorers, including Francisco Vásquez de Coronado in 1540. His group passed deep into the American West, and they became the first Europeans to see the Grand Canyon. However, his failure to find great cities of gold and silver put an end to Spanish designs on the region for the next forty years. In 1583, another Spanish party moved into the north-central part of Arizona. They claimed the territory of the Hopis for Philip II of Spain. Although failing to achieve their immediate goals, these explorers claimed vast territories for Spain that would define its relationship with the Indians and with its European rivals for the next two centuries. Most of the American West remained under Western control until its purchase and seizure by the United States, which had emerged as the major player in western North America.

American Exploration

Although they still wanted to seize control of the vast area, the United States first sought to know the land. Early expeditions, such as that by Lewis and Clark, will be discussed elsewhere. However, by the mid-1800s, the movement that had prioritized natural history at the beginning of the century returned to more scientific roots in an effort to collect and tabulate information about the American West. Most important, the resources of the western United States required the attention of the surveyors who could prepare the land for Anglo habitation and help to bring the region under the better control of the U.S. government. The priority of the nation was to move settlers westward. However, the government wanted to ensure it retained the rights to lands rich in minerals. Surveyors served each of these ends.

The discovery of gold and the need to find the best passages for railroad expansion made the geology of the West a federal priority after 1850. In 1859, for the first time, the value of the products of U.S. industry exceeded the value of agricultural products. Gold was discovered in Colorado, silver was discovered at the Comstock lode in western Nevada to begin the era of silver mining in the West, and the first oil well in the United States was successfully drilled in northwestern Pennsylvania.

On March 2, 1867, Congress for the first time authorized western explorations in which geology would be the principal objective: a study of the geology and natural resources along the fortieth parallel route of the transcontinental railroad, under the Corps of Engineers, and a geological survey of the natural resources of the new State of Nebraska, under the direction of the General Land Office. During the next decade, a series of government-financed surveys took place. The best known of these are the expeditions by Clarence King, John Wesley Powell, and Ferdinand V. Hayden. Legislation stipulated that the Hayden, Powell, and King surveys be discontinued as of June 30, 1879, when Congress would establish the United States Geological Survey (USGS). Its responsibilities would include "classification of the public lands, and examination of the geological structure, mineral resources, and products of the national domain."

For his expedition, Ferdinand V. Hayden carefully chose a team who would help him to bring a visual record of the West back for Americans. Primary among this team were Thomas Moran, a well-known landscape painter, and William Henry Jackson, an accomplished photographer.

Although many of the early survey teams included artists and photographers to aid with documentation of the West, most often the creative minds worked separately. During this expedition, Moran and Jackson collaborated in the selection of views. Their united aesthetic

Thomas Moran's 1874 view of the Lower Yellowstone range presented a unique American landscape that many Americans would never see for themselves. Through his paintings, though, such places became cherished American symbols. Library of Congress.

appreciation came to shape the entire nation's view of the West. Rather than wildlife or specific aspects of the natural environment, each artist was drawn to the magnificent symbols of the West, including landforms and people. Whether intentional or not, their views helped to form the icons that constructed the mythical view of life in the West.

During one of the most influential episodes in the summer of 1871, Moran traveled through what would become Yellowstone National Park. In an effort to express the awe he felt, he made sketches of the Gardiner River, Mammoth Hot Springs, Liberty Cap, and Tower Fall. Then he created *Grand Canyon of the Yellowstone*, which became known as a masterwork of his and of all American art. His paintings, as well as Jackson's photos, circulated through the halls of Congress, where they helped pacify legislators' doubts about the wonders of Yellowstone. In no small measure, the images of Moran and Jackson helped persuade lawmakers to create Yellowstone National Park, the nation's first such site. Seven months after Moran created his images of Yellowstone, the area was designated the nation's first national park.

Moran's impact continued, however. At a price of $10,000, Moran's *Grand Canyon of the Yellowstone* became the first landscape painting hung in the Capitol. Moran's unique view of the importance of American wilderness and its ties to the West now became symbolic of American nationalism. Ultimately, the art in the Capitol would be made to contain both the idealized landscapes of early American painting as well as the romantic images of late-nineteenth-century painting.

Conclusion: A Land Apart

The resources of the American West had finally been appropriated by the close of the nineteenth century. At each juncture in this story, foreign leaders had not requested the right to seize this area and its valuable resources. Were they justified? Was the United States justified in seizing this land from its native occupants? Historians continue to debate this story. Clearly, however, this tradition of exploration demonstrates the unique story of this region and the diverse people who have come to call it home.

Sources and Further Reading: Crosby, *Ecological Imperialism*; Diamond, *Guns, Germs and Steel: The Fates of Human Societies*; Lavathes, *When China Ruled the Seas: The Treasure Fleet of the Dragon Throne 1405–1433*; Marks, *Origins of the Modern World*; National Humanities Center, http://www.nhc.rtp.nc.us/pds/amerbegin/contact/contact.htm.

CAN THE CITY OF BOSTON BE ARTIFICIALLY GROWN?

Time Period: 1700s
In This Corner: Organizers of the city of Boston
In the Other Corner: Competing cities in the Northeast
Other Interested Parties: Producers in surrounding land and traders abroad
General Environmental Issue(s): City planning, watershed development, coastal management

"Where nature has provided some but not all of these geographical requirements," wrote historian Benjamin Labaree, "man has not hesitated to intervene" (Labaree 1999). Each American city provides some example of the use of planning and ingenuity to solve basic problems of development. However, no city was aggressively managed and developed as early as Boston, Massachusetts. Therefore, when residents pondered whether or not they should or could improve the deficiencies of their location, the response was a resounding "Yes!"

Expansion in Early Boston

Founded in 1630, the city of Boston in Massachusetts had a simple reason for being from the start: access to the sea. By 1710, Bostonians had constructed seventy-eight wharves to connect the city center to ships arriving from all over the world. The most impressive of these, Long Wharf, extended 800 feet from the shoreline. The competition of this new era spurred seaport and city growth into the 1800s as other American cities sought to keep up with Boston port.

European settlers arrived in the area of Boston around 1620. The town was founded ten years later by a Protestant religious sect called the Puritans. They named the new town for their former home in Lincolnshire, England. Boston served as the capital of the Massachusetts Bay Colony, but its greatest strength came from its unique geography: bounded on three sides by water. With its dominance of the sea trade, Boston quickly became the largest British settlement on the continent.

Such commerce also quickly compounded other aspects of city development. Although Boston had the natural benefit of open access to the Atlantic trade, nature had not blessed Boston with a great deal of surface area. City leaders faced a simple quandary, however: do we accept our limited supply of land or try to alter the very nature of our location? With its answer to this question in the 1700s, Boston set a national precedent and ushered in an era of urban growth that helped to define the United States.

Ship-building at East Boston, 1855, was made possible by land expansion in the harbor area. By expanding, Boston became one of the leading U.S. ports. Library of Congress.

In short, Bostonians chose not to accept nature's limitations. Using dirt, stone, and any other "fill" material, Boston's leaders expanded the city's real estate many times over. The first "filling" project set the tone for many others that would follow, continuing to the twenty-first century. A hotbed of the Revolution, Boston's proximity to the early New England settlements made it the nation's most active port through the 1700s. This activity, however, was more than matched by its furious efforts at filling and, thereby, expanding itself.

The original site for Boston was the hilly, 800-acre Shawmut Peninsula. Surrounding these hills were salt marshes, mudflats, and inlets of water. As Boston outgrew her site in the 1800s, most of the hills were leveled, and their dirt and stone were used as fill to create Boston's famous Back Bay district. Because of this early filling, the three original hills of the city (Pemberton, Beacon, and Mt. Vernon) no longer dominate the landscape. In addition to creating new land, these surrounding hills were carted to the sea's edge and dumped in the coves to provide more area for building. Typically, such filling joined the areas between existing wharves. Once the land was made contiguous, developers added new wharves extending out from the new coast.

As Boston grew, it became the primary regional port, tying together a hinterland of smaller ports and fishing towns throughout New England. New York, Philadelphia, and Baltimore, conversely, were primarily supported by agricultural regions inland. Each of these ports helped to define the maritime era that helped solidify the U.S. economic base in the early 1800s. The priority for American cities was clearly growth.

Sources and Further Reading: Labaree, *America and the Sea*; Stilgoe, *Alongshore*; Stilgoe, *Common Landscapes of America*.

``BROAD ARROW'' POLICY CONSERVES NEW ENGLAND FORESTS AND FOMENTS REVOLUTION

Time Period: 1600–1800
In This Corner: British Crown, particularly the Navy
In the Other Corner: Timber suppliers in colonies
Other Interested Parties: Competing nations
General Environmental Issue(s): Forest management, resource management

Historian William Carlton wrote, "Masts, in the days of wooden ships, played a far greater part in world affairs than merely that of supporting canvas. They were of vital necessity to the lives of nations. Statesmen plotted to obtain them; ships of the line fought to procure them" (Carlton 1939, 4). During the Age of Sail, mast wood was a matter of national security for Great Britain and the great naval powers of the 1600s. The first recorded shipment of masts from the colonies was sent from Virginia to England in 1609. Other shipments soon followed from what is now the United States and Canada. By the late 1700s, American timbers supplied all the masts for ships in the navies of Spain and France, as well as England.

This is only part of the wood's significance, however. As one of the most important and valuable international commodities of the era, mast wood increased the importance of the British colonies as a trade port. As a commodity, however, mast wood introduced some of the first inclinations of colonists to retain their resources and profits for themselves. Valuable timbers helped to stimulate the American movement for independence.

Broad Arrow Conserves Great Trees

Without white settlers to cut them and European markets to consume them before settlement, North American trees grew until the decay of old age brought them down. In Maine, New Hampshire, and Vermont, white pine and other trees were known to occasionally attain a diameter of four feet or more. Often these trees had no branches until they reached one hundred feet, making them ideal for use as masts.

The greatest ships of this era were found in the English Navy. In such vessels, the mainmast ideally reached forty inches in diameter at the base. Out of necessity, iron bands were often used to tie together smaller trees, for instance Baltic firs that seldom exceeded twenty-seven inches in diameter. Thus, most English ships were made with wood acquired from the Baltic countries and Norway.

Companies involved in the growing commercial trade in the Atlantic often competed with navies in bidding for the best timber. The East India Company as a rival often outbid the navy for the best timber. These ships, however, were smaller and had lower expectations in terms of size and durability. Each navy yard had its "mast pool," which had to be kept full even if the navy had to fight to keep it so, because the fleet that after battles was soonest repaired and returned to service, controlled the seas (Albion 1926).

American timber became crucial to England during the first Dutch War of 1654, when the Dutch and Danes cut off English access to these supplies. As a result, the English Navy set out to develop the colonial supply of mast wood and turned to the New England forests as a primary supplier for the next 120 years.

English Scrutiny of American Lumbering

Sawmills had been established in New England as early as 1629. For instance, early records of Dover, New Hampshire, and other settlements show that prominent families dealt in lumber. To do so, most owned their own ships and shipyards for transporting lumber to the West Indies and England.

The seriousness of this endeavor changed, however, under the scrutiny of the English Navy. For their masts, the Navy demanded a new and specialized type of lumbering. They also required suppliers to construct "mast roads" over the rocky hills and through the low-land swamps. Because the great trees had such length, these roads needed to be quite straight, which was very different from most roads of the era, which held many twists and turns to follow the terrain. Mast ships also had to be specially constructed so that they could be loaded through openings in their sterns. A typical mast ship had to be able to carry between forty and one hundred masts at once.

Mast trees usually were cut in the fall, when they were full of resin. The trees were care-fully felled on prepared beds, limbed, and squared. During the winter, the rough timbers, or baulks, were dragged by brute strength onto sleds and hauled out of the woods by teams of oxen. Because one great tree could weigh as much as eighteen tons, this was a difficult and dangerous process, requiring great skill and the efforts of as many as one hundred oxen.

In 1674, Edward Randolph was sent to see how the trade agreements and navigation acts were being observed. He reported serious violations but stressed that the pine for masts here was the best in the world. This quickened the interest of the London trading community in New England. His findings were clear: white pine (*Pinus strobus*) must be protected in New England to advance the sea power of England; Scotch fir (*Pinus sylvestris*), found in abun-dance in the north European forests, was inferior, less durable, and more difficult to obtain and transport. British inspectors soon began to implement restrictions and guidelines in 1685. These early actions ultimately resulted in more organized laws.

England's Broad Arrow Law

When the value of New England's timber became clear, the king appointed a surveyor gen-eral, under whom certain men, usually surveyors, were assigned to mark the trees to be so re-served. The Broad Arrow mark was made on the base of the tree by three blows with a marking hatchet. It was said to resemble a crow's track more than an arrow. In 1691, the surveyor general enacted the regulations that became known as the "Broad Arrow Law," which included these stipulations:

> … for better providing and furnishing of Masts for our Royal Navy wee do hereby reserve to us … ALL trees of the diameter of 24 inches and upward at 12 inches from the ground, growing upon any soils or tracts of land within our said Province or Territory not heretofore granted to any private person. We … forbid all persons whatsoever from felling, cutting or destroying any such trees without the royal license from us … upon penalty of 100 pounds … for every such tree so felled … without such license …

Prices and sizes were described as follows. In general, there were three size groups for masts: (1) small, eight to twelve inches in diameter; (2) middling, twelve to eighteen inches in diameter; and (3) great, eighteen inches and over in diameter. Most New England mast logs were of the "great" variety. Prices paid for trees delivered in England varied. Some examples of actual contracts include the following: twenty-four-inch diameter at base, twenty-seven yards long, thirty-five pounds; thirty-six-inch diameter at base, thirty-five yards long, 135 pounds; thirty-six-inch diameter at base, thirty-six yards long, 153 pounds (Albion 1926). In 1711, the Broad Arrow Law was extended to include all of New England, New Jersey, and New York. Supervisors intended that the process would follow a basic sequence. The Navy Board first placed its order for an entire year's needs. Typically, they intended only to work with a few contractors whom they trusted. These trusted lumberers received the order from the Navy Board and then sent a copy to his representative or agent on the first vessel bound for New England. The agent hired crews, prepared equipment, and selected the cutting area. Next, the contractor requested a royal license from the Privy Council, which was often a drawn-out process. Once these papers were in order, the surveyor general was notified of each contract by the Navy Board. Part of his job was to assist the holder of the contract and to compare terms of contract with terms of license and approve selection of trees.

In forests of North America, the cutting of an "approved tree" was filled with the care and precaution necessary to fell a tall pine without damaging it when it fell to the ground. New Hampshire historian Compatriot Davison describes the process:

> Normally, a path was cleared from the base of the tree in the direction it was to be felled, for the same distance as the height of the tree. The ground had to be nearly level. All the large branches were cut from the Broad Arrow tree before it was felled. Also, all nearby trees were cut to prevent damage to the mast tree as it fell. The small branches were left on to help reduce the force of the fall. If there was snow on the ground, the path of the fall was thoroughly probed to discover hidden rocks and stumps. One of these under the tree as it fell might break it, especially in the upper part, which had the longest fall and hit the ground with the greatest force. Frequently snow and brush were brought in to smooth out the area where the mast tree was to fall. Every reasonable precaution was taken to see that damage to the tree would be avoided. When the tree was on the ground, the task was less than half done. It had to be trimmed of all branches and transported to a port for shipment to England. Great pines weighed many tons and usually could not be dragged. When possible they were floated down rivers but with great care to avoid rapids and falls. If moved overland, they were laced on several pair of wheels. and pulled by many yoke of oxen at the front and along each side of the mast log. (Davison, *White Pines for the Royal Navy*)

The Broad Arrow laws remained in effect until 1775; however, they did not necessarily have the outcome that the Crown had desired. For the first few years after 1691, New England paid little attention to the preservation clause. Licenses were issued freely on assurance that the king's trees would not be molested. Sawmills continued with even greater activity and with a growing disregard for the protection of "Broad Arrow" tall white pines. Overall, the mast trade with England declined steadily. Surveyor generals and their agents were very unpopular, and many colonists made their enforcement efforts more difficult, as they stepped up the marking of more and more trees with the "Broad Arrow."

It has been estimated that around 4,500 masts were shipped to the Royal Navy between 1694 and 1775. This was only one percent of the trees reserved for Navy use. During July and September of 1763 alone, 6,389 logs were seized in Maine and New Hampshire and in Massachusetts towns on the Merrimac (Albion 1926).

Conclusion: American Trees

Many aspects of colonization led to the revolution of American colonists; however, the Broad Arrow policies must be included on this list. There is a direct confluence of the mast tree on the development of this state and nation and its contribution in precipitating the armed conflict in 1775. Ultimately, wrote Carlton, "The 'Broad Arrow' policy was a failure in New England. Violations were so widespread that the mast supply itself would not have lasted indefinitely. The discontent which the policy provoked was so active that the angry colonies were in no mind to bear the added oppressions of the last decade before the Revolution" (Carlton 1939).

Typically, the mast ships sailed from colonial ports early in the spring, thus being able to return for another cargo later in the year. In the spring of 1775, the ships sailed, unmolested; however, when the news of the April fighting reached the northern ports, the mast supply was shut down. The last shipment of masts reportedly reached England on July 3.

When the news of Concord and Lexington reached northern colonial ports, men were already removing masts from the loading pools and towing them to secluded spots for safe keeping. Eight days before the historic encounters, the Massachusetts Provincial Congress sent word to the shipping points in Maine, instructing them "to make use of all proper and effective measures to prevent" the masts then on hand being supplied to the enemy. In Georgetown, Maine, the local Committee of Safety closed the British dockyard and drove the workers to their boats. At Portsmouth, New Hampshire, mast ships were stopped and any shipping was discontinued.

Without its overseas markets, logs actually became more of a nuisance to New England's expanding population, which was more desirous of open land for farming than for more tall trees. Early policies often treated New England forests as a "limitless resource" that was hampering economic and agricultural growth and development. Consequently, with the resource in such abundance and minimal foreign buyers, the log export market stagnated until late in the twentieth century.

Sources and Further Reading: Adams, *Renewable Resource Policy: The Legal-Institutional Foundation*; Albion, *Forests and Sea Power*; Carlton, *New England Masts and the King's Navy*; Davison, *White Pines for the Royal Navy*, http://www.nhssar.org/essays/Whtpines.html; Williams, *Deforesting the Earth*.

WILLIAM PENN AND ROGER WILLIAMS ESTABLISH A UNIQUE AMERICAN MODEL

Time Period: 1600s
In This Corner: Quakers, American intellectuals
In the Other Corner: Colonial authorities, including Puritans
Other Interested Parties: British Royal Authority
General Environmental Issue(s): Environmental ethics, expansion, Native Americans, colonial settlement

As a nation of settlers, early Americans came to North America for diverse reasons. For some, their idealism did not stop with the decision to set out for new opportunity. Many of the early Americans pursued idealistic, even utopian, efforts to establish communities and to develop towns, cities, and regions. Their efforts were often quite distinct from one another. Although some of these early visionaries were criticized at the time of their efforts, they often became the intellectual forebears of policies and ideas adopted for the entire nation when it was established in 1776. Two such intellectual forefathers are William Penn and Roger Williams. Although each rejected strictures of colonial authority, they did so with particular motivations.

William Penn and the Experience of Penn's Land

In what has become known as Pennsylvania, William Penn used a charter to create a system of government without royal intervention. The government created by Penn reflects some of the earliest conscious attempts at modern democracy. Grounded in his Quaker philosophy, Penn emphasized principles including a representative government, separation of church and state, and the elimination of nobility and ranks. His emphasis was self-rule and peaceful coexistence among peoples of differing cultural and ethnic backgrounds. He used the charter to establish these ideals for Penn's Land. Thomas Jefferson hailed Penn as "the greatest law giver the world has produced."

Penn's visionary opportunity, of course, derived from efforts to colonize the "New World," as Europeans referred to North America. Penn's father served as an admiral of the British Navy in the mid-1600s and used his own funds to support the fighting force. When

In 1681, William Penn's treaty with the Indians when he founded the province of Pennsylvania in North America was one of the best examples of Europeans attempting to find a way of dealing with Native Americans somewhat fairly. Library of Congress.

he died, the crown owed him a debt of approximately £16,000. The younger Penn, who had joined the Religious Society of Friends, or Quakers, three years before his father's death, petitioned King Charles II in May 1680 with a plan: he would forgive the crown its debt to the Penn family in return for land in the New World.

His idea appealed to the crown, which would be forgiven debt for a commodity for which it had paid nothing. The appeal to Penn, of course, was that on this land he would establish a colony for Quakers and persecuted religious groups. King Charles II accepted Penn's drafted charter in 1681 but insisted that the new colony be named Pennsylvania to honor Penn's father. Penn's hope had been to name the place "New Wales" or simply "Sylvania" (a Latin word for forest). However, Penn acquiesced and called the place Pennsylvania (Ries and Stewart, *This Venerable Document*). The text of the Charter of Progress reads as follows:

> William Penn Proprietary and Governour of the Province of Pennsilvania and Territories thereunto belonging To all to whom these presents shall come Sendeth Greeting Whereas King Charles the Second by his Letters Patents under the Great Seale of England beareing Date the fourth day of March in the Yeare one thousand Six hundred and Eighty was Graciously pleased to Give and Grant unto me my heires and Assignes forever this Province of Pennsilvania with divers great powers and jurisdictions for the well Governement thereof And whereas the King's dearest Brother James Duke of York and Albany &c by his Deeds of Feofment under his hand and Seale duely perfected beareing date the twenty fourth day of August one thousand Six hundred Eighty and two Did Grant unto me my heires and Assignes All that Tract of Land now called the Territories of Pennsilvania together with power and jurisdictions for the good Governement thereof And whereas for the Encouragement of all the Freemen and Planters that might be concerned in the said Province and Territories and for the good Governement thereof I the said William Penn in the yeare one thousand Six hundred Eighty and three for me my heires and Assignes Do Grant and Confirme unto all the Freemen Planters and Adventurers therein Divers Liberties Franchises and properties as by the said Grant Entituled the Frame of the Government of the Province of Pennsilvania and Territories thereunto belonging in America may Appeare which Charter or Frame being found in Some parts of it not soe Suitable to the present Circumstances of the Inhabitants was in the third Month in the yeare One thousand Seven hundred Delivered up to me by Six parts of Seaven of the Freemen of this Province and Territories in Generall Assembly mett provision being made in the said Charter for that End and purpose And whereas I was then pleased to promise that I would restore the said Charter to them againe with necessary Alterations or in liew thereof Give them another better adapted to Answer the present Circumstances and Conditions of the said Inhabitants which they have now by theire Representatives in a Generall Assembly mett at Philadelphia requested me to Grant Know ye therefore that for the further well being and good Governement of the said Province and Territories and in pursuance of the Rights and Powers before mencioned I the said William Penn doe Declare Grant and Confirme unto all the Freemen Planters and Adventurers and other Inhabitants in this Province and Territories these following Liberties Franchises and Priviledges soe far as in me lyeth to [be] held Enjoyed and kept by the Freemen Planters and Adventurers and other Inhabitants of and in the said Province and Territories thereunto

Annexed for ever first Because noe people can be truly happy though under the Greatest Enjoyments of Civil Liberties if Abridged of the Freedom of theire Consciences as to theire Religious Profession and Worship. And Almighty God being the only Lord of Conscience Father of Lights and Spirits and the Author as well as Object of all divine knowledge Faith and Worship who only [can] Enlighten the mind and perswade and Convince the understandings of people I doe hereby Grant and Declare that noe person or persons Inhabiting in this Province or Territories who shall Confesse and Acknowledge one Almighty God the Creator upholder and Ruler of the world and professe him or themselves Obliged to live quietly under the Civill Governement shall be in any case molested or prejudiced in his or theire person or Estate because of his or theire Conscientious perswasion or practice nor be compelled to frequent or mentaine any Religious Worship place or Ministry contrary to his or theire mind or doe or Suffer any other act or thing contrary to theire Religious perswasion And that all persons who also professe to beleive in Jesus Christ the Saviour of the world shall be capable (notwithstanding theire other perswasions and practices in point of Conscience and Religion) to Serve this Governement in any capacity both Legislatively and Executively he or they Solemnly promiscing when lawfully required Allegiance to the King as Soveraigne and fidelity to the Proprietary and Governour And takeing the Attests as now Establisht by the law made at Newcastle in the yeare One thousand Seven hundred Intituled an Act directing the Attests of Severall Officers and Ministers as now amended and Confirmed this present Assembly Secondly For the well Governeing of this Province and Territories there shall be an Assembly yearly Chosen by the Freemen thereof to Consist of foure persons out of each County of most note for Virtue wisdome and Ability (Or of a greater number at any time as the Governour and Assembly shall agree) upon the first day of October forever And shall Sitt on the Fourteenth day of the said Month in Philadelphia unless the Governour and Councell for the time being shall See cause to appoint another place within the said Province or Territories Which Assembly shall have power to choose a Speaker and other theire Officers and shall be judges of the Qualifications and Elections of theire owne Members Sitt upon theire owne Adjournments, Appoint Committees prepare Bills in or to pass into Laws Impeach Criminalls and Redress Greivances and shall have all other Powers and Priviledges of an Assembly according to the Rights of the Freeborne Subjects of England and as is usuall in any of the Kings Plantations in America And if any County or Counties shall refuse or neglect to choose theire respective Representatives as aforesaid or if chosen doe not meet to Serve in Assembly those who are soe chosen and mett shall have the full power of an Assembly in as ample manner as if all the representatives had beene chosen and mett Provided they are not less then two thirds of the whole number that ought to meet And that the Qualifications of Electors and Elected and all other matters and things Relateing to Elections of Representatives to Serve in Assemblies though not herein perticulerly Exprest shall be and remaine as by a Law of this Government made at Newcastle in the Yeare One thousand [Seven] hundred Intituled An act to ascertains the number of members of assembly and to Regulate the elections Thirdly That the Freemen [in Ea]ch Respective County at the time and place of meeting for Electing [th]eire Representatives to serve in Assembly may as often as there shall be Occasion choose a Double number of persons to present to the Governour for Sheriffes and Coroners to

Serve for three Yeares if they Soe long behave themselves well out of which respective Elections and Presentments the Governour shall nominate and Commissionate one for each of the said Officers the third day after Such Presentment or else the first named in Such Presentment for each Office as aforesaid shall Stand and Serve in that Office for the time before respectively Limitted And in case of Death and Default Such Vacancies shall be Supplyed by the Governour to serve to the End of the said Terme Provided allwayes that if the said Freeman shall at any time neglect or decline to choose a person or persons for either or both the aforesaid Offices then and in Such case the persons that are or shall be in the respective Offices of Sheriffes or Coroner at the time of Election shall remaine therein untill they shall be removed by another Election as aforesaid And that the justices of the respective Counties shall or may nominate and present to the Governour three persons to Serve for Clerke of the Peace for the said County when there is a vacancy, one of which the Governour shall Commissionate within Tenn dayes after Such Presentment or else the first Nominated shall Serve in the said Office dureing good behaviour fourthly That the Laws of this Government shall be in this Stile Viz "By the Governour with the Consent and Approbation of the Freemen in Generall Assembly mett" And shall be after Confirmation by the Governour forthwith Recorded in the Rolls Office and kept at Philadelphia unless the Governour and Assembly shall Agree to appoint another place fifthly that all Criminalls shall have the same Priviledges of Wittnesses and Councill as theire Prosecutors Sixthly That noe person or persons shall or may at any time hereafter be obliged to answer any Complaint matter or thing whatsoever relateing to Property before the Governour and Councill or in any other place but in the Ordinary courts of justice unless Appeales thereunto shall be hereafter bylaw appointed Seventhly That noe person within this Governement shall be Licensed by the Governor to keep Ordinary Taverne or house of publick entertainment but Such who are first recommended to him under the hands of the Justices of the respective Counties Signed in open Court which justices are and shall be hereby Impowred to Suppress and forbid any person keeping Such publick house as aforesaid upon theire Misbehaviour on such penalties as the law cloth or shall Direct and to recommend others from time to time as they shall see occasion Eighthly If any person through Temptation or Melancholly shall Destroy himselfe his Estate Reall and personall shall notwithstanding Descend to his wife and Children or Relations as if he had dyed a Naturall Death And if any person shall be Destroyed or kill'd by casualty or Accident there shall be noe forfeiture to the Governour by reason thereof And noe Act Law or Ordinance whatsoever shall at any time hereafter be made or done to Alter Change or Diminish the forme or Effect of this Charter or of any part or Clause therein Contrary to the True intent and meaning thereof without the Consent of the Governour for the [time being and] six parts of Seven of the Assembly [mett] But because the happiness of Mankind Depends So much upon the Enjoying of Libertie of theire Consciences as aforesaid I Doe hereby Solemnly Declare Promise and Grant for me my heires and Assignes that the first Article of this Charter Relateing to Liberty of Conscience and every part and Clause therein according to the True Intent and meaneing thereof shall be kept and remaine without any Alteration Inviolably for ever And Lastly I the said William Penn Proprietary and Governour of the Province of Pennsilvania and Territories thereunto belonging for my Selfe my heires and Assignes

Have Solemnly Declared Granted and Confirmed And doe hereby Solemnly Declare Grant and Confirme that neither I my heires or Assignes shall procure or doe any thing or things whereby the Liberties in this Charter contained and expressed nor any part thereof shall be Infringed or broken And if any thing shall be procured or done by any person or persons contrary to these presents it shall be held of noe force or Effect In wittnes whereof I the said William Penn at Philadelphia in Pennsilvania have unto this present Charter of Liberties Sett my hand and Broad Seale this twenty Eighth day of October in the Yeare of our Lord one thousand Seven hundred and one being the thirteenth yeare of the Reigne of King William the Third over England Scotland France and Ireland &c And in the Twenty first Yeare of my Government. And notwithstanding the closure and Test of this present Charter as aforesaid I think fitt to add this following Provisoe thereunto as part of the same That is to say that notwithstanding any Clause or Clauses in the above mencioned Charter obligeing the Province and Territories to joyne Togather in Legislation I am Content and doe hereby Declare That if the representatives of the Province and Territories shall not hereafter Agree to joyne togather in Legislation and that the same shall be Signifyed to me or my Deputy In open Assembly or otherwise from under the hands and Scales of the Representatives (for the time being) of the Province or Territories or the Major part of either of them any time within three yeares from the Date hereof That in Such case the Inhabitants of each of the three Counties of this Province shall not have less then Eight persons to represent them in Assembly for the Province and the Inhabitants of the Towne of Philadelphia (when the said Towne is Incorporated) Two persons to represent them in Assembly and the Inhabitants of each County in the Territories shall have as many persons to represent them in a Distinct Assembly for the Territories as shall be requested by them as aforesaid Notwithstanding which Seperation of the Province and Territories in Respect of Legislation I doe hereby promise Grant and Declare that the Inhabitants of both Province and Territories shall Seperately Injoy all other Liberties Priviledges and Benefitts granted joyntly to them in this Charter Any law usage or Custome of this Governement heretofore made and Practised or any law made and Passed by this Generall Assembly to the contrary hereof Notwithstanding. (American Philosophical Society, *William Penn, Charter of Privileges for the Province of Pennsylvania,* *1701*)

A Model of Openness in Penn's Land

Following his advertisements proclaiming the ideals of his new settlement, large numbers of emigrants began pouring into the province. The year 1682 saw twenty-three ships bring some 2,000 colonists to settle in Pennsylvania. Ninety more ships followed during the next three years, and, by 1715, approximately 23,000 emigrants had relocated there. Most were either Quakers or Quaker sympathizers. By 1750, the Society of Friends was the third largest denomination in Britain's American colonies (Ries and Stewart, *This Venerable Document*).

On one other point, Penn's charter offered an idealism that later Americans would find impractical: toleration and unity with native occupants of the land. In general, Quakers enjoyed a far different relationship with the Indians than did most other colonists. Penn and other Quakers viewed Native Americans as children of God no different from themselves.

For this reason, Penn sought ways in which he could deal more honestly and fairly with them. This included his interest in paying them a fair price for the land to which they claimed ownership. Fortunately for Penn, he dealt with the Lenni Lenape who also wanted to have a peaceful relationship with the settlers (Soderlund 1983).

Roger Williams and Religious Toleration in New England

In a similar vein, Roger Williams guided the United States toward its idea of religious toleration. Unlike Penn, however, to pursue his vision, Williams had to endure being the subject of the exact intolerance against which he argued.

New England of the 1600s, of course, was dominated by the expanding settlements of "Puritans": immigrants from England who hoped to "purify" the Church of England. As Charles I became less tolerant of the Puritans' activities in England, he drove them to the New World in larger numbers: between 1630 and 1640, 20,000 Puritans sailed for New England in hopes of practicing their religion in peace. They wanted to build a holy community in which people would live by the rules of the Bible. They expected their Massachusetts Bay Colony to be an example for all the world.

Although the Puritans sought religious freedom, this was a very different concept from toleration. They had little tolerance or even respect for the Pequot Indians, who lived in nearby Connecticut and Rhode Island. They called them heathens. As more and more Puritan settlers moved into their land, the Pequots got angry and resisted. In 1637, war broke out, and the Puritans, helped by Mohican and Narragansett Indian allies, massacred 600 Pequots in their fort, burning many alive. Their ministers preached that it was wrong to practice any religion other than Puritanism, and those who did so were helping the devil. They believed they followed the only true religion.

Roger Williams, a Puritan minister, disagreed. He once said, "Forced worship stinks in God's nostrils." Williams thought that killing or punishing in the name of Christianity was sinful. He respected the beliefs of others, including the Native Americans. He began to openly proclaim that land should not be forcibly taken from the Indians. For his public, vocal criticism, the Puritans arrested Williams and banished him from their colony.

He fled south, bought land from the Indians, and started a new colony called Providence, which would later become the capital of Rhode Island. In his new colony, Williams put his ideals into practice: he welcomed everyone, Quakers and Catholics, Jews and atheists, even when he disagreed with their religion.

One of the great examples of his idealism is Williams' 1643 book *A Key into the Language of America*, which marked the first English-language study of an Indian language, the Narragansett language. He wrote the following:

> I once traveled to an island of the wildest in our parts, where in the night an Indian (as he said) had a vision or dream of the sun (whom they worship for a god) darting a beam into his breast which he conceived to be the messenger of his death: this poor native called his friends and neighbors, and prepared some little refreshing for them, but himself was kept waking and fasting in great humiliations and invocations for ten days and nights; I was alone (having traveled from my bark, the wind being contrary) and little could I speak to them to their understandings especially because of the

change of their dialect or manner of speech from our neighbors: yet so much (through the help of God) I did speak, of the true and living only wise God, of the creation: of man, and his fall from God, etc. that at parting many burst forth, "Oh when will you come again, to bring us some more news of this God?" ...

Nature knows no difference between Europe and Americans in blood, birth, bodies, etc. God having of one blood made all mankind (Acts, 17), and all by nature being children of wrath (Ephesians, 2).

More particularly,

Boast not proud English, of thy birth and blood
Thy brother Indian is by birth as good.
Of one blood God made him, and thee, and all.
As wise, as fair, as strong, as personal.
By nature, wraith's his portion, thine, no more
Till grace his soul and thine in Christ restore.
Make sure thy second birth, else thou shalt see
Heaven ope to Indians wild, but shut to thee.

Conclusion

The strict Puritan roots of the United States contradict many of the founding principles of documents such as the Constitution and Bill of Rights. Although these ideas would crystallize with the global intellectual enlightenment of the late 1700s, thinkers such as Penn and Williams show us that, despite the difficulties of colonial settlement, there were already clear indications of the great American "experiment" by the end of the 1700s.

Sources and Further Reading: American Philosophical Society, *William Penn, Charter of Privileges for the Province of Pennsylvania, 1701*, http://www.amphilsoc.org/library/exhibits/treasures/charter.htm; Miller, *The Transcendentalists: An Anthology*; Patton, A. *Soderlund, William Penn, and the Founding of Pennsylvania*; Ries and Stewart, *This Venerable Document*, http://www.phmc.state.pa.us/bah/dam/charter/charter.html.

ALEXANDER HAMILTON ENVISIONS AN INDUSTRIAL AMERICA

Time Period: Late 1700s to early 1800s
In This Corner: Hamilton and other Federalists
In the Other Corner: Jeffersonian Democrats, farmers, some political leaders
Other Interested Parties: Working-class members of New England communities
General Environmental Issue(s): River management, city development, industry and technology

Nature, argued Thomas Jefferson, would influence everyday life in America by enabling productive agricultural development over open lands made available with the removal of native occupants. When Alexander Hamilton served as secretary of the treasury for both of George

Washington's terms as president, he also argued that nature must play an important role in the future of American life. In his mind, however, the natural resources would be used to supply and power new industries that would put Americans to work in a new economy of factories and manufactures. Each vision had nature at its core; however, Hamilton's nature was instrumentalized—changed, through the use of human ingenuity—technology.

To settle this discussion once and for all, Hamilton set out to create a community that could serve as a model that others might follow. Hamilton used his leadership to develop American industry and to help the nation gain international respect in manufacturing and trade. Taking an opposite view from Jefferson, Hamilton argued that a strong central government needed to direct the nation's activities and be able to assist economic development without interference from the states. Internal improvements that would benefit the nation as a whole, he argued, should be financed by the federal government. These two intellectuals represented two different paths of national development. To some degree, the United States would not choose a single point of view; instead, nineteenth-century Americans pursued both paths of development.

In 1791, Hamilton released the *Report on the Subject of Manufactures*, which revealed the full range of his plan for industrializing the United States. Rejecting the common assumption that America could prosper with an agricultural base, Hamilton argued that the new republic should concentrate on developing industry. To nurture American industry, he proposed the imposition of protective tariffs and the prohibition of imported manufactured goods that would compete with domestic products. To Hamilton, it did not seem plausible that a nation of farmers could compete against the industrial might of Europe. He argued that the United States could only ensure its political independence by maintaining economic independence.

In his 1791 report, Hamilton set out the framework of the American industrial revolution of the 1800s:

THE SECRETARY OF THE TREASURY, in obedience to the order of the House of Representatives, of the 15th day of January, 1790, has applied his attention at as early a period as his other duties would permit, to the subject of Manufactures, and particularly to the means of promoting such as will tend to render the United States independent on foreign nations, for military and other essential supplies; and he thereupon respectfully submits the following report....

The expediency of encouraging manufactures in the United States, which was not long since deemed very questionable, appears at this time to be pretty generally admitted....

The objections which are commonly made to the expediency of encouraging, and to the probability of succeeding in manufacturing pursuits, in the United States, having now been discussed, the considerations, which have appeared in the course of the discussion, recommending that species of industry to the patronage of the Government, will be materially strengthened by a few general, and some particular topics, which have been naturally reserved for subsequent notice.

There seems to be a moral certainty that the trade of a country, which is both manufacturing and agricultural, will be more lucrative and prosperous than that of a country which is merely agricultural.... There is always a higher probability of a favorable balance of trade, in regard to countries in which manufactures, founded on the basis of a thriving agriculture, flourish, than in regard to those which are confined wholly, or almost wholly, to agriculture....

Not only the wealth, but the independence and security of a country, appear to be materially connected with the prosperity of manufactures. Every nation, with a view to those great objects, ought to endeavor to possess within itself, all the essentials of a national supply. These comprise the means of subsistence, habitation, clothing, and defence.... (Hamilton, Report)

In this concluding portion of his report, Hamilton directly tied industrial development to national independence. In this argument, the natural resources of the nation were, indeed, directly tied to maintaining and perpetuating the nation's independence from European powers. Technology would guide American entrepreneurs in most effectively putting nature to work for the nation.

Politicians, however, were too slow to swing to his perspective for Hamilton's liking. When it became obvious to him that Jeffersonian Democrats and others were resistant to his plan, Hamilton founded the Society for the Establishment of Useful Manufactures (S.U.M.). Whereas Jefferson's dream for the nation relied largely on skills and abilities that were already known to most Americans, Hamilton's vision relied on technology not available yet in the United States. Using private capital, the S.U.M. set out to create a model that would convince the American public of the future in industry. The focal point of the plan was the energy of the seventy-seven-foot-high Great Falls of the Passaic River, which was second only to the Mississippi in total volume in the eastern United States. Using this prime mover, Hamilton and the S.U.M. planned an industrial community that would serve as the "Cradle of American Industry."

Paterson, as the industrial town was called, used the power of the Passaic to run woolen and textile mills. Pierre L'Enfant, who also designed the original plan for Washington, DC, designed Paterson. Soon after water from the falls was brought to the district around 1794, developers erected the first cotton-spinning factory to manufacture yarn. The expansion of the raceway system in 1807 helped increase the textile production (flannel, silk ribbons, woolens, and cotton), as well as the number of mills that produced metal goods such as kettles, spades, pans, and nails. Before moving to Hartford, Connecticut, in the early 1840s, the Colt Gun Mill also manufactured its well-known revolvers in Paterson.

Hamilton's vision for an industrial economy created cradles of development in many different areas. Rapid expansion of technology created entirely new industries, such as coal mining and steelmaking. However, in other cases, the texture and spirit of the industrial age took existing technology, such as the water-powered mill, and expanded the undertakings' scale and scope to make them almost unrecognizable compared with those of the eighteenth century. Unlike later industry, however, none of the endeavors of the early industrial era grew too distant from the natural resources that made them possible.

Although Hamilton's ideas continued and would shape much of the American future, his outspoken public persona created significant difficulties for him. Both Jefferson and Hamilton resigned from their cabinet positions before Washington's resignation from the presidency. For many Americans who favored his view of an industrial future, Hamilton was a favorite to become president. A personal scandal, however, erupted in 1797 and derailed any such aspirations.

A pamphlet published that year revealed Hamilton's affair with a woman named Maria Reynolds and linked him to a scheme by Reynolds' husband to illegally manipulate federal securities. To prove his innocence, Hamilton resorted to publishing love letters he wrote to Maria Reynolds. This cleared Hamilton of financial impropriety but badly damaged his reputation. In 1800, Hamilton's old enemy, Aaron Burr, obtained and published a confidential document Hamilton wrote that was highly critical of Federalist John Adams (who now served as president). Publication of the article created a rift in the Federalist party, helping Republicans Thomas Jefferson and Aaron Burr win the race for presidency in 1801.

In the New York gubernatorial race of 1804, Hamilton again clashed with Aaron Burr, who was running as an independent. Hamilton feared that Burr would eclipse him in the Federalist leadership. He spoke out against the vice president, and New York Republicans George and DeWitt Clinton led a brutal media campaign against Burr. The Clintons, not Hamilton, were responsible for Burr's defeat.

Still, after reading in a newspaper that Hamilton had expressed a "despicable opinion" about him during the campaign, Burr challenged Hamilton to a duel. They met on the dueling grounds at Weehawken, New Jersey, on July 11, 1804. Both men fired their pistols; only Hamilton was hit. He died of his wounds the next day. The sitting vice president killed his and Jefferson's greatest political rival and also the torchbearer of America's industrial future.

Sources and Further Reading: Conzen, *The Making of the American Landscape*; Opie, *Nature's Nation*; Trachtenberg, *The Incorporation of America*.

NAVIGATING THE CHESAPEAKE BAY DURING THE EARLY REPUBLIC

Time Period: Late 1700s
In This Corner: Federalists, Mid-Atlantic developers, George Washington
In the Other Corner: Republicans, representatives of other regions
Other Interested Parties: Federal government, state and local leaders
General Environmental Issue(s): Coastal management, trade corridors

During colonial times, dangerous shoals, fog-enshrouded capes, and other dangers made navigation near land treacherous for ship captains. In the area of Virginia, Maryland, and Delaware, however, it was essential for the nation that the Chesapeake Bay be traveled safely. As the volume of shipping increased, governments in these areas began considering their options. First, however, they needed to deal with critics who argued that governments had no right to meddle in areas such as building aids to commercial interests.

In the early 1700s, the colonial governments of Virginia and Maryland agreed to allocate funds to build a lighthouse that would serve this common need. Work began almost immediately with the delivery of 4,000 tons of Aquia sandstone from Brooks Quarry on the Rappahannock River. The first effort at "public works," however, was not meant to be. Escalating costs and increasing conflict with Great Britain forced the colonies to abandon the project. The valuable stone was left in its place.

During the first session of the first U.S. Congress in 1789, Jacob Wray, a collector of Customs at Hampton, Virginia, reported to Secretary of the Treasury Alexander Hamilton that the absence of a light on the shores of Cape Henry was a danger to all who passed in the Bay. He made the specific claim that such dangers had claimed fifty-seven vessels because of problems navigating the treacherous waters.

By August, the House and the Senate were convinced that such a project merited some of the first effort and funding of the young federal government. Later that year, the Congress sent a bill to President George Washington titled *The Act for the Establishment and Support of Lighthouses, Beacons, Buoys, and Public Piers*. It read as follows:

Section 1. Be it enacted by the Senate and House of Representatives of the United States of America in Congress assembled, That all expenses which shall accrue from and after the fifteenth day of August one thousand seven hundred and eighty-nine, in the necessary support, maintenance and repairs of all lighthouses, beacons, buoys and public piers erected, placed, or sunk before the passing of this act, at the entrance of, or within any bay, inlet, harbor, or port of the United States, for rendering the navigation thereof easy and safe, shall be defrayed out of the treasury of the United States: Provided nevertheless, That none of the said expenses shall continue to be so defrayed by the United States, after the expiration of one year from the day aforesaid, unless such lighthouses, beacons, buoys and public piers, shall in the mean time be ceded to and vested in the United States, by the state or states respectively in which the same may be, together with the lands and tenements thereunto belonging, and together with the jurisdiction of the same.

Sec. 2. And be it further enacted, That a lighthouse shall be erected near the entrance of the Chesapeake Bay, at such place, when ceded to the United States in manner aforesaid, as the President of the United States shall direct.

Sec. 3. And be it further enacted, That it shall be the duty of the Secretary of the Treasury to provide by contracts, which shall be approved by the President of the United States, for building a lighthouse near the entrance of the Chesapeake Bay, and for rebuilding when necessary, and keeping in good repair, the lighthouses, beacons, buoys, and public piers in the several States, and for furnishing the same with all necessary supplies; and also to agree for the salaries, wages, or hire of the person or persons appointed by the President, for the superintendence and care of the same.

Sec. 4. And be it further enacted, That all pilots in the bays, inlets, rivers, harbors and ports of the United States, shall continue to be regulated in conformity with the existing laws of the States respectively wherein such pilots may be, or with such laws as the States may respectively hereafter enact for the purpose, until further legislative provision shall be made by Congress. (Lighthouse Act)

Entrusted with attempts to ensure the safety of commercial traffic in the Bay, the federal government took over operations of state facilities and specifically of the abandoned building project of the lighthouse to be housed on the southern shore of the Chesapeake Bay. Without funds to pay for the project, the federal government passed an act establishing a system of duties on ships and vessels as well as on the imported goods they carried.

Next, the Virginia General Assembly in November 1789 provided conveyance of the land "lying and being in the County of Princess Anne at the place commonly called the head land of Cape Henry" to the new government "for the purpose of building a lighthouse." Alexander

Hamilton contracted with John McComb Jr. of New York on March 31, 1791. McComb was the designer of the Government House, the planned residence for the president, in New York City. He contracted with him to create an octagonal structure with three windows in the east and four windows in the west, rising seventy-two feet from the water table to the top of the stone work. The agreement also stipulated the design and construction of a two-story house to be a residence for the keeper and for safe storage of the oil to be used for the light.

In early October 1792, George Washington again focused his interest on this most historic project taking place in his native region. He personally requested a list of applicants for the keeper. After presidential review, Laban Goffigan, who is believed to have hailed from Norfolk, became the first keeper of the Cape Henry Lighthouse. A primary task of the keeper was to light the fish-oil-burning lamps. The new government completed its first federal work project and fulfilled its obligation to the sea travelers of the Virginia coast. The final cost of $17,700 exceeded the first estimate by $2,500.

Cape Henry Light guided use and development of the Chesapeake Bay area. It still stands today.

Sources and Further Reading: APVA Preservation Virginia, *Old Cape Henry Lighthouse*, http://www.apva.org/capehenry/origin.php; Hamilton, http://history.sandiego.edu/gen/text/civ/1791manufactures.html; Labaree, *American and the Sea*; Lighthouse Act, http://www.lighthousefoundation.org/museum/natllighthouseday_info.htm; New Jersey Lighthouse Society, *An Act for the Establishment and Support of Lighthouse, Beacons, Buoys, and Public Piers*, http://www.njlhs.org/historicdocs/act.htm; Stilgoe, *Alongshore*; Stilgoe, *Common Landscapes of America*.

INTERNAL IMPROVEMENTS IN THE EARLY REPUBLIC

Time Period: Late 1700s
In This Corner: Federalists, Midwestern developers
In the Other Corner: Republicans, Jeffersonian Democrats
Other Interested Parties: Federal government, state and local leaders
General Environmental Issue(s): Resource management, trade corridors, roads

The strength and appropriate powers and responsibility of the federal government are a much-debated portion of the American legacy. From the earliest phases of U.S. history, political leaders have vied to add new responsibilities to the federal government or to simplify and streamline the federal authority to place more responsibility on individuals or state and local authorities. One of the first times that the nation wrestled with this problem concerned large-scale construction projects, particularly those aiding internal trade and commerce. Even today, internal improvements and who should build them stir debate.

However, these concerns over financing and support of internal improvements were even more pronounced when the fledgling United States needed improving in almost all regards. What was the responsibility of the federal government? Should taxes be used for such projects? Were gargantuan projects such as roads and canals even practical for a nation of such sprawling size? Some scholars argue that the responses to such questions revolved around the larger concerns of nationalism and sectionalism. However, the reality was that the nonexistence and impassibility of avenues of transportation made many fear political disintegration even before the nation had solidified.

The political division on the issue of internal improvements largely followed party lines. The early Federalists, led by Alexander Hamilton, argued that the Constitution gave the federal government no explicit authority to make internal improvements. At best, they argued, that power was implied. On the other side, Thomas Jefferson's Republican party, concerned about giving too much power to the federal government, took a dim view of that interpretation. Initially, the Republicans opposed the use of federal power to make internal improvements such as roads and canals, yet when Jefferson gained the presidency in 1801, his party actively pursued the idea of federal authority overseeing and even financing such projects. Jefferson's secretary of the treasury, Albert Gallatin, became a key figure in favor of federal financing for such projects. He developed the first systematic proposal for a national network of roads and canals, which will be discussed below.

In December 1791, Secretary of the Treasury Alexander Hamilton, a rival of Jefferson, encouraged the nation to establish a "comprehensive plan" for nationwide internal improvements, which were "improvements which could be prosecuted with more efficacy by the whole than by any part or parts of the Union" (American State Papers, *Gallatin's Report on Roads and Canals*, 1981), but no comprehensive plan emerged at the national level until after the Jeffersonians came to power in 1800. Even at that point, not all Republicans agreed on the efficacy or method of a national internal improvements system. Historian Lee Fornwalt wrote the following:

> While conservative Jeffersonians or Old Republicans considered the idea of federal expenditures for canals and roads to be repugnant, more moderate Republicans, including Jefferson himself, approved of government expenditure of surplus funds if a constitutional amendment allowing such action was passed. Still other moderate Republicans saw no constitutional impediment at all to federal government support of internal improvements. (Fornwalt 1981, 3–4)

Gallatin led the latter group. When he became Jefferson's secretary of the treasury in 1801, he set out to commit the federal government to planning and funding support of internal improvement works throughout the country.

The American Way

Early efforts to create internal improvements, even if they were performed by private groups, were hampered by various difficulties, including the lack of capital and the absence of technical or engineering expertise and equipment. Fornwalt wrote, "Americans were wealthy in ships and land, but as late as 1800 only three corporations in the United States had a capital of a million dollars—the Bank of the United States, the Bank of North America, and the Bank of Pennsylvania" (Fornwalt 1981, 5).

During the nation's first few decades of existence, Americans debated the federal government's relationship to the states. One of the first tasks related to this issue was the surveying of western lands and waterways. Even in these early days, the drive for internal improvements had one esteemed proponent: at the Constitutional Convention, Benjamin Franklin became the principal advocate of federal sponsorship for internal improvements. His interests, however, were left out of the document. The Constitution gave the new federal government the power to "provide for the common Defense and general Welfare of the United States," but it limited explicit construction authority to military structures such as arsenals and fortifications, lighthouses, and post roads, post offices, dockyards, and "other needful buildings."

In 1802, Congress initiated the practice of appropriating money for specific internal improvement projects within the states. For instance, in that year, it authorized payment of $30,000 for the repair and erection of public piers in the Delaware River. However, Jefferson and Congress retained doubts about the constitutionality of such financial appropriations.

National Road

The first well-known publicly funded project was the construction of the National Road, which was also known as the Cumberland Road. In 1802, the same year that Ohio was admitted as a state, Congress passed the bill authorizing construction of the passage westward. However, the actual construction did not begin until 1815. In the interim, small stretches of turnpike and informal roads steadily grew.

The earliest roads stretched from Baltimore to Frederick and Hagerstown. In fact, the portion between Boonsboro and Hagerstown was the first piece of macadamized road in the United States. No turnpikes, however, reached west to Wheeling. To extend the development in this direction, in March 1806, Congress passed a law providing for the construction of the road from Cumberland to the Ohio River. Great care went into selecting the point at which the road would strike the Ohio, where it seemed certain a great city would take shape. Operations on the road were commenced forthwith, and, up to 1817, it cost $1.8 million and had, moreover, in some portions become so worn out as to need extensive repairs.

When it was completed, the National Road covered approximately 620 miles and provided a connection between the Potomac and Ohio Rivers. As such, it also served as a gateway westward for many settlers. The first project supported by federal funds, the National Road, has remained an important corridor; today, it exists as U.S. Highway 40.

Gallatin Report

In the case of the National Road, the federal government facilitated the construction of one important thoroughfare. It did not necessarily suggest a wholesale shift in the role of the federal government. This shift, however, occurred almost simultaneously with the work of Gallatin. Thanks to a close friendship with Senator Thomas Worthington of Ohio, Gallatin was able to manage a congressional resolution on March 2, 1807, that directed the treasury secretary to prepare and report to the Senate, at their next session, a plan for the application of such means as are within the power of Congress to the purposes of opening roads and making canals. They would also need to provide a statement of the efforts underway to create objects of public improvement and what, if any, assistance was needed from the federal government.

In July 1807, Gallatin sent questionnaires to customs collectors throughout the United States to collect information about the canals and turnpike roads that had been built, were being constructed, or were still in the planning stages in their respective districts. In addition, he began consulting with Robert Fulton and Benjamin Henry Latrobe for technical engineering advice. Fornwalt reports that Latrobe became Gallatin's informal chief technical expert from the summer of 1807 to April 1808, when the latter submitted his comprehensive plan to the Senate. He wrote, "Not only did Latrobe compose a thirty page essay on canals and river improvements in America but he met with Gallatin at least every other day during March 1808 when the secretary was sorting out the masses of information that had poured into Washington from canal and turnpike representatives" (Fornwalt 1981, 105).

On April 4, 1808, Gallatin submitted his historic document to the Senate. John Quincy Adams' Senate committee ordered that 1,200 copies of the report be printed, including the essays and maps by Latrobe and Fulton. He proposed federal financing (approximately $20 million over ten years) to support canals from the coastal plain to the interior. He argued that transportation improvements could stimulate additional projects connecting to a region, ultimately benefiting the general population even if the initial investors failed to make a profit. The text of *Gallatin's Report on Roads and Canals* is as follows (American State Papers 1808, 724):

The Secretary of the Treasury, in obedience to the resolution of the Senate of the 2d March, 1807, respectfully, submits the following report on roads and canals.

The general utility of artificial roads and canals, is at this time so universally admitted, as hardly to require any additional proofs. It is sufficiently evident that, whenever the annual expense of transportation on a certain route in its natural state, exceeds the interest on the capital employed in improving the communication, and the annual expense of transportation (exclusively of the tolls) by the improved route; the difference is an annual additional income to the nation. Nor does in that case the general result vary, although the tolls may not have been fixed at a rate sufficient to pay to the undertakers the interest on the capital laid out. They indeed, when that happens, lose; but the community is nevertheless benefited by the undertaking. The general gain is not confined to the difference between the expenses of the transportation of those articles which had been formerly conveyed by that route, but many which were brought to market by other channels, will then find a new and more advantageous direction; and those which on account of their distance or weight could not be transported in any manner whatever, will acquire a value, and become a clear addition to the national wealth. Those and many other advantages have become so obvious, that in countries possessed of a large capital, where property is sufficiently secure to induce individuals to lay out that capital on permanent undertakings, and where a compact population creates an extensive commercial intercourse, within short distances, those improvements may often, in ordinary cases, be left to individual exertion, without any direct aid from government.

There are however some circumstances, which, whilst they render the facility of communications throughout the United States an object of primary importance, naturally check the application of private capital and enterprise, to improvements on a large scale.

The price of labor is not considered as a formidable obstacle, because whatever it may be, it equally affects the expense of transportation, which is saved by the improvement, and that of effecting the improvement itself. The want of practical knowledge is no longer felt: and the occasional influence of mistaken local interests, in sometimes thwarting or giving an improper direction to public improvements, arises from the nature of man, and is common to all countries. The great demand for capital in the United States, and the extent of territory compared with the population, are, it is believed, the true causes which prevent new undertakings, and render those already accomplished, less profitable than had been expected.

1. Notwithstanding the great increase of capital during the last fifteen years, the objects for which it is required continue to be more numerous, and its application is generally more profitable than in Europe. A small portion therefore is applied to objects which offer only the prospect of remote and moderate profit. And it also happens that a less sum being subscribed at first, than is actually requisite for completing the work, this proceeds slowly; the capital applied remains unproductive for a much longer time than was necessary, and the interest accruing during that period, becomes in fact an injurious addition to the real expense of the undertaking.

2. The present population of the United States, compared with the extent of territory over which it is spread, does not, except in the vicinity of the seaports, admit that extensive commercial intercourse within short distances, which, in England and some other countries, forms the principal support of artificial roads and canals. With a few exceptions, canals particularly, cannot in America be undertaken with a view solely to the intercourse between the two extremes of, and along the intermediate ground which they occupy. It is necessary, in order to be productive, that the canal should open a communication with a natural extensive navigation which will flow through that new channel. It follows that whenever that navigation requires to be improved, or when it might at some distance be connected by another canal to another navigation, the first canal will remain comparatively unproductive, until the other improvements are effected, until the other canal is also completed. Thus the intended canal between the Chesapeake and Delaware, will be deprived of the additional benefit arising from the intercourse between New York and the Chesapeake, until an inland navigation, shall have been opened between the Delaware and New York. Thus the expensive canals completed around the Falls of Potomac, will become more and more productive in proportion to the improvement, first of the navigation of the upper branches of the river, and then of its communication with the western waters. Some works already executed are unprofitable, many more remain unattempted, because their ultimate productiveness depends on other improvements, too extensive or too distant to be embraced by the same individuals.

The general government can alone remove these obstacles.

With resources amply sufficient for the completion of every practicable improvement, it will always supply the capital wanted for any work which it may undertake, as fast as the work itself can progress, avoiding thereby the ruinous loss of interest on a dormant capital, and reducing the real expense to its lowest rate.

With these resources, and embracing the whole union, it will complete on any given line all the improvements, however distant, which may be necessary to render the whole productive, and eminently beneficial.

The early and efficient aid of the federal government is recommended by still more important considerations. The inconveniences, complaints, and perhaps dangers, which may result from a vast extent of territory, can no otherwise be radically removed, or prevented, than by opening speedy and easy communications through all its parts. Good roads and canals, will shorten distances, facilitate commercial and personal intercourse, and unite by a still more intimate community of interests, the most remote quarters of the United States. No other single operation, within the power of government, can more effectually tend to strengthen and perpetuate that union, which secures external independence, domestic peace, and internal liberty....

It must not be omitted that the facility of communications, constitutes, particularly in the United States, an important branch of national defence. Their extensive territory opposes a powerful obstacle to the progress of an enemy. But on the other hand, the number of regular forces, which may be raised, necessarily limited by the population, will for many years be inconsiderable when compared with that extent of territory. That defect cannot otherwise be supplied than by those great national improvements, which will afford the means of a rapid concentration of that regular force, and of a formidable body of militia, on any given point.

Amongst the resources of the union, there is one which from its nature seems more particularly applicable to internal improvements. Exclusively of Louisiana, the general government possesses, in trust for the people of the United States, about one hundred millions of acres fit for cultivation, north of the river Ohio, and near fifty millions south of the state of Tennessee. For the disposition of those lands a plan has been adopted, calculated to enable every industrious citizen to become a freeholder, to secure indisputable titles to the purchasers, to obtain a national revenue, and above all to suppress monopoly. Its success has surpassed that of every former attempt, and exceeded the expectations of its authors. But a higher price than had usually been paid for waste lands by the first inhabitants of the frontier became an unavoidable ingredient of a system intended for general benefit, and was necessary in order to prevent the public lands being engrossed by individuals possessing greater wealth, activity or local advantages. It is believed that nothing could be more gratifying to the purchasers, and to the inhabitants of the western states generally, or better calculated to remove popular objections, and to defeat insidious efforts, than the application of the proceeds of the sales to improvements conferring general advantages on the nation, and an immediate benefit on the purchasers and inhabitants themselves. It may be added, that the United States, considered merely as owners of the soil, are also deeply interested in the opening of those communications, which must necessarily enhance the value of their property. Thus the opening of an inland navigation from tide water to the great lakes, would immediately give to the great body of lands bordering on those lakes, as great value as if they were situated at the distance of one hundred miles by land from the sea coast. And if the proceeds of the first ten millions of acres which may be sold, were applied to such improvements, the United States would be amply repaid in the sale of the other ninety millions....

The manner in which the public monies may be applied to such objects, remains to be considered.

It is evident that the United States cannot under the constitution open any road or canal, without the consent of the state through which such road or canal must pass. In order therefore to remove every impediment to a national plan of internal improvements, an amendment to the constitution was suggested by the executive when the subject was recommended to the consideration of Congress. Until this be obtained, the assent of the states being necessary for each improvement, the modifications under which that assent may be given, will necessarily control the manner of applying the money. It may be however observed that in relation to the specific improvements which have been suggested, there is hardly any which is not either already authorized by the states respectively, or so immediately beneficial to them, as to render it highly probable that no material difficulty will be experienced in that respect.

The monies may be applied in two different manners: the United States may with the assent of the states, undertake some of the works at their sole expense or they may subscribe a certain number of shares of the stock of companies incorporated for the purpose. Loans might also in some instances be made to such companies. The first mode would perhaps, by effectually controlling local interests, give the most proper general direction to the work. Its details would probably be executed on a more economical plan by private companies. Both modes may perhaps be blended together so as to obtain the advantages pertaining to each. But the modifications of which the plan is susceptible must vary according to the nature of the work, and of the charters, and seem to belong to that class of details which are not the immediate subject of consideration.… (American State Papers)

Although it was never realized in its entirety, Gallatin's was the first plan for a national highway and canal system.

While Gallatin's proposals were being debated, the War of 1812 approached and soon stopped all thought of the projects. After the war, they were brought up again, and four roads were built but no canals. In the end, Gallatin's report was prophetic for its strategic design of resource use, whether projects were financed by the federal government (such as the Intracoastal Waterway) or by the states (such as the Erie Canal). The subject of internal improvements became increasingly divisive during the antebellum period, pitting Whigs, who generally supported federal funds for transportation improvements, against Democrats, who did not.

Conclusion

The debate over the extent of federal involvement in internal improvements continued after the War of 1812 and for decades afterward. Despite the nationalist fervor following the War of 1812, politicians, especially in the agrarian South where national planners had proposed few major transportation improvements, spoke most about the need to restore states' political power. Large-scale internal improvements (first, canals and then later, railroads) were undertaken with government aid during the antebellum period but that assistance emanated from the state legislatures. Not until several years before the Civil War did the federal government begin to assist railroad construction but only in a haphazard manner.

The ongoing debate clearly extended into the federal government's very right to plan and carry out such projects. In 1817, President Madison vetoed the *Bonus Bill* of John C. Calhoun to use profits from the Second Bank of the United States for internal improvements. Madison recognized the value of canals and roads but thought federal financing would lead to too much federal power:

The legislative powers vested in Congress are specified and enumerated in the eighth section of the first article of the Constitution, and it does not appear that the power proposed to be exercised by the bill is among the enumerated powers, or that it falls by any just interpretation within the power to make laws necessary and proper for carrying into execution those or other powers vested by the Constitution in the Government of the United States....

I am not unaware of the great importance of roads and canals and the improved navigation of water courses, and that a power in the National Legislature to provide for them might be exercised with signal advantage to the general prosperity. But seeing that such a power is not expressly given by the Constitution, and believing that it can not be deduced from any part of it without an inadmissible latitude of construction and a reliance on insufficient precedents; believing also that the permanent success of the Constitution depends on a definite partition of powers between the General and the State Governments, and that no adequate landmarks would be left by the constructive extension of the powers of Congress as proposed in the bill, I have no option but to withhold my signature from it.... (Fornwalt 1981)

By the 1820s, President James Monroe had led the country in quite a different direction. In 1822, President Monroe issued his celebrated internal improvement message and argued for the general improvement policy of the country to enlarge on the propriety of the government. During his presidency, the federal government appropriated funds for the following projects: (1) a survey of the Maine-New Hampshire coast; (2) $2,500 to repair seawalls and build lighthouses; (3) $6,000 to remove obstacles in Gloucester Harbor; and (4) $150 to survey the harbor entrance at Presque Isle, Pennsylvania.

In 1824, President James Monroe signed the *General Survey Bill*, which expanded authority beyond shipping routes to create surveys for the routes of roads and canals "of national importance, in a commercial or military point of view, or necessary for the transportation of public mail." Shortly after, Congress passed and the president approved the first true rivers and harbors bill. This act appropriated $75,000 to improve navigation on the Ohio and Mississippi Rivers by removing sandbars, snags, and other obstacles. After 1824, federal programs on rivers and harbors increased, although states and private interests still carried the greater financial burden. Most of the federal focus was on the "public highways"—that is, the great rivers such as the Mississippi, Ohio, Missouri, and Arkansas—and on river and harbor work close to or on ocean ports.

By the 1850s, the issue of federal involvement in internal improvements composed one of the most clearly partisan issues in American politics. In 1852, the Whig Party championed the cause of federal assistance by proclaiming Congress's power to improve and maintain all navigable rivers for either defense or the protection of commerce. In contrast, the Democrats maintained the position that they established in the 1840s: namely, that Congress had no power to carry on a general system of internal improvements. When the Republican party weighed in in 1856, it did so by declaring that "appropriations by Congress for the improvement of rivers and harbors of a national character, required for the accommodation and security of our existing commerce, are authorized by the Constitution, and justified by the obligation of the government to protect the lives and property of its citizens" (Fornwalt 1981).

The Civil War sidetracked this debate; however, at the same time, the events of the early 1860s demonstrated the capabilities of federal planning and organization in the industrial era that was emerging. By the end of the 1860s, the necessity of government involvement was uncontested.

Sources and Further Reading: Conzen, *The Making of the American Landscape*; Ewing, *America's Forgotten Statesman: Albert Gallatin*; Fornwalt, *Benjamin Henry Latrobe and the Revival of the Gallatin Plan of 1808*; Gifford, http://onlinepubs.trb.org/onlinepubs/trnews/trnews244newvision.pdf; Opie, *Nature's Nation*; Kushida, "Searching for Federal Aid: The Petitioning Activities of the Chesapeake and Delaware Canal Company"; American State Papers, *Gallatin's Report on Roads and Canals*, http://www.union.edu/PUBLIC/ECODEPT/kleind/eco024/documents/internal/internal_calendar.htm; Virginia Places, http://www.virginiaplaces.org/transportation/canals.html.

WHAT IS THE IMPORTANCE OF AMERICA'S NATURAL HISTORY?

Time Period: Late 1700s and early 1800s
In This Corner: Naturalists, including Audubon, Jefferson, Peale, and Bartram
In the Other Corner: Religious or royal leaders who do not want symbols
Other Interested Parties: American colonists/citizens
General Environmental Issue(s): Natural history, species management

In the early 1800s, many Americans believed that their nation was turning a corner from being a settler society to becoming a more civilized nation to rival those of Europe. In trying to stimulate such development, many Americans made extensive comparisons between the United States and the long-standing European societies. The young American nation compared unfavorably in many categories, especially arts and culture. However, in its natural wonders, there could be no disputing the majesty of the United States. For this reason, some Americans sought new ways to highlight the natural splendor that distinguished the United States from Europe. They came to believe that, although the United States had little history compared with European nations, it could, instead, offer natural history. By getting to know the continent's nature better, argued early naturalists, the United States could gain its identity.

For a few scientists, chronicling North America's everyday nature as well as its natural history became an effort of both science and patriotism. Charles Wilson Peale worked with the Philosophical Society of Philadelphia to initiate this process in 1784 when he established the first natural history museum in the United States (Rhodes 2004, 34–35). His efforts to preserve and catalog the species of North America were shared by Thomas Jefferson. Referred to as "natural history," this effort to know the continent through the creatures living on it spurred at least one of the young nation's first unified, federal undertakings: Peale's effort to excavate a mastodon skeleton from New York state in the late 1700s.

Natural History and National Meaning

When a skeleton was excavated by Peale from a Hudson River Valley farm in 1801, he could not yet verify what type of creature it was. After studying it in Philadelphia, Peale identified the skeleton as that of a mastodon. In Europe and elsewhere, the mastodon had been the focus of a debate about the prehuman past as well as about the concept of extinction. Because of this heightened level of interest, Peale's mastodon discovery became important world news. The bones came to serve as an international puzzle for scientific-minded people everywhere.

The ability to excavate and reassemble the skeleton also became an important symbol for the stability of the young United States. For many Americans, the animal's symbolic meaning far outweighed its scientific significance as evidence of extinct species or a prehuman past. "Indeed," wrote historian Paul Semonin, "while Lewis and Clark were exploring the western wilderness, Peale had remounted his skeleton with its tusks pointing downward to magnify its ferocity." Most historians view this as a representative moment of scientific naiveté, and yet, Semonin suggests that this understandable lapse, instead, demonstrates that the mastodon was "the nation's first prehistoric monster" used by the nation's founders as "a symbol of dominance in the first decades of the new republic" (Semonin 2000).

Also a product of the zeal for natural history, the Lewis and Clark Expedition will be discussed in another essay, and yet it is important to note that President Thomas Jefferson had Meriwether Lewis travel to Philadelphia to receive advice from Peale, the nation's leading naturalist. In particular, Jefferson hoped not that the expedition would create more bones; instead, he actually hoped the explorers would find a living mastodon!

Symbolizing the United States

If they had found such a beast roaming the American West, America's national emblem may have turned out very differently. As it was, most interested Americans focused on fowl to serve as a national emblem. However, there was a bit of debate over which bird was best. The bald eagle was chosen because of its long life, great strength, and majestic looks and also

The Great Seal of the United States immortalized the nation's symbol, the bald eagle, for the ages. Library of Congress.

because it was then believed to exist on this continent only. The national bird of the United States, the bald eagle (*Haliaeetus leucocephalus*), is the only eagle unique to North America. The common name dates from a time when "bald" meant "white," not hairless. The bald eagle ranges over most of the North American continent, from the northern parts of Alaska and Canada to northern Mexico.

As a national symbol, the eagle eventually was placed on the backs of coins. In addition, the Great Seal of the United States became the outstretched eagle with a shield covering his breast. On the shield, one finds thirteen perpendicular red and white stripes surmounted by a blue field with the same number of stars. The eagle clutches a bundle of thirteen arrows in his left talon, an olive branch in his right, and, finally a scroll in his beak inscribed with "E Pluribus Unum."

There were some dissenters to this imagery and selection. Benjamin Franklin wrote that the eagle was "a bird of bad moral character, he does not get his living honestly, you may have seen him perched on some dead tree, where, too lazy to fish for himself, he watches the labor of the fishing-hawk, and when that diligent bird has at length taken a fish, and is bearing it to its nest for the support of his mate and young ones, the bald eagle pursues him and takes it." Franklin favored the humble turkey as a symbol of America.

Other symbols of strength and permanence emanated from North America's natural elements. Jefferson decided that one of the preeminent examples was the Natural Bridge near his home in Virginia. Historian Charles Miller hypothesizes that Jefferson likely saw the bridge for the first time in 1767 and then became its first American owner in 1774 when his family purchased it with the adjoining 150 acres (Miller 1988, 105–7). Jefferson proceeded to enlist artists to paint the bridge and erected a small cabin for tourist guests. The Natural Bridge is just one example showing that, from the earliest days of the 1800s, natural wonders often embodied many of the ideals that Americans wanted to present to the world.

Creating an American Natural History

Finally, this passion for natural history inspired young naturalists to record for posterity what they found. John James Audubon began using his painting talents to preserve each species of bird that he could find (and kill) in North America. His collection, *The Birds of North America*, first appeared in 1824.

The hunter-artist Audubon represents well the mixed motives of most collectors who were impressed with the natural wonders of North America. Audubon, wrote biographer Richard Rhodes, "engaged birds with the intensity (and sometimes the ferocity) of a hunter because hunting was the cultural frame out of which his encounter with birds emerged. In early nineteenth-century America, when wild game was still extensively harvested for food, observation for hunting had not yet disconnected from observation for scientific knowledge" (Rhodes 2004, 75). Audubon observed American fowl extensively, but he also killed samples (at least six and as many as hundreds) of each species. The catch was skinned and stuffed with frayed rope. Audubon then posed each sample in positions he had observed in the wild. He used the samples as puppets or manikins to create the illusion that the bird sample was still alive and then he would paint it. Audubon wrote, "By means of threads I raised or lowered a head, wing, or a tail and by fastening the threads securely I had something like life before me" (Rhodes 2004, 12–15).

Audubon's efforts in painting were mirrored in the form of the written word by the Bartrams. The American tradition in nature writing grew out of the efforts of John Bartram and his third son, William. Keeping journals during their extensive travels throughout the southeastern United States, the Bartrams gave many American and European readers their first understanding of the details of American nature. Most of their trips took place in the late 1700s; however, the writings that they published inspired the writers who followed during the 1800s.

Although the details of their exploration brought new understanding and appreciation to their readers, the real value of the Bartrams' accounts derived from the basic aesthetic appreciation of the nature with which the father and son approached the landscapes of North America. John is given credit, for instance, for being the first naturalist to use the term "sublime" to describe nature. Charged with the meaning of romantic nature, sublimity valued nature for completely nonutilitarian reasons. He also wrote about the great virtues of native peoples whom he contacted, particularly their relationship with and appreciation of natural surroundings. In each of these cases, the Bartrams countered the accepted approach of most of American culture (Nash 1982, 54–56). John's description of a mountain thunderstorm in Rabun County, Georgia, near the North Carolina border, provides a good example:

> It was now after noon; I approached a charming vale, amidst sublimely high forests, awful shades! darkness gathers around, far distant thunder rolls over the trembling hills; the black clouds with august majesty and power, moves slowly forwards, shading regions of towering hills, and threatening all the destructions of a thunderstorm; all around is now still as death, not a whisper is heard, but a total inactivity and silence seems to pervade the earth … the face of the earth is obscured by the deluges descending from the firmament, and I am deafened by the din of thunder; the tempestuous scene damps my spirits, and my horse sinks under me at the tremendous peals, as I hasten for the plain. (Bartram 1983)

Accounts such as this one formed the foundation of American Romanticism, which took shape as a multimedia effort in the early to mid-1800s. The first realm for the aesthetic appreciation of nature was on the canvas of oil painting.

Conclusion: A National Commitment to Natural History

Inviting the nation to the opening of his museum, Charles Wilson Peale wrote the following:

> Mr. Peale respectfully informs the Public, that having formed a design to establish a MUSEUM, for a collection, arrangement and preservation of the objects of natural history and things useful and curious, in June 1785 … he began to collect subjects, and to preserve and arrange them in Linnaean method … the museum having advanced to be an object of attention to some individuals … he is there for the more earnestly set on enlarging the collection with a greater variety of birds, beasts, fishes, insects, reptiles, vegetables, minerals, shells, fossils, medals, old coins … with sentiments of gratitude, Mr. Peale thanks the friends of the Museum, who have beneficially

added to his collection a number of precious curiosities, from many parts of the world;—from Africa, from Indies, from China, from the Islands of the great Pacific Ocean, and from different parts of America. (Semonin 2000)

To establish his museum, Peale relied heavily on the help of his sons: Rubens, Franklin, Titian II, Rembrandt, and Raphaelle. Together, the Peales accepted donations of trophy animals shot all over the world from many Americans, including George Washington. Other American collectors donated insects, shells, and plants collected internationally. Finally, Lewis and Clark presented Peale with many specimens taken during their exploration of the American continent, including a prong-horned antelope (the only known specimen of its kind), Lewis's woodpecker, Clark's crow, a western tanager, and a large California condor.

By April 1799, Peale listed his holdings as including more than 100 quadrupeds, 700 birds, 150 amphibians, and thousands of insects, fishes, minerals, and fossils. Peale also began to collect and catalog various specimens of unknown creatures and biological oddities. One of his earliest unidentified specimens was a lizard from Louisiana presented by President Thomas Jefferson. There was also a Ripley's-Believe-It-Or-Not feel to specimens such as cows with additional heads and tails.

The commitment to this record of the nation's natural history represented an important watershed to the United States. Through it, the young nation discerned itself from every other nation through the celebration of its unique and bountiful natural resources.

Sources and Further Reading: Bartram, *Travels*; Nash, *Wilderness and the American Mind*; Opie, *Nature's Nation*; Sellers, *Mr. Peale's Museum*; Semonin, *American Monster: How the Nation's First Prehistoric Creature Became a Symbol of National Identity*; Steinberg, *Down to Earth*.

MODELING PUBLIC WORKS IN PHILADELPHIA

Time Period: 1700s to 1820
In This Corner: Private developers
In the Other Corner: Public works sympathizers
Other Interested Parties: Residents of Philadelphia, political leaders
General Environmental Issue(s): City planning, water management

Creating safe and healthy communities may seem to be in everyone's best interest, but public works and city planning required enormous financial and civic organization. Without a central authority such as a king or emperor, American communities were not immediately able to organize such projects. Typically, problems forced American communities to find solutions. First, in so many ways, Philadelphia was a national leader in confronting problems of city life and also of creating solutions.

The sounds could not be heard in any other locale in America during the 1820s: mechanical pumps moving frothing water into different ponds and through pipes. The futuristic scene attracted onlookers from around the world and clearly defined Philadelphia as one of the world's most advanced cities, but this was just the beginning. Upstream from the Waterworks, the city had also set aside the land directly along some of the rivers that fed the city's water system. Converted into Fairmount Park, these open spaces offered some of the nation's first urban green spaces. Of course, this corridor was only made possible by the great technology of what Americans called the Waterworks.

Philadelphia's attention had been brought to its water supply by health concerns. The 1790s were known as the "plague decade" for Philadelphia, which was then the nation's largest city and capital. In 1793, between 10 and 15 percent of the city's population of 45,000 died from an outbreak of yellow fever. Similar epidemics preceded the 1793 outbreak and recurred annually throughout the 1790s. The panicked public searched for possible causes of the disease. The city's water supply quickly came to the attention of citizens as both a possible culprit and solution. A more effective mode for acquiring the city's water became a priority.

Most Philadelphians blamed general filth for the health problems and called for the city to be flushed. City officials regularly washed the streets with fresh water, but this was an expensive process that wasted the city's limited supply of fresh water. Like other major cities of the 1790s, Philadelphia derived its waters from a disorganized system of wells, cisterns, and springs. Worried about the issue of clean water, Benjamin Franklin's 1789 will bequeathed £1000 to the city to be invested in developing a new water system. Interested parties began studying neighboring waterways, particularly the Schuylkill River.

During the plague of 1793, a "Committee of Health" was added to the city's docket of commissions. Although this group of doctors and citizens could help solve immediate difficulties, they also became convinced that long-term solutions required the planning, funds, and authority of the city government. They hoped new technology would deliver better health to the city. In an era when the field of city planning had not yet been created, the proposal came from a Philadelphia resident who knew the city very well. Of course, Benjamin Henry Latrobe also happened to be the country's most famous architect and engineer. He wrote in 1798, "The great scheme of bringing the water of the Schuylkill to Philadelphia to supply the city is now become an object of immense importance, though it is at present neglected from a failure of funds. The evil, however, which it is intended collaterally to correct is so serious and such magnitude as to call loudly upon all who are inhabitants of Philadelphia for their utmost exertions to complete it" (University of Virginia, *The Diseased City*).

Latrobe's plans to create the Philadelphia Waterworks were fleshed out by other engineers in the early 1800s who extended the plans into what is now known as the Fairmount Park and Waterworks. The ingenuity of the water pumps was matched only by the splendor of the architecture.

In addition to mastering the engineering to provide the city with a safe and clean water supply, Latrobe believed that, if properly designed, the technology could provide a symbol for Americans. If wondrous enough, such mechanisms might supplant the natural wonders of the New World as symbols of the young American nation. In the early 1800s, the Philadelphia Waterworks was one of the first sites to demonstrate this symbolic possibility. Latrobe designed the Waterworks to function in a revolutionary way but also to appear as a symbol for a new age. When Leo Marx described the impulses of the machine and the garden, he also suggested the possibility of a "middle ground" in which these contradictory forces exist in balance. The Waterworks is one of the first examples of the middle ground that Marx described.

The technology to supply and purify Philadelphia's water placed this city ahead of nearly every city in the nation. However, in prioritizing the symbolic value of this accomplishment—one of the first public works in American history—the technology was carefully wrapped in an aura of classical art. The wheels of the primary work, for instance, were contained in a building resembling a Greek temple. By veneering the cutting-edge technology, the works linked the classical styles with the modern age of technology.

In contrast, the grounds only functioned to embellish the majestic appearance of the works. The heavily designed buildings contrasted with the rustic, natural surroundings of the brown-gray fieldstone, trees, and the flowing river. However, the designers did not attempt to blend the Waterworks in with their natural surroundings; the instruments of the Waterworks were intended to stand as monuments of a new era. Even so, the open land around the Waterworks still became part of Fairmount Park, the nation's first planned recreational natural environment. Even without the public grounds related to the Waterworks, Philadelphia already had a ready-made park landscape in the early 1800s. The banks of the Schuylkill River held some of the city's most stately aristocratic country homes. Modeled after the London area where aristocrats commuted to their estates along the Thames, Philadelphia's elite often spent summers at these river estates.

The evolution of the park was assisted by the river in two important ways. First, to escape the city's epidemics and summer heat, estates and villas had sprung up on the banks of the Schuylkill River. As suburbanization made the era of villas come to a close, these carefully manicured landscapes provided raw material that could become a park. In fact, unlike New York City's Central Park with carefully designed man-made landscapes, Fairmount Park evolved through the absorption of older estates without significant alterations.

The park took form during the following decades as grand estates were purchased by the city to keep the Schuylkill River, which fed the Waterworks, as clean as possible. The nation's first example of "watershed conservation" did not use either of these terms; however, the Fairmount Park, which ensured clean runoff into the city's water system, was officially founded in 1855 when the Lemon Hill estate was dedicated as a public park and renamed Fairmount Park. In late 1858, the city council's Committee on Public Property invited "Plans for the Improvement of Fair Mount Park." They sought to create a park corridor that connected the interior park to some of the newly acquired upriver estates. Initially including approximately 130 acres, the committee urged entries into the design competition that would create a unified park similar to Central Park under development in New York City.

With the field of landscape architecture still relatively undeveloped in the United States, eight different applicants submitted designs. The winner was the firm of James Clark Sidney, which proposed a few eye-catching elements, including a Grand Avenue and carriage drive, an open parade ground, and a terraced garden woven into a network of serpentine paths. The Sidney plan shared a basic approach with Olmsted's Central Park design in that the entire layout was intended to accentuate the contrast of the city's artificiality with the organic natural environment. By both Olmsted and Sidney, the park was intended to provide a buffer from the city and at all times to resist reminding visitors of the urban environment in which they lived. It was an escape. Therefore, each designer used underpasses and bridges to screen the major roads that, out of necessity, had to bisect the park.

Each designer also kept a clear regard for the preferences of the parks' primary users: upper-class leisure seekers. Sidney created a great entry for the park using a prominent carriage drive. This grand entry, Sidney intended, formed the transition from the geometric regularity of the city to the irregular landscape beyond. It commenced at the southern entrance to the park in a graceful sweeping curve and then straightened into a grand avenue ninety-six feet in width lined with American lindens; strips of grass separated the carriage drive, sixty feet in width, from the pedestrian paths on either side. This promenade led toward the river, where it terminated in a roundel, beyond which it turned into a meandering romantic drive along the river. In length, the Grand Avenue extended half a mile. Other carriage drives measured thirty to forty feet in width and meandered more picturesquely through the wilder sections of the park. Together, the whole park encompassed more than three miles of carriage drives and three miles of walks (The Fairmount Water Works Interpretive Center).

Ultimately, the park's design was completed by a young German architect, Hermann J. Schwarzmann, who was appointed park engineer in 1869.

Housing a Vision of America

In 1876, this landscape, which had been originally planned for reasons of health and urban renewal, became the site of the Centennial Exposition, a seminal event in the emergence of the United States. Philadelphia was the natural site for America's first entry into the culture of World's Fairs and Expositions. Centered on a vast machinery hall, which contained thirteen acres of new devices and gadgets, the Exposition presented a view of America's future in technology.

One of the most popular exhibits in the Machinery Hall was a prototype slice of the cable that Roebling Brothers would use for the Brooklyn Bridge. They would end up using 6.8 million pounds of these first galvanized cables, covered with zinc and with a strength of 160,000 pounds per square inch (double that of the iron wire used at Niagara). The Machinery Hall also featured other novelties, such as the first typewriter and a telephone. The real attraction, however, was the gigantic Corliss steam engine that powered all the

The main building of the 1876 U.S. Centennial Exposition held at Fairmount Park, Philadelphia, presented the technological wonders of the era to the world. Library of Congress.

machinery in the great hall, one of the world's first artificially lit environments. The 1,500-horsepower double Corliss steam engine connected to five miles of shafting used to move this power throughout the vast Machinery Hall (Hunter and Bryant 1991, 207–8).

Even before the surrounding park housed this great view into the nation's future in 1876, the Waterworks—this first great public works project—attracted tourists from all over the world. One tourist, Caroline Gilman, describes the Waterworks in her collection *The Poetry of Traveling in the United States*:

> I visited Fair Mount, and rejoiced like another Undine in its waterfalls and fountains, and felt how the river was like God's spirit, spreading somewhere at first in unattainable beauty, then carried through the dark channels of human life, seemingly lost until man inquires and strives for it, and then breaking out in new modifications, pouring its blessings on all who ask, and they are glad.
>
> I am grateful for beauty in all its forms. Had I been carried blindfolded to the machinery at Fair Mount, and then permitted to behold it alone, I should have been agreeably excited by its singular combination of simplicity and power; its wheels would have rolled on a while in my memory, I should have paid the usual tribute of wonder to man's ingenuity, and have dreamt of those iron arms that seem so human in their operations; but now that I have gazed on the placid river, marked the shaded green of its beautiful borders, seen the sculptured images awakening graceful associations, stood by the clear basin and felt a longing like a youth to rush in and stand under its showery fountain, heard the roar of the giant Art contending with and counteracting the giant Nature, climbed the precipitous eminence, and watched the setting sun throwing his golden smile on all, this leaves a deeper stamp—the stamp of the beautiful.... (Gilman 1838, 32)

Similar to the great construction projects of Ancient Rome, Latrobe meant for Fairmount to provide inspiration for Americans to seize a new era of using technology to elevate the standard of living in cities and towns. From the symbol of Fairmount, other American cities followed.

Sources and Further Reading: Gordon and Malone, *The Texture of Industry*; Hunter and Bryant, *A History of Industrial Power in the United States, 1780–1930. Vol 3: The Transmission of Power*; Lewis, *The First Design for Fairmount Park*, http://www.historycooperative.org/journals/pmh/130.3/lewis.html; The Fairmount Water Works Interpretive Center, http://www.fairmountwaterworks.com; Nash, *Wilderness and the American Mind*; Stilgoe, *Common Landscapes of America*.

INSTRUMENTALIZING THE RIVERS OF NEW ENGLAND

Time Period: 1700s to 1860
In This Corner: River developers
In the Other Corner: Other users of the rivers in question, states or regions with fewer rivers or less capital
Other Interested Parties: Laborers, political leaders
General Environmental Issue(s): Rivers, industry, energy

Many commodities of colonial New England were simple to place a value on. A tall white pine or smooth beaver pelt was worth a specific range of money to Europeans and others, and, therefore, its value could be assessed and considered rather specifically. In one of the most confounding instances, however, New Englanders made a commodity of a specific portion of the landscape: rivers. The uncertain legal and economic nature of rivers made them a consistent focus of debate and uncertainty in the 1700s to 1800s, when such rivers served as a primary energy source for industry.

U.S. riparian law took form during this time period; however, the most frequent response to the question "Who owns the river?" was whoever first controlled it. Most often this meant damming streams to divert it through waterwheels contained in an industrial establishment, but this, of course, meant that entrepreneurs or farmers downstream from the dam did not have the same options. This was one of the most important issues to be solved by riparian law in New England.

Throughout much of the 1700s, the American colonies had defined themselves as the suppliers of raw materials to the industry of Europe. In an era when rivers were the most popular prime mover for industrial use, England housed most of the knowledgeable businessmen. By the late 1700s, however, England had few remaining opportunities for river development. Many skilled engineers left to take on the undeveloped streams and rivers of New England in the United States. To protect its place as the world's industrial leader, England passed laws forbidding the export of machinery or the emigration of those who could operate it. Despite these laws, one of the world's first "brain drains" occurred when laborers in the British textile industry secretly emigrated to the United States.

Samuel Slater, who was born in England, became involved in the textile industry at fourteen years of age when he was apprenticed to Jedediah Strutt, a partner of Richard Arkwright and the owner of one of the first cotton mills in Belper. Slater spent eight years with Strutt before he rose to oversee Strutt's mill. In this management position, Slater gained a comprehensive understanding of Arkwright's machines.

Slater's business acumen was bold. He believed that the textile industry in England had peaked. With his expertise, however, Slater was restricted from leaving England. To make his move, Slater posed as a farm laborer to secretly emigrate to America in 1789. Slater was the first immigrant to arrive with such expertise in building and operating textile machines. With funding from Providence investors and the help of local skilled workers in Pawtucket, Slater opened the first successful water-powered textile mill in 1793. His mill was primarily staffed by women and children, aged seven to twelve years of age. The laborers worked with machines to spin yarn, which local weavers then turned into cloth. To attract poorer families to work in the mills, Slater added housing. Eventually, as Slater established company stores and paid the workers in credit that could only be used at these stores, Slater's mills grew into small industrial towns.

The millwrights and textile workers who trained under Slater contributed to the rapid proliferation of textile mills throughout New England in the early nineteenth century. His own mill grew to employ one hundred workers by 1800. However, his model of small, rural spinning mills became known as the "Rhode Island System" and truly set the tone for early industrialization in the United States. By 1810, sixty-one cotton mills in the United States used water power to turn more than 31,000 spindles. The manufacturing centers remained primarily in the Rhode Island and the Philadelphia regions. By the time other firms entered the industry, Slater's organizational methods had become the model for his successors in the Blackstone River Valley.

Although Rhode Island is viewed as the cradle of river development in the United States, the Merrimack River in Massachusetts possessed the raw power to surpass the Passaic and the Blackstone as an industrial center. Located just outside of Boston, the Merrimack became the next center of American industry when the businessman Francis Cabot Lowell used Slater's idea but exploded the scale to create an entire industrial community entirely organized around turning the power of the river into textile manufacture. The workable power loom and the integrated factory, in which all textile production steps take place under one roof, made Lowell the model for future American industry.

The city's brick mills and canal network were, however, signs of a new human domination of nature in America. Urban Lowell differed significantly from all of the farms and villages in which the vast majority of Americans lived in the early nineteenth century. Farming represented human efforts to work with and accommodate natural patterns, but Lowell followed more of a bulldozer approach: mill owners prospered by regimenting the natural world. To make the factory processes as profitable as possible, Lowell's factory owners imposed a regularity on the workday radically different from the normal agricultural routine, which followed seasons and sunlight. The typical mill ran an average of twelve hours per day, six days per week, totaling more than 300 days per year. Mill owners also used the light of whale-oil lamps to resist seasonal rhythms and to set their own schedule. Through their management, Lowell's mills might remain productive year round.

The power behind the factory began with the river. Simply damming the existing waterway did not create enough power to run the mills. Lowell's industrial life was sustained by naturally falling water that was engineered and repositioned by the town's designers. At Pawtucket Falls, just above the Merrimack's junction with the Concord, the river drops more than thirty feet in less than a mile. This continuous surge of kinetic energy provided the mills with more than 10,000 horsepower. Without the falls, Lowell's success would have been impossible, but nature was not perfect. In addition, Lowell relied on the construction of canals to better position the Merrimack's water for mill use. To increase efficiency, mill owners added dams and allowed ponds to form overnight for use the next day. Anticipating seasonal dry spells, planners turned the river's watershed into a giant millpond. In addition, they were aggressive in purchasing water rights in New Hampshire, storing water in lakes in the spring and releasing it into the Merrimack in the summer and fall (Steinberg 1991, 3–14).

The rise of Lowell in the second quarter of the nineteenth century became a national symbol of economic success. The poet John Greenleaf Whittier described Lowell as "a city springing up ... like the enchanted palaces of the Arabian Tales, as it were in a single night—stretching far and wide its chaos of brick masonry.... [The observer] feels himself ... thrust forward into a new century" (Steinberg 1991). As a symbol of modern industry, the city became an important attraction for Europeans touring the United States.

Conclusion: Reconstruing Rivers

In each case, the rights to the water in the river needed to be considered in a very different manner from any time previously. In an economy grown on the power of rivers, this potential energy was a commodity. The river often came to represent a community's greatest opportunity for economic success.

Sources and Further Reading: Gordon and Malone, *The Texture of Industry*; Hunter and Bryant, *A History of Industrial Power in the United States, 1780–1930. Vol 3: The Transmission of Power*; Steinberg, *Nature Incorporated: Industrialization and the Water of New England*.

THE LAND ORDINANCE OF 1785–1787 CONSTRUCTS THE AMERICAN GRID FOR LAND "DISPOSAL"

Time Period: Late 1700s

In This Corner: Jefferson, Democrats

In the Other Corner: Federalists, Republicans

Other Interested Parties: Native peoples, settlers, foreign nations, particularly France, Spain, Britain, and Russia

General Environmental Issue(s): Land dispersal, agriculture, cultural diffusion

How big was too big for the young United States? To many political leaders, size mattered: if the nation became too large, it would compromise its stability and security. Very likely, expanding the nation's borders, they argued, would be the nation's downfall.

Passed on May 20, 1785, and updated in 1787, the Ordinance for Ascertaining the Mode of Disposing Lands in the Western Territory, which later became known as the Northwest Ordinance, constructed the mechanisms to transfer vast amounts of lands into private ownership. Primarily, the disposal of western lands would be performed at public auction with a standard price of $1 per acre. This ordinance represented a victory for Thomas Jefferson and his dream of America as a nation of farmers.

Without a doubt, one of the most unique resources of the United States was vast amounts of "open" land. Covering immense territory, however, the efficient use of this land resource required centralized authority and centralized ideology. Although the American Revolution organized the nation around central ideas, it was Thomas Jefferson who assiduously wound land into the nation's ideology. Many critics, however, argued against Jefferson's plans. They argued that too expansive a nation could be the downfall of the United States.

During his political career, Jefferson had unparalleled opportunity to influence the policies of the young nation. During the Revolutionary War, Jefferson was a statesman and diplomat to France, securing that nation's financial and military support for the colonies. After the war, Jefferson served as the secretary of state to President George Washington, who was a fellow Virginia plantation owner. This common heritage, however, did not necessarily mean that Washington and Jefferson viewed the natural environment in the same fashion. In fact, Washington often favored the utilitarian ideas of his secretary of the treasury, Alexander Hamilton. Jefferson's career continued, however, and he was elected vice president in 1800 and then president in 1804 and 1808.

Jefferson's hopes for the nation grew from his wariness of a strong central government. He believed that individual owners working a plot of land provided the most stability for the future of the republic. He spoke and wrote often of the dangers that accompanied giving one group too much central control. Jefferson believed that the rapid and aggressive industrial growth favored by Hamilton and others would likely do just that. Jefferson put his trust in an agrarian republic filled with sturdy, independent farmers.

Before the ordinances of the 1780s, property had traditionally been cordoned off through the system known as "metes and bounds." Deeds and other documents in this system described the borders of property by using the details of the land, including trees, rocks, and streams. As the federal government debated the quickest way to dispense lands in the "Northwest" (the term used to describe the Ohio Valley and beyond), the metes and bounds system was criticized as imprecise and the root of many lawsuits, but indiscriminate location would not result in a coherent settlement progressively moving westward. Jefferson was determined to create a new system that would abolish feudal ideas such as hereditary rights and peasant tenancy that had been carried over from England. He first suggested that the land not be sold but given in long-term leases. When critics ignored his interest in leases, Jefferson followed the popular desire for low-cost purchase as an American right. Jefferson, of course, dreamed of a nation of yeoman farmers on small tracts of land (40–160 acres) who would make poverty practically nonexistent. Jefferson chaired the congressional committee to draft a distribution plan for the western lands.

The political debate over this new system resulted in a system of squares that is today described as the grid. The geographer of the United States supervised a surveyor from each state who used thirty-two-and-one-half-foot chains to make north and south measurements. Townships were reduced to thirty-six square miles, which were subdivided into squares of 640 acres numbered from one to thirty-six beginning at the southeast corner and proceeding left to right and then right to left. In 1786, surveyors began measuring on the north shore of the Ohio River. The sale of these squares began in the following year. The surveyors made detailed notes, which remain a valuable record of original vegetation along the compass lines. The vast majority of the continental United States was measured out using the grid system (Opie 1998, 103–5). The only exceptions to the grid system were those areas that had been settled before the national survey. Spanish and French areas along the Mississippi River used other systems. Similarly, coastal colonies also used separate standards for property measurement.

The grid system that makes up the classic American rural landscape spread out residents on farms that typically ranged between 40 and 160 acres. The use of squares inadvertently separated homes by great distances compounding the isolation of rural life. The overall federal district would be separated into three to five territories. When the population reached 5,000, the territory could elect a legislature to share power with a council of five who had been chosen by the governor and Congress. A nonvoting delegate could then be sent to the U.S. Congress. To apply for statehood, however, territories needed to have a total population in excess of 60,000 (Conzen 1990, 136).

Sales of land in northwestern Ohio began in 1829. By 1900, 221,897 farms totaling nearly six million acres were carved out of the former Public Domain of Indiana (Conzen 1990, 136). The system of development grew from the land dispersal system: four schoolhouses per township, cemeteries, and small churches were dispersed at fairly equal intervals throughout the township. Whereas the population of the trans-Appalachian West was approximately 0.1 million in 1790, it had risen to 1.1 million in 1810, 3.7 million in 1830, and 9.9 million in 1850. Thirteen new states were created during the years between Kentucky and Vermont (1792) and Wisconsin (1848). Texas was annexed in 1845 (Williams

1992, 111). As historian John Opie wrote, the land ordinance "changed the entire national domain west of Pennsylvania and north of the Ohio River from a formless wilderness into a national geometry of gigantic squares and rectangles" (Opie 1998, 102).

With the acceptance of the ordinance and the grid that it placed on the landscape, America had the raw material for growth. Called the agrarian ideal on other mythological terms, the agricultural policies of the United States during the 1800s prioritized settlement on western lands.

Sources and Further Reading: Conzen, *The Making of the American Landscape*; Opie, *Nature's Nation*.

NATIVE RESISTANCE TO EUROPEAN SETTLEMENT

Time Period: 1700s to 1800s
In This Corner: American politicians, settlers, military leaders
In the Other Corner: Native leaders, some Christian leaders
Other Interested Parties: Western settlers
General Environmental Issue(s): Native rights, land use

Americans faced many difficulties in expanding and settling the United States. Challenges related to climate and region tested each settler. In addition, Americans needed to face up to the challenge that people already lived in many of the unsettled regions. Moving native peoples could have been purely a military undertaking of the U.S. government. Although the cavalry was one tool in this displacement process, American ideals demanded a different model of expansion. The difficulty came in establishing what exactly that model should entail. Through trial and error, federal efforts to make way for settlers created a legacy of inconsistent and unfair policies and treaties.

Although many American politicians in the early 1800s agreed that they wanted to possess all of the North American continent, few believed a nation priding itself on the founding principle of individual freedom and democracy should simply force native peoples from the land of their heritage. When public sentiment fueled a national policy of displacement by force during the 1800s, American politicians faced an unforeseen problem: resistance among native people. Although many Americans of the nineteenth century gave little credit to the intellectual capabilities of native peoples, there is a clear record of organized efforts to resist displacement among many tribes and peoples.

Historians argue that this movement began in 1799 when a member of the Seneca named Handsome Lake lay near death. In his semiconscious state, Handsome Lake had a vision in which the Creator directed him to awaken a new religion and way of life for the Seneca. He developed "the Longhouse Religion," which combined traditional beliefs with Christian additions. Although Handsome Lake proposed many changes in traditional Indian ways, his vision provided hope for many Seneca. In 1805, the Shawnee Prophet Tenskwatawa also began preaching his vision of a new life. His vision, however, reached beyond Shawnee culture to influence a half dozen additional tribes. Followers joined him in Prophetstown on the Tippecanoe River in Indiana. Another Shawnee, Tecumseh, traveled throughout the American South preaching the need for collective native resistance to the American efforts.

Such efforts were among the first by native groups to overcome tribal distinction and to unite to negotiate and combat the American onslaught (Calloway 1999, 217–20).

The activities of these native leaders during the first decade of the 1800s created serious concern among American leaders. Most American leaders saw the nation's future in the expansion westward. The "Indian problem" stood in the way of this vision of the nation's future. After the War of 1812, Tennessean Andrew Jackson led a movement to implement Thomas Jefferson's idea for how to settle the "problem": moving the tribes west of the Mississippi River. In 1830, President Jackson approved the *Indian Removal Act*, which required that all tribes be moved west. Native resistance resulted in a series of wars, including the Black Hawk War (1832), the Creek War (1835–1836), and the Second Seminole War (1835–1842). Overall, the federal forces crushed the native resistance. Between 1831 and 1839, members of the Five Civilized Tribes (Choctaw, Chickasaw, Creek, Seminole, and Cherokee) exchanged their lands for an area between Missouri and Arkansas known as Indian Territory.

The removal process of these tribes was characterized by the brutal removal of the Cherokee, which became known as the "Trail of Tears." When 14,000 Cherokee walked from Georgia to Oklahoma in the early 1830s, 4,000 died from dysentery and starvation. By the 1840s, only the Western and Great Plains tribes remained beyond the control of the federal government. In 1850, Commissioner of Indian Affairs Luke Lea proposed a system of reservations that would place all tribes on specific areas of land where they could be controlled, educated, Christianized, and taught to farm. The captive people would be deprived of their culture and made into Americans at a series of schools, including the Carlisle School in Carlisle, Pennsylvania.

A burlesque parade led by Andrew Jackson, this cartoon satirizes various aspects of his administration. Inside the cage a forlorn Indian sings "Home! Sweet home!" This no doubt refers to Jackson's controversial Indian resettlement program, whereby thousands of Cherokees, Seminoles and other natives of the eastern United States were uprooted and moved to less desirable lands farther west. Library of Congress.

A crucial part of making western territories settled and inhabited for American settlers was to rid them of the animals and other extreme influences. By the early 1800s, it was clear that, to ensure westward expansion, American leaders had to interpret native peoples as one of these influences that had to be either tamed or exterminated. For the native groups who survived, they resided on reservations, often far from home. The sociological and cultural consequences were severe.

Sources and Further Reading: Calloway, *First Peoples*; Conzen, *The Making of the American Landscape*; Krech, *The Ecological Indian: Myth and History*; McNeil, *Something New Under the Sun: An Environmental History of the Twentieth-Century World*; Miller, *The Transcendentalists: An Anthology*; Opie, *Nature's Nation*.

INDIAN EXPROPRIATION AND THE TRAIL OF TEARS

Time Period: Early 1800s
In This Corner: President Andrew Jackson
In the Other Corner: U.S. Supreme Court, native peoples
Other Interested Parties: Western settlers
General Environmental Issue(s): Indian policy, land use

The advance westward necessitated that the current residents of the region not remain. The U.S. government attempted to reconcile its founding principles with the need to conquer the land occupied by native peoples and to remove them. This led to nearly a century of debate, a retraction of existing policy initiatives, and the adoption of new approaches. Uncertainty and inconsistency dominated native people–United States relations for decades in the mid-1800s, but it began with the initiation of the "removal" policy adopted by some American leaders in the 1830s. Ill feelings and visceral debate over this federal policy divided politicians as well as entire branches of the federal government.

Initially, the rapidly growing United States expanded into areas of the lower South in the first decades of the 1800s. Already home to the Cherokee, Creek, Choctaw, Chicasaw, and Seminole nations, most Americans perceived these people as figuratively and literally blocking their progress to converting the land for large-scale production of cotton.

Tennessean Andrew Jackson led the American army as the conflict became violent. In 1814, he commanded the U.S. military forces to victory over one portion of the Creek nation. In their defeat, the Creeks lost twenty-two million acres of land in southern Georgia and central Alabama. The United States acquired more land in 1818 when, spurred in part by the motivation to punish the Seminoles for their practice of harboring fugitive slaves, Jackson's troops invaded Spanish Florida.

The U.S. willingness to displace the native peoples by force was an intimidating act. From 1814 to 1824, eleven treaties by the United States forced the southern tribes of the East to give up their land in exchange for lands in the west. Of these, Jackson played an important role in negotiating nine. As a result of the treaties, the United States gained control over three-quarters of Alabama and Florida, as well as parts of Georgia, Tennessee, Mississippi, Kentucky, and North Carolina. This was a period of voluntary Indian migration, however, and only a small number of Creeks, Cherokee, and Choctaws actually moved to the new lands. In 1823, the Supreme Court handed down a decision that stated that Indians could occupy lands within the United States but could not hold title to those lands. In response to the great threat this posed, the Creeks, Cherokee, and Chicasaw instituted policies of restricting land sales to the government. They wanted to protect what remained of their land before it was too late.

Nonviolent attempts at resistance had already been used by some of the five nations. In one example, traditional cultures altered themselves to adopt Anglo-American styles of farming, education, and language. Some had even taken up slave holding. The federal government recognized these efforts and designated them the "Five Civilized Tribes." With these groups, the United States adopted a policy of assimilation.

Of the Civilized Tribes, however, the most unique case of resistance was the Cherokee. Learning about the Americans' language and legal system, they set out to file suit to protect their interests from land-hungry white settlers, who continually harassed them by stealing their livestock, burning their towns, and squatting on their land. To clarify their interests in advance of the suit, in 1827, the Cherokee adopted a written constitution declaring their status as a sovereign nation. The Cherokee hoped to use this status to aid in their negotiation; however, the state of Georgia did not recognize their sovereign status. Instead, the state perceived the Cherokee as tenants living on state land. The Cherokee lost their case in the U.S. Supreme Court.

Filing a new suit in 1831, the Cherokee based their appeal on an 1830 Georgia law that prohibited whites from living on Indian territory after March 31, 1831, without a license from the state. The state legislature wrote this law to justify removing white missionaries who were helping the Indians resist removal. The Court this time decided in favor of the Cherokee. The Supreme Court's decision stated that the Cherokee had the right to self-government and declared Georgia's extension of state law over them to be unconstitutional.

Revealing the overriding cultural view of the era, the state simply refused to abide by the Court decision. Having made his name in wars such as those in the South, Andrew Jackson now led the entire nation as president. He refused to enforce the law. Next, in 1830, Jackson pushed through the "Indian Removal Act" that granted the president power to negotiate removal treaties with Indian tribes living east of the Mississippi. Any tribes that refused would become citizens of their home state. When nations resisted leaving their land voluntarily, Jackson forced them to leave. The act reads as follows:

> Be it enacted by the Senate and House of Representatives of the United States of America, in Congress assembled, That it shall and may be lawful for the President of the United States to cause so much of any territory belonging to the United States, west of the river Mississippi, not included in any state or organized territory, and to which the Indian title has been extinguished, as he may judge necessary, to be divided into a suitable number of districts, for the reception of such tribes or nations of Indians as may choose to exchange the lands where they now reside, and remove there; and to cause each of said districts to be so described by natural or artificial marks, as to be easily distinguished from every other.

> SEC. 2. And be it further enacted, That it shall and may be lawful for the President to exchange any or all of such districts, so to be laid off and described, with any tribe or nation within the limits of any of the states or territories, and with which the United States have existing treaties, for the whole or any part or portion of the

territory claimed and occupied by such tribe or nation, within the bounds of any one or more of the states or territories, where the land claimed and occupied by the Indians, is owned by the United States, or the United States are bound to the state within which it lies to extinguish the Indian claim thereto.

SEC. 3. And be it further enacted, That in the making of any such exchange or exchanges, it shall and may be lawful for the President solemnly to assure the tribe or nation with which the exchange is made, that the United States will forever secure and guaranty to them, and their heirs or successors, the country so exchanged with them; and if they prefer it, that the United States will cause a patent or grant to be made and executed to them for the same: Provided always, That such lands shall revert to the United States, if the Indians become extinct, or abandon the same.

SEC. 4. And be it further enacted, That if, upon any of the lands now occupied by the Indians, and to be exchanged for, there should be such improvements as add value to the land claimed by any individual or individuals of such tribes or nations, it shall and may be lawful for the President to cause such value to be ascertained by appraisement or otherwise, and to cause such ascertained value to be paid to the person or persons rightfully claiming such improvements. And upon the payment of such valuation, the improvements so valued and paid for, shall pass to the United States, and possession shall not afterwards be permitted to any of the same tribe....

SEC. 7. And be it further enacted, That it shall and may be lawful for the President to have the same superintendence and care over any tribe or nation in the country to which they may remove, as contemplated by this act, that he is now authorized to have over them at their present places of residence. (Calloway)

Jackson knew the culture of many native groups quite well by this point. However, similar to many Americans, he viewed them in a manner that was clearly paternalistic and patronizing. In this point of view, Indians were less developed, almost like children in need of assistance. Jackson and many other Americans expected that the United States would never extend beyond the Mississippi; therefore, disposing of the Indians there would permanently rid the United States of its Indian problem.

Although the United States was barely fifty years old, activists were ready to voice outrage if its founding ideals were interpreted to be violated. Legal challenges followed as humanitarian groups took up the Native Americans' cause. Ultimately, two cases—*Worcester v. Georgia* (1832) and *Cherokee Nation v. Georgia* (1831)—challenged the constitutionality of the Removal Act. Despite these challenges, President Jackson pressed forward with his personal plan for American expansion.

With this policy at least temporarily in place, Jackson defiantly enforced it. The U.S. government used the Treaty of New Echota in 1835 to justify the removal. The treaty, signed by about one hundred Cherokees and known as the Treaty Party, relinquished all lands east of the Mississippi River in exchange for land in Indian Territory and the promise of money, livestock, and various provisions and tools.

Their protests did not save the southeastern nations from removal, however. The Choctaws were the first to sign a removal treaty, which they did in September 1830. Some chose to stay in Mississippi under the terms of the Removal Act, but, although the War

Department made some attempts to protect those who stayed, it was no match for the land-hungry whites who squatted on Choctaw territory or cheated them out of their holdings. Soon, most of the remaining Choctaws, weary of mistreatment, sold their land and moved west.

Under orders from President Jackson, the U.S. Army began enforcement of the Removal Act. Around 3,000 Cherokees were rounded up in the summer of 1838 and loaded onto boats that traveled the Tennessee, Ohio, Mississippi, and Arkansas Rivers into Indian Territory. Many were held in prison camps awaiting their fate. In the winter of 1838–1839, 14,000 Native Americans were marched 1,200 miles through Tennessee, Kentucky, Illinois, Missouri, and Arkansas into rugged Indian Territory. An estimated 4,000 died from hunger, exposure, and disease. The journey became an eternal memory as the "trail where they cried" for the Cherokees and other removed tribes. Today, it is remembered as the Trail of Tears (Calloway 1999, 215).

The U.S. government struggled for the next twenty-eight years to force relocation of the southeastern nations. Although a group of Seminoles signed a removal treaty in 1833, most refused to leave. The resulting struggle was the Second Seminole War, which lasted from 1835 to 1842. Scholars estimate that this war cost the federal government between $40 and $60 million. The remaining Seminole fought again in the Third Seminole War (1855–1858), when the U.S. military attempted to drive them out. Finally, the United States paid the remaining Seminoles to move west.

The Creeks also refused to emigrate. By 1836, the secretary of war ordered the removal of the 15,000 Creeks as a military necessity. The Chickasaws signed a treaty in 1832 stating that the federal government would provide them with suitable western land. They migrated there in the winter of 1837–1838.

Finally, the Cherokee may have had the most difficult experience. In 1833, a small group agreed to the Treaty of New Echota. The leaders of this treaty group were not the recognized leaders of the Cherokee nation; Chief John Ross and 15,000 Cherokees signed a petition in protest of this false treaty. The Supreme Court ignored their demands and ratified the treaty in 1836. The Cherokee were given two years to migrate voluntarily, at the end of which time they would be forcibly removed.

By 1838 only 2,000 had migrated; 16,000 remained on their land. The forced removal became known as the infamous "Trail of Tears." The U.S. government sent in 7,000 troops, who forced the Cherokees into stockades like cattle. The troops then led the forced march to the western lands. Scholars estimate that 4,000 Cherokee people died of cold, hunger, and disease during the Trail of Tears. On October 1, 1838, one observer of the journey gave her recollection:

> At noon all was in readiness for moving. The teams were stretched out in a line along the road through a heavy forest, groups of persons formed about each wagon. The day was bright and beautiful, but a gloomy thoughtfulness was depicted in the lineaments of every face. In all the bustle of preparation there was a silence and stillness of the voice that betrayed the sadness of the hearts. At length the word was given to move on. Going Snake, an aged and respected chief whose head eighty summers had whitened, mounted on his favorite pony and led the way in silence, followed by a number of younger men on horseback. At this very moment a low sound of distant thunder

fell upon my ear ... a voice of divine indignation for the wrong of my poor and unhappy countrymen, driven by brutal power from all they loved and cherished in the land of their fathers to gratify the cravings of avarice. The sun was unclouded—no rain fell—the thunder rolled away and seemed hushed in the distance. (Cerritos College, http://www.cerritos.edu/soliver/American%20Identities/Trail%20of%20Tears/quotes.htm)

As a nation, the United States had clearly chosen to view the native inhabitants of North America as a portion of nature that could and should be eliminated. During the 1840s, many settlers settled on Native American land. The native groups of the west were, for the most part, beyond the control of the federal government until after the Civil War. This was particularly true of the nomadic tribes of the Great Plains.

In 1850, Commissioner of Indian Affairs Luke Lea proposed a system of reservations that placed all tribes on reservations where they could be controlled and educated. Part of this education was an effort to Christianize the Indians and to teach them to farm in the European model. The reservations would be a permanent area in which native peoples could live in stationary homes and own their own land. The organized, federal expulsion of natives from lands east of the Mississippi continued until after the Civil War.

Conclusion

The U.S. government had no easy answer to the Indian problem once it had established the national priority to move westward and to fill the land with Anglo settlers. Proponents argued that the future of the nation depended on the displacement of native peoples. Even so, from a humanitarian standpoint, the Trail of Tears remains one of the darkest portions of American history.

Sources and Further Reading: Calloway, *First Peoples*; Conzen, *The Making of the American Landscape*; Krech, *The Ecological Indian: Myth and History*; McNeil, *Something New Under the Sun: An Environmental History of the Twentieth-Century World*; Miller, *The Transcendentalists: An Anthology*; Opie, *Nature's Nation*; White, *It's Your Misfortune and None of My Own*.

WAS A WHALE WORTH THE EFFORT TO NANTUCKETEERS?

Time Period: Late 1700s
In This Corner: No whale conservation groups in 1700s, native peoples
In the Other Corner: Businessmen of New England
Other Interested Parties: Illumination consumers worldwide
General Environmental Issue(s): Species management

The bounty of the sea produced some of America's first full-blown industrial endeavors in New England. When early settlers commented on the plentiful whales close to shore along the New England coast, fishermen from Long Island northward began using the technology of Basque sailors and others to pursue, kill, and bring whales to shore. From these kills, some coastal towns were producing whale oil by the early 1700s. This very limited hunt, however, was soon to give way to a new era off Nantucket Island. In the limited supply of whales, we find a superb example of early industrialization in the United States. This era was defined by reliance on human labor for seemingly small gains. For many observers, the danger of the whale hunt was not worth the products acquired from the largest mammal.

Before the European habitation of New England, native peoples residing in coastal villages closely followed the migrations of humpback and right whales. They told stories of butchering beached whales and even intentionally beaching whales found near shore. When European colonists established fishing communities along the northeastern coast of North America, some residents also struck out after the plentiful cetaceans. Between 1645 and 1655, the settlers of Southampton, Long Island, fitted boats for whaling along shore and employed native peoples, paying them in shares of the oil gotten. Other sites rapidly followed, including Martha's Vineyard in 1652, Cape Cod before 1670, and Nantucket in 1672.

In Nantucket during the 1670s, James Loper became the first-known individual to gain full-time employment as a whaler. As an inducement for performing the hunt for two years, the Quaker community granted Loper ten acres of land. Despite the concentrated effort of Loper and others, the Nantucket whale fishery still faltered as a result of a lack of sailors and a lack of interest among Quakers, who were disinterested in becoming involved in external markets. The absence of able-bodied sailors became particularly acute when the Wampanoags, who were native to the island, became less and less inclined to subject themselves to the danger of the whale hunt. Around 1690, Nantucket inhabitants learned that Cape Cod sailors had become very proficient at taking the whales and collecting oil. The Nantucketers hired Ichabod Paddock, one of the successful Cape Cod whalers, to come to Nantucket and instruct them. The hunt grew rapidly, and, in 1726, Nantucket reported eighty-six whale captures.

A variety of characteristics distinguished the Nantucket whale fishery from its predecessors. Clearly, however, the primary distinction in the new fishery was its prey. In 1712, the boat of Christopher Hussey, a Nantucket whaler, blew far from shore in a gale. In the deeper waters of the Atlantic, Hussey took the first known sperm whale and brought it back to Nantucket. The sperm whale oil was superior to standard whale oil, but its pursuit required refitting the Nantucket fleet to travel greater distances and to stay longer at sea. By 1715, Nantucket had six small sloops of thirty to forty tons, each plighing the waters for sperm whale.

The general process of American whaling soon took shape. Large sailing vessels were soon outfitted with six small whale boats. From the main mast of the sailing ship, the lookout called "There she blows!" on spotting a whale's blow or spout. Six sailors piled into each small boat and paddled after the whale. If they were able to kill it with harpoons and knives, they would then tow it back to the sailing ship. Once lashed along the side of the main ship, the whale's blubber was peeled off and then boiled in the ship's try works. Once the oil was separated off, it was placed in barrels below deck.

From this point until the American Revolution, the American whale fishery grew rapidly in size and number of ships. Most impressive, Quakers introduced technologies to make deep-water whaling more practical. Although seaports developed all along the New England coast, Nantucket remained the center of American whaling. In 1730, the Nantucket fleet of twenty-five vessels (of thirty-eight to fifty tons each) brought in 3,700 barrels of oil; in 1775, the fleet of 150 vessels (of 90–180 tons each) brought in 30,000 barrels of oil. From a curious adventure, whaling had rapidly become a pursuit that involved the rendering of blubber and pockets of oil and spermaceti into candles and oil for illumination. A few merchants developed small-scale sites on the island to process the oil, but most of the oil was shipped to foreign markets.

After the Revolutionary War, Britain realized that the world was being fed whale products now being harvested almost exclusively by its former subjects on Nantucket. Britain made extensive efforts to recruit the whalemen to become British subjects and operate from outposts in Halifax and Nova Scotia. France, in retaliation, made a similar effort to recruit the whalemen from Nantucket in 1785, establishing an outpost for them in Dunkirk. In the United States, John Adams and Thomas Jefferson noted the industry's critical importance to the United States at the close of the eighteenth century. As Jefferson considered the nation's trade difficulties in 1789, he ranked the importance of whale oil as second only to tobacco. Proving him correct, the ensuing decades marked a bidding war for the fishery and those on Nantucket Island who had perfected the pursuit.

The necessity of such negotiation demonstrates the global importance of the commodity in the late eighteenth century, but it also suggests an ongoing source of conflict within the industry: the supply of whales could not magically increase without new technologies. Whales exist as a common resource with neither the control nor reliability of an owned or land-based resource. No matter how adept Americans got at catching whales, they would never provide enough energy to meet the needs of the growing nation.

Sources and Further Reading: Black, "Organic Planning: Ecology and Design in the Landscape of TVA"; Creighton, *Rights and Passages*; Gordon and Malone, *The Texture of Industry*; Hohman, *The American Whaleman*.

DEFINING A UNIQUE SOUTHERN STYLE OF AGRICULTURE

Time Period: 1670s to 1850
In This Corner: Landed gentry of the South, political leaders
In the Other Corner: African laborers
Other Interested Parties: Americans living in the North, France, Britain, Spain
General Environmental Issue(s): Regionalism, agriculture, the South

There is no doubt that the use of slavery plantations composes a tragic portion of the American story; however, the use of slaves composed one portion of an economic and ecological system. Southern economic development followed a vision of land use and productivity that was uniquely honed to the region's climate and soil. By understanding such an entity as a mechanism within a larger agricultural system, we better comprehend its complexity and also the ongoing debate over the American future for which it was the single best example.

Historians have pieced together many descriptions of the processes of agriculture on the plantation, but one of the best sources remains Frederick Douglass's description contained in *My Bondage and My Freedom* when he wrote the following:

> Public opinion in such a quarter, the reader must see, was not likely to be very efficient in protecting the slave from cruelty. To be a restraint upon abuses of this nature, opinion must emanate from humane and virtuous communities, and to no such

opinion or influence was Col. Lloyd's plantation exposed. It was a little nation by itself, having its own language, its own rules, regulations, and customs. The troubles and controversies arising here were not settled by the civil power of the State. The overseer was the important dignitary…. There were, of course, no conflicting rights of property, for all the people were the property of one man, and they could themselves own no property….

Old master's house, a long brick building, plain but substantial, was centrally located, and was an independent establishment. Besides these houses there were barns, stables, store-houses, tobacco-houses, blacksmith shops, wheelwright shops, cooper shops; but above all there stood the grandest building my young eyes had ever beheld, called by every one on the plantation the great house. This was occupied by Col. Lloyd and his family. It was surrounded by numerous and variously-shaped out-buildings. There were kitchens, wash-houses, dairies, summer-houses, green-houses, hen-houses, turkey-houses, pigeon-houses, and arbors of many sizes and devices, all neatly painted or white-washed, interspersed with grand old trees, ornamental and primitive, which afforded delightful shade in summer and imparted to the scene a high degree of stately beauty. The great house itself was a large white wooden building with wings on three sides of it. In front, extending the entire length of the building and supported by a long range of columns, was a broad portico, which gave to the Colonel's home an air of great dignity and grandeur. It was a treat to my young and gradually opening mind to behold this elaborate exhibition of wealth, power and beauty. (Douglass 1987, 44–47)

Historians Theodore Steinberg and Jack Temple Kirby have suggested that the Civil War was precipitated by distinct land-use choices made early in the colonization of North America. By the late 1700s, debates over land use largely took the form of discourse over the impropriety of slavery. These debates, however, were moot; in fact, by the mid-1700s, southern agriculturalists had committed themselves to forms of agriculture that required extensive use of laborers. Plantation agriculture perpetuated some of the most troubling ideas of the era, but none more than the rationalization of forced servitude, otherwise known as slavery. Although these workers, of course, did not need to be slaves, southern states suffered from a shortage of available labor. By the early 1800s, the southern states were active participants in the worldwide trade in African slaves.

Atlantic System and Trade Network

In the logic of economic expansion that drove European powers abroad, New World resources became sources of military and political power at home. Efforts to collect and develop these resources allowed many leaders and merchants between 1400 and 1800 to suspend consideration of the rights of native peoples or of the human rights of laborers. By far, however, the most glaring example of this was the traders' ability to see humans, specifically Africans, as another commodity of value to be moved, sold, and collected.

Africans had been traded as slaves for centuries, reaching Europe via the Islamic-run, trans-Saharan trade routes. Between 1450 and the end of the nineteenth century, slaves were obtained from along the west coast of Africa with the full and active cooperation of African kings and merchants. In return, the African kings and merchants received various trade goods, including beads, cowrie shells (a type of money), textiles, brandy, horses, and guns.

By the mid-1400s, Portugal had a monopoly on the export of slaves from Africa. It is estimated that, during the four and a half centuries of the trans-Atlantic slave trade, Portugal was responsible for transporting more than 4.5 million Africans (approximately 40 percent of the total). However, during the eighteenth century when more than six million Africans were traded; Britain was the worst transgressor, estimated to have been responsible for trading almost 2.5 million Africans.

The system of trade that took shape in the Atlantic Ocean is often referred to as the "triangle trade." This term derives from a basic movement that currents and goods often followed. In this self-contained system, the eastward wind pattern, which blows on the southern part, came to be known as the "trade winds" because they enabled ships to cross the Atlantic. The westward wind pattern, blowing on the northern part, came to be known as the "westerlies."

Sailing ships were highly constrained by dominant wind patterns, and, therefore, the trading system followed this pattern. Manufactured commodities were exported from Europe to go in two directions: toward the African colonial centers and toward the American colonies. Slaves then left Africa bound primarily for Central and South American colonies (Brazil, West Indies). Tropical commodities (sugar, molasses) flowed from these colonies to the American colonies or to Europe. Within this Atlantic system, North America also exported tobacco, furs, indigo, and lumber to Europe.

Middle Passage

The leg of the trade system that carried Africans into slavery was often referred to as the "Middle Passage." European ships were loaded with groups of six people chained together with neck and foot shackles. On board, they were put below the decks, placed head to foot, still chained in long rows. Most Africans suffered from seasickness and additionally lost hydration caused by vomiting. Additionally, poor food and fearful conditions contributed to diarrhea. The ensuing conditions below deck, of course, led to the outbreak of disease, including typhoid fever, measles, yellow fever, and smallpox.

The unhealthy conditions were made worse by the common practice of overcrowding a ship to maximize profit. The longer the ship was at sea, the higher the slave mortality rate. Shorter voyages were expected to result in a 5–10 percent mortality rate; on longer voyages, however, traders expected between 30 and 50 percent of the slaves to perish.

The extreme human degradation of the Middle Passage exerted severe psychological shock on nearly every African. This was compounded by a common fear among the Africans that they had been taken by the Europeans to be eaten, to be made into oil or gunpowder, or that their blood was to be used to dye the red flags of Spanish ships. Of course, traders were instead attracted to the Africans' abilities as agricultural laborers and their adaptability to tropical climates.

Since the beginning of colonial settlement in the southern United States, agriculture was defined by an imbalance between population and land. Because of climate considerations and the preferences of some of the initial European settlements, southern planters focused on crops such as tobacco, rice, cotton, and sugar. Each of these crops required large tracts of

This nineteenth-century cotton plantation on the Mississippi River was one example of the prevailing agricultural method of the southern United States. Although this Currier and Ives image romanticizes the scene, one can see the diversity of activities that composed life on this more-than-a-farm. Library of Congress.

land as well as many laborers. For these reasons, many planters organized their land into plantations. These vast agricultural colonies presented a very different model of economic development and use of nature than the United States followed in other regions.

By definition, plantations are large agricultural estates cultivated by bonded or slave labor under central direction. After being used on the islands of the Caribbean, slavery on plantations was introduced into North America in the British colonies of Virginia, the Carolinas, and Georgia. By the late seventeenth century, slavery was firmly established in Virginia and the Carolinas. The plantation was not the only agricultural model of the South, but it was widespread enough to cause many to identify it with the region. Geographer Sam Hilliard describes six essential elements to a plantation: it should generally have more than 250 acres of property, a distinct division of labor and management, specialized production of one or two products or monoculture, a location in the South with a tradition of plantation agriculture, distinctive spatial organization to reflect centralized control, and particularly intensive use of human labor (Conzen 1990, 106).

Although most farmers in the antebellum South did not own slaves, those who did dominated agricultural production. Planters who owned slaves also possessed power not just to dominate other human beings and profit from their labor but also over the difficult environment. Slavery and exploitation of the environment went hand in hand.

One's mental picture of the plantation economy that dominated the antebellum South does not do justice to the diverse agriculture that dominated the subtropical climates of North America. Southern life and trade were largely organized by rivers. From the original tobacco sites along the James River in Virginia, growers of "the noxious weed" spread to the north, south, and west. Even as tobacco growers moved inland to the Piedmont Mountains, most plantations were located on or near rivers. By 1800, however, growers had expanded so far inland that about three-quarters of all tobacco was sold first within the United States before the remainder was sold at ports to companies that would ship to England (Conzen 1990, 110).

Historian Lewis Cecil Gray wrote that many southerners "bought land as they might buy a wagon—with the expectation of wearing it out" (Steinberg 2002, 74). Through their land-use practices and crop selection, many southerners mined the fertility out of the land. Typically, such farmers would then move farther inland to begin anew. Partly for this reason, the initial push westward came over the Appalachians from the Mid-Atlantic and South. Geographer Terry Jordan refers to this early push as the "backwoods frontier," which was a settlement process that also established a fairly consistent cultural pattern (Jordan and Kaups 1986, 2–3).

The infrastructure of the young nation did not function equally. Regional distinctions became apparent during the 1800s. Primarily, whereas the Northeast and Midwest moved more deeply toward industrialization, the South resisted infrastructural development and increased the use of slaves to expand the growth of cotton through the region and to the Mississippi River. Although the scale and scope of southern agriculture grew massively, rivers remained the primary courses of trade. Historian Ted Steinberg wrote, "There is nothing the least bit natural about slave labor, but in the antebellum South, at least, it owed its rise to a climate that favored the growth of short-staple cotton. The development of the Cotton Belt rested on a set of climatic conditions; without them it is hard to imagine slavery taking on the role that it did in southern political culture" (Steinberg 2002, 87).

Despite a similar westward migration, the development of cities and towns in the South took a different path than that of the North and West. As the Cotton Belt grew in the interior South, new towns were created and the population increased in existing towns. This growth, however, was slower than corresponding urban expansion in the North. A clear differentiation evolved between North and South in the nineteenth century. The influence of the planter culture on the growth of cities in the South and on the types of cities that did emerge are evident in both the social and economic spheres.

In fact, however, plantations and not the city were seen as the location of opportunity. Early on, in Virginia and the Piedmont, the plantations had little need of town services, because the excellent river transportation facilitated trade between plantations and England with relative ease. Later, when American ports became primary destinations of planter production, each plantation formed a fairly self-contained unit of operation and life, separated from other plantations, often more closely tied to the port of destination of its crop than to local towns. Plantations generally produced many of the items that in other areas of the country were available only in towns.

Land Conversion

In the American South, the logic of capital came to the tidal South and demanded changes to the naturally occurring ecosystems. The tool to perform this change, of course, was slave labor. Slaves leveled the ground and built embankments around squared-off areas of land. Typically, these squares were one-quarter acre. Once the ground was leveled, slaves constructed trunk lines to carry the water from the river to the fields. The hydraulic machine was organized and carried out through a system of labor called the task system.

Whereas slaves on tobacco and cotton plantations worked in gangs to complete specific jobs, rice workers were given a task each day. Typically, one slave would have responsibility for performing whatever task was needed on a specific quarter-acre plot. Once that task was completed, the rice worker was free to do as he wanted. Often, the rice plantation required less oversight than other plantations. Tasks such as digging were easy to measure.

In the Tidewater region, this system of rice cultivation gave way to tidal irrigation during the early 1800s. Partly through the know-how of West African slaves, the rice landscape was reengineered to make use of the ebb and flow of ocean tides for flooding and draining rice fields. This tidal movement killed weeds, which previously needed to be hoed out by slaves. In addition, it also lessened the amount of hoeing of any type that slaves needed to perform. Whereas inland rice plantations produced approximately 600–1,000 pounds of rice per acre, yields on tidal plantations reached 1,200–1,500 pounds (Stewart, 1996).

Different plantations were designed and orchestrated to produce a variety of crops. In each case, the physical organization of the structures varied. One of the primary reasons for this variation was the different tasks required to produce and harvest each crop.

Tobacco

To a much greater extent than in the North or Mid-Atlantic, southern farmers prioritized profit-making cash crops. The entire settlement of the South was predicated on the tropical climate and the unique crops that could be grown there and sold in Europe. Tobacco grown in Maryland, Virginia, and the Carolinas propelled settlement of new lands farther inland. Known as "the noxious weed," tobacco dictated the layout of farms as well as the political and economic considerations of its growers. The tobacco landscape created a permanent imprint on southern states and sustained the system of slavery for two centuries.

In the Chesapeake Bay region, one found growers producing the Sweet Scented variety of tobacco. In Virginia, growers focused on the Oronoko variety. In either case, seeds were sown first in beds. After rain in May, the plants were moved to hills and set approximately four feet apart in a cleared field. From this point forward, the tobacco fields needed to be consistently weeded. Next, the top of the plant needed to be cut, allowing approximately a dozen large-sized leaves to remain. In August, the ripe plants were cut low to the ground and hung on pegs in a ventilated drying house. They remained there for five to six weeks, after which the leaves were cut from the plant and tied into bundles.

With little labor available, early planters grew only a few acres of tobacco. African laborers began arriving in Jamestown in 1620. Identified as a primary source for the labor needed to expand southern agriculture, African laborers grew in number. Their status also shifted from indentured servants to slaves. By 1790, there were 660,000 southern slaves. From the original tobacco regions along Virginia's James River, tobacco cultivation expanded. Between 1800 and 1860, tobacco culture spread throughout Virginia, Maryland, and North Carolina.

In Virginia, the typical plantation ranged in size from 150 to 250 acres. Many tobacco plantations possessed a big house that was less grand than those found on some of the more profitable plantations growing rice and cotton. Beyond the manor house, however, the

Virginia plantation comprised many outbuildings, including a dairy, a cellar, a stable, a hen-house, a kitchen, slave cabins, possibly a schoolhouse, and other smaller homes for white workers and the overseer.

Rice

The coastal regions of South Carolina and Georgia were settled by financiers from Britain with a specific crop in mind: rice. In no other agricultural region of the South was the staple crop as intensely and singularly produced. In the Georgia-Carolina rice coast, settlers managed the natural environment in an ingenious effort to reform it as an ideal environment for rice production. Throughout the antebellum period, the southern Atlantic coast supplied nearly 90 percent of the national rice production.

Rice plantations were located on river floodplains, which aided irrigation. Plantations were most prevalent on rivers that were influenced by the Atlantic tides, including the Savannah, Ogeechee, Altamaha, Satilla, PeeDee, Waccamaw, Cooper, and Combahee. Tidewater planting became popular during the 1800s. Unlike inland swamp cultivation, tidewater planting used freshwater directly from the rivers rather than water reservoirs. The power of the tides lifted the level of the streams high enough for their water to spread through the fields. If the salinity of the water rose too high, the area was no longer good for rice cultivation.

The most successful rice plantations had large workforces. Clearing the land and constructing the streambeds required technical knowledge as well as many laborers. Because of the capital required to set up a rice plantation, most rice growers concentrated on producing only this staple crop. Seeds were planted in the spring, and the fields were flooded and drained throughout the growing season. In September, laborers harvested the rice by hand with a sickle. Most often, rice plantations employed significant numbers of slaves, who performed each stage of the labor.

Conclusion: The Problems of Making Cotton King of the South

The spread of cotton production in the early 1800s stemmed directly from the introduction of the cotton gin in the late 1700s. Cotton was grown from Virginia to Texas and to the north in Ohio, Illinois, and Missouri. The floodplains in the Deep South, however, in Alabama, Mississippi, and Louisiana, marked the highpoint of cotton culture. This area possessed the essential necessities for cotton growing: fertile land, convenient transportation routes mainly along the river, and massive amounts of available labor through the use of slaves.

Eli Whitney's invention of the cotton gin in 1793 was originally believed to have made slaves less necessary. The device eliminated the need for many slaves to pick through the cotton to remove seeds. Instead, the device ran the floss through brushes and combs. Instead of lessening the need for slaves, however, the gin enabled planters to grow short-staple cotton, which was more adaptable to diverse soils. This fueled massive expansion of cotton planting into the inland portions of the South.

The plantation landscape included a number of outbuildings. Among these, the cotton gin replaced the tobacco drying houses or the rice mill. Many cotton plantations spanned 1,000 acres or more. In addition to cotton fields, many plantations grew corn to feed livestock and residents. Other garden areas were tended by slaves, providing crops for their own use or sale. The sheds and barns were typically placed in the center of the property, with equal access to any portion of the land.

In addition to these difficulties, the emphasis of the plantation agriculture on singular crops left the entire southern economy susceptible to a collapse in the market. For cotton, the difficulty came in the form of a mite or bug.

The boll weevil entered the United States near Brownsville, Texas, in 1892. The first boll weevil found in Mississippi was on September 20, 1907, by W. D. Hunter. By 1915, the boll weevil covered the entire state. By 1920, all areas to the east were infested. The boll weevil covered 600,000 square miles in approximately thirty years.

As the weevil advanced across the South, land values plummeted and large numbers of farm workers migrated to northern industrial states. Since the boll weevil entered the United States a century ago, estimates of cost to U.S. cotton producers is estimated to be approximately $13 billion.

In 1958, the National Cotton Council passed a resolution addressing the economic hardship caused by the boll weevil. In 1960, the Boll Weevil Research Laboratory was established at Mississippi State University to develop trapping, suppression, and reproduction control techniques. In 1977, a pilot boll weevil eradication project was conducted in south Mississippi. This pilot project was successful and prepared the way for a full eradication program. This program was started in 1983 in North Carolina and South Carolina.

The National Boll Weevil Eradication Program ranks close to Eli Whitney's invention of the cotton gin as one of the greatest advancements ever for the U.S. cotton industry. This federal-state-grower cost share program has helped thousands of U.S. cotton growers become more competitive and has been a plus for the environment.

Today, the boll weevil has been eradicated in the southeast states of Virginia, North Carolina, South Carolina, Georgia, Florida, and Alabama and the far west states of Arizona and California. The final step, one already under way, is eliminating the weevil as an economic pest from the other two Cotton Belt regions: Louisiana, Mississippi, Arkansas, Tennessee, and Missouri in the mid-South and Oklahoma, Texas, and New Mexico in the Southwest.

Sources and Further Reading: Conzen, *The Making of the American Landscape*; Douglass, *My Bondage and My Freedom*; Kirby, *The American Civil War: An Environmental View*; Stewart, *"What Nature Suffers to Groe": Life, Labor and Landscape on the Georgia Coast, 1680–1920.*

JEFFERSON ARGUES FOR THE LOUISIANA PURCHASE AND AMERICAN EXPANSION

Time Period: 1801–1806
In This Corner: Expansionist Democrats
In the Other Corner: Federalists, those resistant to expansion
Other Interested Parties: Native peoples, Spain, France, Britain
General Environmental Issue(s): West, expansion, aridity

As the chief executive in 1801, Thomas Jefferson had a remarkable opportunity to use his own philosophies and ideals to organize the future of the young republic. As the system for land dispersal became organized with the Land Ordinance of 1787, it was possible for the nation to contemplate expansion into additional areas. Since American independence, Louisiana held a special place in the young nation's expansionist dreams. Louisiana defined the western border of the United States along the Mississippi from the Gulf of Mexico to present-day Minnesota. This sprawling acreage, however, also shaped the dreams of the leaders of other nations, including native groups who already resided there. Finally, many American politicians completely disagreed with Jefferson's expansionist ideas. To them, expanding the nation could be the end of the precarious experiment that was America.

The land in question was essentially what France lost at the end of the French and Indian Wars in 1763. Extending from New Orleans up the Missouri River to modern-day Montana, the French intended the region to supply raw materials, including flour, salt, lumber, and food, for the empire's sugar plantations in the Caribbean. In 1763, the Treaty of Fontainbleau ceded this vast area to Spain (eastern portions of the French colony went to the victorious British). After the United States established its independence from Great Britain in 1783, many leaders of the young nation looked to its Spanish neighbor with concern. For some American politicians, the entire western border seemed insecure. In addition, many American traders, thwarted by the Appalachian Mountains, found an easier trade route down by flatboat and down the Ohio and Mississippi Rivers to the port of New Orleans, from which goods could be put on ocean-going vessels. The difficulty, of course, was that the area, and particularly New Orleans, was under Spanish control.

Treaties temporarily allowed American traders to access New Orleans. By 1802, U.S. farmers, businessmen, trappers, and lumbermen were bringing more than $1 million worth of produce through New Orleans each year. Simultaneously, U.S. settlement crept closer to Spanish territory. Whoever controlled the port of New Orleans thus had a hand on the throat of the American economy. Americans preferred Spanish control of Louisiana to control by France, a much stronger power.

Rumors of the transfer of Louisiana back to France proved true in 1802. France acquired the territory in secret pacts with Spain in 1800 and 1801, but the United States learned of the transfer only in 1802, when Napoleon Bonaparte threatened to rebuild the French empire in the New World. In 1800, Emperor Napoleon signed the secret Treaty of Ildefonso with Spain. Their agreement stipulated that France would provide Spain with a kingdom for the son-in-law of Spain's king if Spain would return Louisiana to France. This plan, however, collapsed with the successful revolt of slaves and free blacks in the French colony of Saint-Domingue. This defeat forced French troops to return to France, which prevented them from reaching Louisiana. As Napoleon's New World empire disintegrated, Louisiana became unnecessary.

Dumping Louisiana

Spain was eager to divest itself of Louisiana, which was a drain on its financial resources. On October 1, 1800, Napoleon Bonaparte, First Consul of France, concluded the Treaty of San Ildefonso with Spain, which returned Louisiana to French ownership in exchange for a Spanish kingdom in Italy. New Orleans was, thus, a pawn in a global chess game. Napoleon's ambitions involved creating a great new empire centered on the Caribbean sugar trade. He viewed Louisiana as a depot for these sugar islands and as a buffer to U.S. settlement.

However, these plans rapidly began to collapse as a slave revolt in Santo Domingo stalled his overall progress. Finally, when Spain refused to sell Florida, Napoleon's attention left the New World and returned to the "Old World," Europe. Louisiana now served no purpose for him in larger plans and he needed funds for fighting his neighbors.

Meanwhile, in the United States, Thomas Jefferson, the third president, was disturbed by Napoleon's plans to reestablish French colonies in America. With the possession of New Orleans, Napoleon could close the Mississippi to United States commerce at any time. In October 1802, when Spain's King Charles IV signed a decree transferring the territory to France and the Spanish agent in New Orleans, Jefferson needed to figure out his next move swiftly. Although he and Secretary of State James Madison worked to resolve the issue through diplomatic channels, some factions in the West and the opposition Federalist Party called for war and advocated secession by the western territories to seize control of the lower Mississippi and New Orleans.

Behind the scenes, Jefferson authorized Robert R. Livingston, U.S. Minister to France, to negotiate a deal. Jefferson hardly entertained thoughts of obtaining all of the Louisiana territory. Instead, he wanted to acquire the area near New Orleans to ensure the safety of Mississippi trading. He sent Robert Livingston to France to offer Napoleon $2 million for a small tract of land on the lower Mississippi on which Americans could build their own seaport. Soon, Jefferson grew impatient with the lack of decision from France, so he sent James Monroe to Paris to increase the offer to $10 million for New Orleans and West Florida.

Napoleon's spokesman, Talleyrand, in April 1803 asked Livingston how much the United States was prepared to pay for Louisiana. Livingston was confused, because his instructions only covered the purchase of New Orleans and the immediate area, not the entire Louisiana territory. Nervous that they would lose such an opportunity by taking time to ask Jefferson, Livingston and Monroe immediately opened negotiations. By April 30, they closed a deal for the purchase of the entire 828,000-square-mile Louisiana territory for 60 million francs (approximately $15 million).

Even before the sale was complete, many members of Congress expressed fear that the land purchase made the nation too large. Other critics charged that the Constitution did not provide the president the right to purchase territory without congressional consent. Jefferson's political opponents in the Federalist Party argued that Louisiana was a worthless desert. Despite these complaints, the Senate ratified the treaty on October 20 by a vote of 24 to 7. France officially transferred the territory to the Americans on December 20. Jefferson had nearly doubled the size of the country for $11,250,000 and assumed claims of its own citizens against France up to $3,750,000, for a total purchase price of $15 million. This expansion provided Jefferson with the land to pursue his dreams for the country. However, there still remained a few questions.

Contemporary scholars have traced Jefferson's thinking about Native Americans as an important portion of his rationale for the Louisiana Purchase. Through his intense study of native culture, Jefferson became convinced that separation was the only possible solution to the "Indian problem." He wrote, "After the injuries we have done them, they cannot love us, which leaves us no alternative but that of fear to keep them from attacking us. But justice is what we should never lose sight of, and in time it may recover their esteem" (Thomas Jefferson to Benjamin Hawkins, Indian agent, August 13, 1786). Although Jefferson publicly argued for philanthropic efforts to educate and assist natives, he moved for expansion to provide new areas in which to deposit Eastern tribes.

Historian Bernard Sheehan offers that, "The acquisition of Louisiana gave removal a feasibility it had not possessed before … the ownership of the territory by the United States and the information about the region supplied by Lewis and Clark provided ground for a new and positive policy" (Sheehan 1973). With the Louisiana Purchase a possession of the United States, removing the Indians across the Mississippi became possible. They would still be within American jurisdiction and could still be cared for by the government but would be far enough away from the frontier that very few white settlers would ever have contact with the displaced tribes. The removed Eastern tribes would, as a secondary consideration, also function as a buffer between the United States and European encroachment. Jefferson wrote the following:

> As a means of increasing the security, and providing a protection for our lower possessions on the Mississippi, I think it also all important to press on the Indians, as steadily and strenuously as they can bear, the extension of our purchases on the Mississippi from the Yazoo upwards; and to encourage a settlement along the whole length of that river, that it may possess on its own banks the means of defending itself, and presenting as strong a frontier on our western as we have on our eastern border…. I have confident expectations of purchasing this summer a good breadth on the Mississippi…. You will be sensible that the preceding views … are such as should not be formally declared. (Little)

When he launched the Lewis and Clark Expedition, Jefferson urged that the explorers collect information about indigenous Indians who already lived in the West. In particular, he wanted to know how they got along with other tribes and how they differed from the Eastern Indians. Historian Christian Keller observed, "Answers to these questions would be critical in assessing how Eastern and Western Indians could cohabit. Jefferson had a plan in mind for Louisiana before it was even American territory." Thus, in addition to the other reasons for the Louisiana Purchase, Jefferson saw it as a key to solving the Indian problem. Once removed to the West, these Indians could live peacefully while white settlers took their former lands in the East.

Conclusion: Jefferson's Dream Becomes America's Reality

Although Jefferson later admitted that he had stretched his power almost to the breaking point, he said that opportunity required that theoretical ideas sometimes be set aside. Ultimately, the majority agreed. Jefferson went on to complete the largest single land purchase in American history: 600 million acres at a cost of less than 3¢ an acre in what today is the better part of thirteen states between the Mississippi River and the Rocky Mountains. He now had the raw material to help make America the agrarian republic of which he dreamed.

An official transfer ceremony was held in New Orleans on November 29, 1803. It was all a bit confusing, however. The Louisiana territory had never officially been turned over to the French; therefore, the first step was for the Spanish to remove their flag and the French to raise theirs. After one day, U.S. General James Wilkinson accepted possession of New

Orleans for the United States. A similar ceremony was held in St. Louis on March 9, 1804. The following day, Captain Amos Stoddard of the First U.S. Artillery raised the American flag. The transfer of Louisiana was official, and the region was placed under the oversight of Meriwether Lewis.

Sources and Further Reading: Conzen, *The Making of the American Landscape*; Keller, *Philanthropy Betrayed*; Little, *Jefferson's Nature*, 1988; Sheehan, *Seeds of Extinction*.

LEWIS AND CLARK SEEK TO KNOW THE UNKNOWN CONTINENT

Time Period: 1800–1806
In This Corner: Jefferson and expansionist interests
In the Other Corner: Anti-expansionists, Spain, France, Britain
Other Interested Parties: Native Americans, American citizens
General Environmental Issue(s): Exploration, west

For as long as Thomas Jefferson had been pressing to open up settlement of the western lands, he had also been yearning to know the details of this unknown landscape. As a member of Congress, Jefferson argued that federal funds should be used to support this exploration, which, ultimately, would be in the national interests. Others disagreed. As Jefferson gained more influence and seniority over federal funding, however, he was able to initiate the exploration that he so desired, even if he could not accompany the explorers.

He attempted to send three different expeditions, the first two of which were under the auspices of the American Philosophical Society. After the arrival of the *Columbia* in modern-day Washington state in 1792, however, Jefferson's urgency to explore the West intensified. The primary quest for any of the excursions was to ascertain the extent of the Missouri River Trade Route (Miller 1988, 239–40).

To not arouse suspicion of critics of the expedition, Jefferson sent a secret message to Congress. In it, he requested $2,500 to finance an expedition that would search out a land route to the Pacific, strengthen American claims to Oregon, and gather precise data about flora and fauna, as well as information about Native Americans and the terrain of the west. This time, the president trusted the expedition to his private secretary, Meriwether Lewis. The group was called the "Corps of Discovery" because its aim was to explore, scout, survey, and map the land of the Louisiana Purchase.

The scientific emphasis of the company was evident from Lewis's personal training: although he had long studied botany at his family's farm, Lewis now endured a battery of instruction by some of the most eminent scientists in Philadelphia. Jefferson clearly wanted to be part of the expedition. Of course, his political responsibilities made this impossible. Therefore, he oversaw Lewis's preparation as a way of participating vicariously in the journey (Miller 1988, 239–40). Despite his schooling, however, Lewis lacked experience in Indian warfare. His former commander in the army, William Clark, would join him to overcome his lack of experience. Although this was not an officially sanctioned military trip, Clark's inclusion quieted critics of Jefferson's scientific band.

On August 31, 1803, the trip began when the band floated down the Ohio River to arrive in St. Louis. Here, the team underwent additional training. Their tour finally began in

May 1804 when they launched a keelboat and two pirogue boats filled with supplies and approximately thirty men (Opie 1998, 161). This group included Touissant Charbonneau, who brought one of his Indian wives, Sacajawea. Historians agree that her chance inclusion in the expedition was one of the single greatest reasons for its success. For instance, many of the Indian tribes did not think this was a military expedition because a woman and her child (whom she gave birth to on February 11, 1804) accompanied the group. In addition, Sacajawea knew a great deal about Indian culture, including their medicines and the uses and names for many local plants and animals foreign to the Easterners.

During their travels, the captains and four other men kept diaries, which today serve as catalogs of the region's nature and climate. The explorers recorded their observations of many animals, ranging from the channel catfish to the prairie rattler, and landscapes, including the raging western rivers and the dry expansive desert. Lewis, Clark, and their companions met members of more than fifty native tribes. The diaries also record the first official council between people from the United States and the Indians inhabiting the western regions, which took place on August 3, 1804.

The expedition extended across the Continental Divide at Lemhi Pass. Then they went through the Lolo Pass to the Bitterroot Mountains and followed the Clearwater River down to the Snake River. After following the Columbia River, the expedition spent the winter of 1805–1806 at a fort on the Pacific coast, called Fort Clatsop. The journey had reached its goal. In the spring, the expedition returned to the east; however, in March 1806, the team split. Lewis and a portion of the team went to explore the Marias River. The rest joined Clark to scout from the Yellowstone River to the Missouri River. On August 12, 1806, the two teams reunited at the mouth of Yellowstone and arrived at St. Louis on September 23, 1806.

Once in St. Louis, Lewis drafted the first of the letters that served as a preliminary report of the expedition's findings. In one, Lewis wrote, "In obedience to your orders we have penetrated the continent of North America to the Pacific Ocean, and sufficiently explored the interior of the country to affirm with confidence that we have discovered the most practicable route which does exist across the continent by means of the navigable branches of the Missouri and Columbia Rivers." This was Jefferson's primary goal for the expedition, of course, but Lewis's letters also offered a glimpse of the region's plentiful wildlife.

To Jefferson and to his political adversaries, wildlife translated into great opportunity for hunting, particularly for fur. Such possibilities combined with wariness over the presence of foreign nations in the West to spur new negotiations to acquire the Western claims of other European nations.

The expedition lost one member en route and cost the nation $39,000. Clearly, Lewis and Clark had not only fulfilled Jefferson's scientific expectations but also reaffirmed his faith in the future economic prosperity of the United States. Its success is also measured by the ability to create a new era of mountain men, explorers, stagecoaches, pioneers, and the transcontinental railroad.

Lewis returned to new celebrity in Washington where he was awarded double pay for his time on the expedition and a warrant for 1,600 acres of land. In addition, he was named Governor of the Territory of Upper Louisiana, effective in early March 1807, after the purchase had been completed. The young explorer sought publishing opportunities for his journals and basked in the moment, actually not returning to St. Louis to take up his appointment as governor until March 1808. Displeased with the flood of opportunists and the lack of organization from the federal government, Lewis fled his responsibilities. Ultimately, he committed suicide.

Jefferson's forecast of the importance of their exploration, of course, proved to be correct. In fact, such exploration would become an emphasis of the federal government, particularly through the U.S. Geological Service, throughout the nineteenth century.

Sources and Further Reading: Ambrose, *Undaunted Courage: Meriwether Lewis, Thomas Jefferson, and the Opening of the American West*; DeVoto, *The Journals of Lewis and Clark*; Jones, *The Essential Lewis and Clark*; Lamar, *The New Encyclopedia of the American West*; Opie, *Nature's Nation*.

ROMANTICS CLAIM THAT NATURE'S VALUE DERIVES FROM BEAUTY

Time Period: 1800s
In This Corner: Writers, artists, transcendentalists
In the Other Corner: Settlers, developers
Other Interested Parties: Government
General Environmental Issue(s): Preservation, Romanticism, ethics, aesthetics

Ideas of beauty vary significantly between different cultures and are rarely consistent even within specific societies. Although it contrasted with much of the aesthetics appreciated by European sensibilities, many American thinkers and artists of the early 1800s made one theme a primary portion of their emerging tastes: nature. Grouped with the aesthetic movement known as Romanticism, this aesthetic turned typical views of nature on their head. Based in biblical teachings as well as in the experience of struggling against nature for survival, most of Western thought before 1800 emphasized the need to civilize and alter nature. In the revisionist thinking, nature could be beautiful without being civilized, settled, or used by human influence.

The Hudson River School

As wealthy Americans longed for their young nation to form a unique cultural tradition that would stand out among the nations of the world, it seems unsurprising that artists would focus on the natural wonders of North America. Known as the Hudson River School, the first internationally recognized genre of art to be initiated in the United States grew between the 1820s and the late nineteenth century. Initially, their paintings were organized around scenes of the Hudson River Valley and the adjoining mountains of New York and Vermont. Eventually, as a view of nature and not a region was identified as the primary organizational device for the genre, Hudson River School artists would paint natural wonders from all over the world (Novak 1980, 18–20).

Hudson River painters conveyed their natural wonders with a fairly clear and unapologetic ideology. They painted with an almost religious reverence for the magnificence of the American wilderness. Their effort to recreate the unique beauty of the American landscape for the public can be viewed as one of the first expressions of the American desire for preservation. For the Hudson River painters, of course, instead of national parks or policies, they set aside locations and moments in time within the boundaries of a frame. In this view, the artists could make nature appear aesthetically wonderful, even exaggerating the existing beauty of a site.

Unlike the orderly landscapes of European artists, Thomas Cole and the artists that followed him created vast, awe-inspiring scenes that were designed to convey threatening characteristics of the American wilderness. Cole hoped to use his influence to encourage more people to love, enjoy, and protect nature. His work inspired a new class of wealthy American and international tourists to the areas he painted. In fact, the irony is that, by inspiring visitation, Cole's work increased development.

These larger societal changes created a border between developed and undeveloped parts of America that also inspired art. Maybe the best example is George Innes's *Lackawana Valley*. This canvas, which was completed in 1855, provides a dramatic visual demonstration of the "machine and the garden" paradigm that was described in the introductory remarks. Unapologetically depicting stumps in the foreground, the scene centers on a train and one of the technological wonders of the rail age, a round house. Overall, however, the scene highlights the vibrant colors of a natural environment that seems to be accepting such changes (Novak 1980, 171–74).

Thomas Cole's *The Oxbow*

Another painting that demonstrated the vicious dichotomy of America's view of nature was Thomas Cole's painting of 1836 entitled *The Oxbow* (The Connecticut River near Northampton). In the painting, the tension between wilderness and garden, savagery and civilization, is recorded visually as European conventions of landscape painting are used to comment on the state of the physical place of America. The savagery of the storm clouds over the wilderness retreats from the advancing cultivated landscape of civilization. Art historian Barbara Novak wrote, "Cole used the storm-ravaged tree, a palimpsest of time associations, almost as a signature. This natural picturesque is of a totally different order from what we might call the man-made or unnatural picturesque, the cut stump of the wood-chopper, utilitarian man feeding his ten thousand fires" (Novak 1980, 161–62).

Cole seemed to express a concern of a growing number of nineteenth-century Americans: was wilderness being lost? His paintings dramatically left no doubt that the answer was affirmative.

Transcendentalism

Early in the nineteenth century, the aesthetic appreciation for nature had no intellectual foundation. After the 1820s, however, writers and intellectuals began knitting together ideas and influences from other parts of the world with sensibilities such as those of visual beauty expressed by Cole. The literary and intellectual movement that grows out of this increased interest in nature is referred to as transcendentalism. This realm of belief became a portion of American Romanticism, ultimately combining spirituality and religion (Nash 1982, 85–86).

Writers and reformers, including Ralph Waldo Emerson, Henry David Thoreau, Margaret Fuller, and Amos Bronson Alcott, developed this line of thinking in New England between 1830 and 1850. Their actions helped to transform transcendentalism, at least partly, into an intellectual protest movement; however, it continued to carry with it a new appreciation for nature. Most often, transcendentalists connected to the ideas of philosophical idealism that derived from German thought, either directly or through the British writers Samuel Taylor Coleridge and Thomas Carlyle.

Emerson emerged as the intellectual leader of this group when he connected Romanticism with Unitarianism. By 1825, Unitarianism had many followers in Massachusetts, where they openly attacked the orthodoxy of the Puritans who dominated New England. In place of Puritan thinking, the Unitarians offered a liberal theology that stressed the human capability for good. Four years after resigning as pastor at Boston's Second Church, Emerson published *Nature* in 1836. Emerson directly challenged the materialism of the age, and his writing was adopted as the centerpiece of transcendentalism (Nash 1982, 86–89).

In the Boston area, the Transcendental Club began to meet to refine and disseminate the ideas that Emerson voiced in his writing. This group of intellectuals also created the famous Brook Farm experiment in communal living (1840–1870), in West Roxbury, Massachusetts. Young Henry David Thoreau became active with the club and began working with its publication, the *Dial*. Thoreau's writing emphasized the role of nature in Americans' lives. Thoreau published his greatest work, *Walden*, in 1854. This book was Thoreau's account of transcendentalism's ideal existence of simplicity, independence, and proximity with nature. In *Walden*, Thoreau extended Emerson's ideas of replacing the religion of early-nineteenth-century America with the divine spirit out of nature. In this paradigm, the natural surroundings took on spiritual significance. More than ever, Thoreau created a model of transcendentalist thought connected to nature.

Thoreau's message from Walden Pond urged Americans to escape from mechanical and commercial civilization to be immersed in nature, even if only for a short time. Although few Americans in 1850 either read *Walden* or immediately came to see nature differently, Thoreau and other transcendentalists laid the foundation for a new way of viewing the natural environment. No longer simply raw material for industrial development, nature possessed aesthetic or even spiritual value.

Through the efforts of transcendentalists, writers, poets, and artists argued for America to be "nature's nation." The symbolic nature of the United States was not necessarily the manicured beauty of the French and British gardens but also the raw wilderness not found in Europe. In April 1851, Thoreau lectured at Concord Lyceum in Massachusetts. After beginning by saying that he "wished to speak a word for nature," he answered proponents of developments and civilization. Finally, he shared a timeless insight when he stated, "In Wildness is the preservation of the world" (Nash 1982, 84).

With such a statement, Thoreau forged a connection between the intellectual approach of transcendentalism to wilderness to American ideals of democracy, independence, and beauty. The attraction of nature would eventually also include an interest in primitivism, one interpretation of Thoreau's "wildness." As society became more industrialized, developed, and urban, a contrary impulse attracted some Americans to seek innocence in raw nature.

Ironically, this interest in primitivism also occurred when the civilizing forces of the United States were mobilized against the primitive native occupants of North America. Clearly, in literature and in art, this interest in primitivism influenced a new appreciation and romanticization of native culture (Nash 1982, 100–103). Whereas many Americans saw the native peoples as a national "problem" and a "nuisance," some romantic viewers began to see them as a people uniquely and admiringly in tune with the natural environment.

Wild nature thus became a source of national pride as the root of character traits for a unique national identity. This imagery was most notable in paintings, which, through their reconfigured realities, allowed artists to create a new creation myth for America that emphasized the primacy of the white settlers. Artists of this genre granted a privileged role for an American elite and enabled the white discovery and settlement of the wilderness by evoking images of classical heroes. Nature was also given a reinterpreted role: rather than presenting nature as an obstacle to the establishment of a civilization, American authors and painters alike upheld nature as the source of the animating spirit behind the American character. Excerpts from two of the genre's most famous authors follow.

Ralph Waldo Emerson, *Essay VI from Nature*

"There are days which occur in this climate, at almost any season of the year, wherein the world reaches its perfection, when the air, the heavenly bodies, and the earth, make a harmony, as if nature would indulge her offspring; when, in these bleak upper sides of the planet, nothing is to desire that we have heard of the happiest latitudes, and we bask in the shining hours of Florida and Cuba; when everything that has life gives sign of satisfaction, and the cattle that lie on the ground seem to have great and tranquil thoughts. These halcyons may be looked for with a little more assurance in that pure October weather, which we distinguish by the name of the Indian Summer. The day, immeasurably long, sleeps over the broad hills and warm wide fields. To have lived through all its sunny hours, seems longevity enough. The solitary places do not seem quite lonely. At the gates of the forest, the surprised man of the world is forced to leave his city estimates of great and small, wise and foolish....

These enchantments are medicinal, they sober and heal us. These are plain pleasures, kindly and native to us.... Nature is always consistent, though she feigns to contravene her own laws. She keeps her laws, and seems to transcend them. She arms and equips an animal to find its place and living in the earth, and, at the same time, she arms and equips another animal to destroy it."

Henry David Thoreau, *Walden*

"After a still winter night I awoke with the impression that some question had been put to me, which I had been endeavoring in vain to answer in my sleep, as what—how—when—where? But there was dawning Nature, in whom all creatures live, looking in at my broad windows with serene and satisfied face, and no question on her lips. I awoke to an answered question, to Nature and daylight. The snow lying deep on the earth dotted with young pines, and the very slope of the hill on which my house is placed, seemed to say, Forward! Nature puts no question and answers none which we mortals ask. She has long ago taken her resolution. 'O Prince, our eyes contemplate with admiration and transmit to the soul the wonderful and varied spectacle of this universe. The night veils without doubt a part of this glorious creation; but day comes to reveal to us this great work, which extends from earth even into the plains of the ether.'" (*Walden*, chap. 16)

William Henry Jackson and Romantic Landscape Photography

During the mid-1800s, the novel technology of photography became a new way of expressing the romantic beauty of the American landscape. The efforts of these early photographers became an important portion of the growing American interest in western lands and their eventual use as national parks. William Henry Jackson picked up where the romantic thinkers and painters left off and created actual views that spoke to the same passionate beauty in natural forms.

Jackson formed his sensibility beginning at the age of ten when he received his first formal artistic training. His emphasis was drawing, and his first job came in 1858 when he was hired as a retoucher for a photographic studio in Troy, New York. In this job, his duty was to enhance or warm up black and white portraits by tinting them with watercolors and India ink. While working at this job, Jackson learned to use cameras and darkroom techniques. He began to apply his artistic sensibility to this new technology. In 1869, he moved to Omaha, Nebraska, and established his own photographic studio. From this base, Jackson made the first subject of his photography American Indians from the nearby Omaha reservation and the construction of the Union Pacific Railroad.

In travel pamphlets such as this one, nineteenth-century railroad companies provided travelers with packages of exotic natural wonders such as the geysers of Yellowstone. The first national parks were clearly in the railroads' interest as well. Library of Congress.

Images such as this stereograph of the lower falls of Yellowstone introduced Americans to the great wonders of the first national parks in the West. Library of Congress.

He established a reputation with these early images and soon was contacted by Dr. Ferdinand Hayden, who was organizing an expedition that would explore the geologic wonders along the Yellowstone River in Wyoming Territory. Hayden realized that a photographer would provide a valuable record of the sights of the exotic western landscape. In addition, Jackson became quite capable of managing the complexities of early photography on the move.

The visual record that Jackson created provided important verification to legends of geysers and waterfalls in the territory. Jackson's images helped to arouse public interest and fueled the push for Congress to officially designate Yellowstone National Park in 1872. Although the well-known painter Thomas Moran also created images based on the travels of the Hayden expedition, Jackson's images struck even more Americans as genuine and unembellished. In fact, during the expedition, Moran often helped Jackson locate the best views and then used the images afterward to compose his paintings. Because no member of Congress had seen Yellowstone, Hayden and his colleagues brought Jackson's photos, along with Moran's watercolors, to Capitol Hill. A year later, in 1873, the Department of the Interior compiled thirty-seven of Jackson's photographs into a portfolio, which was presented to Congress in lobbying funds for future expeditions to the West.

Jackson and Hayden worked with the Department of the Interior to release these and later images as stereo views, which increased their general appeal. These views were made with a special double camera that used two horizontal lenses. Each lens recorded the image as seen by each human eye. The double prints were then placed on a card that could be viewed in a device that allowed the viewers' eyes to combine the images into one scene. More than any other image, stereo views allowed Americans to place themselves into the scene. The stereo views figured significantly in Yellowstone National Park's growing popularity. For some, this was their first and only view of Yellowstone.

Jackson and Hayden joined forces for the U.S. Geological Service (USGS) for almost another decade. Their subject was almost entirely the landscape of the western United States. After his work at the USGS, Jackson continued to work in the West, opening a studio in Denver, Colorado. He traveled widely to photograph railroad construction to min-ing towns in the Rockies. Many of his images were viewed by the millions of visitors to the World Columbian Exposition in Chicago in 1893.

Conclusion: Inspiring a New View of Nature

Through the efforts of these writers and painters, among others, a new paradigm became part of American culture. In this new mindset, nature was granted worth on its own right, particularly for its aesthetic beauty. Although, the majority of Americans maintained a utili-tarian view of nature, the intellectual construction had begun of what would develop into a conservation ethic in the later 1800s.

Sources and Further Reading: Hales, *William Henry Jackson and the Transformation of the American Landscape*; Jackson, *Time Exposure: The Autobiography of William Henry Jackson*; Nash, *Wilder-ness and the American Mind*; Novak, *Nature and Culture*.

BRINGING AMERICA'S NATURE AESTHETIC TO LIFE IN PARKS

Time Period: 1800s
In This Corner: Supporters of public parks, landscape artists
In the Other Corner: Citizens resistant to public works
Other Interested Parties: City residents, political leaders
General Environmental Issue(s): Climate, agriculture, environmental history

Although numerous intellectuals and artists conceived of the romantic in writing and on can-vas, a few also began to do so on the landscape. The first examples of such planning were part aesthetic creation and part utilitarian. Disposal of the dead had become a serious prob-lem for growing urban areas in the early 1800s. Thus, the need for new cemeteries became one of the first applications of the romantic view of nature.

Mount Auburn Cemetery

The landscape that began the rural cemetery movement, Mount Auburn Cemetery opened in Cambridge, Massachusetts, in 1831. Unlike previous places of internment, Mount Auburn prioritized the natural beauty of the surroundings, not just the utilitarian storage of dead bodies. By doing so, it marked an important change in American community planning as well as in Americans' relationship with nature.

With few parks or public areas, the American landscape of the early 1800s still was organized entirely by the everyday needs of residents. There was little interest in the frivolity of aesthetic beauty, and what was considered beautiful by Americans was rarely natural. Mount Auburn began to change this by presenting a real landscape that contained ideas of beauty seen in paintings, particularly those of romantic artists.

In Mount Auburn, tombstone markers were not lined up tightly to bury as many people as possible. Instead, the landscape was designed as a unit by Andrew Jackson Downing. He wove the markers in with the hills of the site and interspersed them with miniature species of exotic shrubs and trees. He used winding paths and benches to create private areas for repose or mourning. However, simultaneously winding wagon trails allowed visitors relaxing for the day to interact with natural beauty. The patterns and choices of his design became the typical characteristics of the rural cemetery movement, which would see similar areas created in many cities by the 1850s.

At Mount Auburn, Americans learned that the landscape could be art. A primary component of this canvas, of course, was the natural details that Americans saw as horrible just a few decades previously. Following the success of rural cemeteries, Americans also became interested in public parks. Each of these developments helped to spur the growth of landscape architecture in the United States (Sloane 1995, 57–60).

The Rural Cemetery Movement

Before 1830, most communities considered cemeteries to be necessary for disposal of the dead but of little aesthetic importance. Normally, urban cemeteries were found adjacent to churches, in churchyards, as the area was often called, or in central common areas of a town or city. Clearly, from the first European settlement in New England, Americans carried on the European tradition of burying the dead amid the living. By the turn of the century, many of these churchyards became overcrowded. Many residents believed that this created a genuine health threat to the living. This was particularly problematic in cities that had come to use their cemeteries as multipurpose areas for common activities including playing, relaxing, and mourning. As a result, cemeteries became one of the first planned landscapes in the United States.

In the young republic of the United States, the "rural" cemetery movement was inspired by romantic perceptions of nature, art, national identity, and the melancholy theme of death. It drew on innovations in burial ground design in England and France, particularly the Père Lachaise Cemetery in Paris, established in 1804 and developed according to an 1815 plan. Between 1830 and 1855, new ideas for interring the dead were grouped under the term "rural" or "garden" cemetery. Nearly a complete contrast from its predecessor, the rural cemetery was most concerned with aesthetics. Its appearance was designed to provide peaceful surroundings by using the beauty of nature, including ornamental shrubs and trees, and romantic layouts, such as winding carriage trails and ornamental monuments. Although the rural cemetery movement represented a social change in American life, it also marked an important moment in the American relationship with nature.

Whereas nineteenth-century community cemeteries typically were organized and operated by voluntary associations that sold individual plots to be marked and maintained by private owners according to individual taste, the memorial park was comprehensively designed and managed by full-time professionals. Often these cemetery parks were operated as a business venture, prioritizing economy of price and natural beauty. In other cases, nonprofit corporations ran the cemetery. Regardless, the natural beauty of cemetery sites continued to be enhanced through landscaping, often incorporating picturesque hills with an overall design of rolling terrain. The cemetery clearly took on an overall design through initial planning and ongoing maintenance.

The "rural" cemetery movement clearly grew out of European trends in gardening and landscape design. However, the cemetery form acted as a conveyor of this aesthetic into the

This aerial view of Laurel Hill Cemetery, Philadelphia, provides an excellent example of the layout and design of a rural cemetery. Library of Congress.

lives of many Americans who had not yet considered the time, expense, and effort needed to sculpt their home landscapes with such beauty. The English roots, of course, grew from traditions of the 1700s that had seeped out of the tastes of English nobility who had urged their gardeners to use classical landscape paintings as their models. The greatest names in English garden designers of this era included Lancelot "Capability" Brown, William Kent, Sir Uvedale Price, Humphrey Repton, and John Claudius Loudon. In an effort to imitate natural forms, these designers built gracefully curving pathways and streams into their rolling land forms. They created a "picturesque" model of design by combining these elements with contrast and variation by massing of trees and plants and incorporating ornamental features. By the end of the eighteenth century, the picturesque landscape was a fairly concrete, well-known form, including open meadows of irregular outline, uneven stands of trees, naturalistic lakes, accents of specimen plants, and, here and there, incidental objects such as an antique statue or urn on a pedestal to lend interest and variety to the scene.

Concentrated around urban areas, these changes in cemetery layout emanated from the Boston area. Ultimately, however, rural cemeteries sprang up throughout the United States. Typically, these cemeteries resembled what would later be called parks. Their construction, therefore, required open tracts of land. For this and aesthetic reasons, the new cemeteries were most often sited on the outskirts of cities and towns. Their location also marked an important step in suburbanization, when human communities began to spread out of urban areas. Well-to-do visitors treated the rural cemeteries less as a place for somber reflection and more for outdoor leisure in beautiful surroundings. A new appreciation of nature began. People began to see nature as something to be enjoyed as well as tamed (Sloane 1995, 32–41).

Mount Auburn was followed by the formation of Laurel Hill Cemetery in Philadelphia in 1836, Green Mount in Baltimore in 1838, Green-Wood Cemetery in Brooklyn, Mount Hope Cemetery in Rochester, New York, in 1839, and ultimately many others. Later in the nineteenth century, the design format would change to that of the perpetual care lawn cemeteries or memorial parks of the twentieth century. The lawn plan system shifted from the rural design by deemphasizing monuments in favor of unbroken lawn scenery or open space.

In the rural design, however, open space would never be a priority. Balance and overall design grew from hilly, wooded sites, even if ground needed to be moved to create the appropriate scene. Such settings, it was thought, stirred an appreciation of nature and a sense of the continuity of life. By their example, the popular new cemeteries started a movement for urban parks that was encouraged by the writings of Andrew Jackson Downing and the pioneering work of other advocates of "picturesque" landscaping.

Andrew Jackson Downing

At Mount Auburn and many other sites, Andrew Jackson Downing introduced a new tastefully designed nature to the American public. A landscaper, Downing wrote widely about his ideas and philosophies of gardening. Ultimately, he created guidebooks and design manuals that made him a definer of taste, similar to the contemporary persona of Martha Stewart. His influence took off in 1846 when Downing founded the magazine *The Horticulturist* to disseminate his ideas to upper-class consumers. In addition to stylistic concerns, the magazine spread Downing's interest in scientific agriculture. The primary readers, of course, were the very wealthy who had leisure to consider such things. By presenting them with factual information, however, Downing helped to create a class of upper-class "gentlemen farmers" and Victorian women most appreciative of beauty in the landscape (Schuyler 1996, 105–7).

Where, however, were country farmers to live? Most wealthy Americans lived in cities and very few had vacation homes. Picking up on English traditions, Downing sought to interest his readership in a new type of countryside. Collaborating with Alexander Jackson Davis in 1842, Downing published *Cottage Residences*, which was a pattern book of country houses. In his drawings, the homes picked up the architecture of the English countryside and blended in aspects of romantic landscape design seen at Mount Auburn and elsewhere. Sounding similar to Jefferson decades before, Downing argued the worth of living close to the land in rural areas in which one could interact with nature. Escape from the urban confusion interested many Americans who were wealthy enough to consider such a move. Soon, Downing had legions of followers.

Downing's priorities and his interest in helping Americans to interact with nature contributed to the American park movement. Ultimately, the spirit of his country homes would inspire the first American suburbs

Although Downing's home designs were directed to the wealthy, he also hoped to create places that would be enjoyed by all classes of society. This desire led Downing to begin advocating large inner-city parks. Downing saw a civilizing aspect of open spaces and wanted to bring one to nearby New York City. Finally, after many years, New York City set aside land for Central Park. Downing and his partner, Calvert Vaux, devised preliminary plans for the park (Schuyler 1996, 202–3).

At roughly the same time, Downing was asked to design the Public Grounds in Washington, DC. This included the Mall and land around the White House. These stunning designs contrast markedly with the monumental landscape that would eventually be built in the nation's capital. Downing used his romantic sensibilities to intermingle landscape features with the monuments. Unfortunately for his Central Park and Washington plans, in 1852, Downing died in a steamboat accident on the Hudson River. He had, however, spurred a version of American taste with the landscape as a vital component.

Conclusion: Planning Beauty in Landscapes

With the rapid growth of urban centers later in the nineteenth century, landscape design and city planning merged in the work of Frederick Law Olmsted, the country's leading designer of urban parks. Olmsted and his partners were influential in reviving planning on a grand scale in the parkways they created to connect units of municipal park systems. Although Olmsted was more closely tied to the naturalistic style of landscape planning, his firm's work with Daniel H. Burnham in laying out grounds for the World's Columbian Exposition of 1893 in Chicago conformed to the classical principles of strong axial organization and bilateral symmetry.

Evolving over a century, the tradition of design that began with rural cemeteries demonstrated the value of planning. By fusing romantic ideas of the picturesque with the engineering of landscape planning, many of the parks became symbols of the possible "middle ground" that could take shape with nature at its core.

During the twentieth century, the work of John C. Olmsted and Frederick Law Olmsted Jr., successors of the elder Olmsted and principals of the Olmsted Brothers firm, expanded these ideas throughout the country.

Sources and Further Reading: Downing, *A Treatise on the Theory and Practice of Landscape Gardening*; Novak, *Nature and Culture*; Schuyler, *Apostle of Taste: Andrew Jackson Downing, 1815–1852*; Sloane, *The Last Great Necessity*.

CONSTRUCTING THE CANAL AGE: WHO SHOULD PAY?

Time Period: 1810–1860
In This Corner: Business and transportation developers

In the Other Corner: Citizens resistant to public works, some political leaders
Other Interested Parties: American citizens, political leaders
General Environmental Issue(s): Rivers, economic development, the federal government's role

Although history proved that every American in the Mid-Atlantic region would benefit from canal development in the early 1800s, there was no plan for how such "internal improvements" could be financed. Public funding seemed reasonable to help construct trade networks that would help entire regions develop; however, there was no precedent for such funding, and the founding fathers of the United States were extremely reluctant to focus such responsibility on the young federal government. As this debate raged, private development took shape to develop canal corridors in three important regions: upstate New York and the Erie Canal; Pennsylvania and the Main Line Canal; and the lower Mid-Atlantic with the Chesapeake and Ohio Canal.

For engineers, there was only a slight leap between managing rivers through a mill and creating roadways of water, called canals. Roads and rivers provided passage for nearly all early trade. Each form of transportation relied on existing passage through varied terrain. Most historians agree that canals marked the first form of transportation that could be sufficiently regularized to create and support inland trade networks. Additionally, the construction of canals marked an early example of tying together large amounts of investment capital to finance the great undertakings. Because of the need for significant sums of money to finance canal construction, the early 1800s included an ongoing political debate over whether the federal government should finance such projects, which were referred to as internal improvements. Nevertheless, the combination of public and private capital, increasingly sophisticated engineering techniques, and a rapidly expanding immigrant labor force made canal building one of the most promising endeavors in the United States by 1820.

To determine where canal passage was practical, designers read the landscape with an eye on details such as elevation and grade. If the canal could follow existing land contours as often as possible, builders saved great costs in excavation and the construction of expensive structures such as earthworks and lockage. Often, however, such routes required twists and turns that demanded the construction of additional miles of trenching. Conversely, a shorter, straight-line route might demand heavy usage of earthworks and locks. At the beginning of the canal age, engineers had minimal experience with constructing infrastructure such as tunnels, and, therefore, their routes were long and heavily reliant on technologies such as locks with which they did have experience. As early as the late 1700s, the interest in digging canals swept through areas between the Chesapeake Bay and rivers such as the Delaware, Susquehanna, and Ohio, as well as areas around Boston. By the 1810s, however, national attention emphasized expanding these short-line canals westward through New York State and Pennsylvania (Shaw 1990, 190).

New York led the way in the era of expansion based on canals. DeWitt Clinton led the way in arranging and financing this historic trade network. He believed a canal connecting the Hudson River with Lake Erie would benefit the economic development of the entire nation. After approval by the New York State Legislature in 1817, the canal was built in short sections by the local workers in each region. The first section, which runs from Utica to Rome, began operating in 1819, and the entire canal was completed on October 26, 1825. Clinton used a corps of engineers who had been trained in England.

The completion of the Erie Canal changed the economic standing of many New Yorkers and reconfigured trade patterns throughout the nation. Freight rates for the 364-mile trip from the Lake Erie port of Buffalo to Albany, near the head of the Hudson River, dropped to an astonishing extent:—from $100 per ton to $10 per ton. In five years, tolls exceeded $1 million a year, with the volume of shipments increasing in both directions. The canal, as part of a vital water network of communication and trade, facilitated the transportation of grain from western lands to the East Coast for domestic consumption and, in the later decades, for export. By 1854, 83 percent of the nation's shipment of grain went by this route. In addition, passenger travel on the Erie Canal stimulated the flow of migration to undeveloped western land. By 1821, more than 1,000 immigrants per day made the trip from New York City to Buffalo.

Although canals were costly to build, their operation was fairly inexpensive. Additionally, the canals lent themselves to technological innovation. Steamboats, which were adapted for river travel in the early 1800s, also used the Erie Canal. These faster vessels made it a much quicker trip for settlers to travel from the Hudson River, through the Erie Canal and into Lake Erie before extending into the Ohio and Mississippi Rivers. By reaching to the Great Lakes, the Erie Canal became a major portion of the American movement westward (Opie 1998, 131–32).

The success of the Erie waterway spurred action to the south. Private funding was rallied in Pennsylvania, and engineers began wrestling with the primary obstacle of its own canal: the Allegheny Mountains. Forming part of the Appalachian Mountain system of the eastern United States, the Allegheny Mountains extend southwest from central Pennsylvania through western Maryland, eastern West Virginia, and western Virginia. The Alleghenies vary in height from about 2,000 feet in the north to more than 4,800 feet in the south. The mountains form a divide between streams that flow into the Atlantic Ocean and those that empty into the Gulf of Mexico. Although the mountains are one of the most thinly populated regions of the eastern United States, the Alleghenies helped to define much of the surrounding human history.

Among native peoples, the Delaware and Seneca called portions of the Alleghenies home. Practicing limited agriculture, their villages were often constructed in small forest clearings at the base of the mountains. The only use of the mountain regions was for hunting. As European settlers moved westward from the Mid-Atlantic, the Alleghenies often became the site of native-European skirmishes and conflict. These altercations culminated in the French and Indian War of the 1750s, which included the important battle of Fort Duquesne near Pittsburgh and on the western side of the mountains.

With the extension of European settlement, the Alleghenies were major barriers to transportation. In 1755, the British general Edward Braddock built a road through a mountain pass called Cumberland Narrows, near Cumberland, Maryland. Part of this road, called Braddock's Road, became part of the National Road, which linked the East Coast and the Ohio River Valley. The pass became known as the Cumberland Gap, one of the first gateways to the American West. Today, the pass is the heart of major transportation routes through the Alleghenies. Only a few highways and railroads run through the mountains at other points because of the need to build tunnels to avoid steep grades.

Construction across the Allegheny Mountains posed far greater problems than had the Erie Canal with its water-level route through northern New York State. To cross the

Appalachian Mountains, the Pennsylvania project required a series of inclined planes, on which canal boats were hoisted on rails, which made the system unusually expensive.

Although canals followed the contours of the land, there remained mountainous terrain that was impenetrable to the technology of the canal age. Often, the innovations that engineers made to circumvent such obstacles foreshadowed the next frontier of technological innovation. The best example of this might be the inclined plane, which was used to circumnavigate many of the smaller peaks of the Appalachians. In the Allegheny Mountains, for instance, engineers cleared the Mainline Canal to reach across the state to Pittsburgh by using fixed steam technology to pull loads up mountains. This technology clearly hinted at the railroads that would dominate American industry by 1850. Pennsylvania followed shortly with the Main Line Canal, the first of its States Works projects, which opened fully in 1834.

Pennsylvania's Main Line of Public Works connected Philadelphia and Pittsburgh to provide the state with a competitor to the Erie Canal. It only reached its destination with the help of the Allegheny Portage Railroad, which used steam railways and inclined planes operated by stationary engines. The power can be men or animals, or even a waterwheel. In the case of the Allegheny Portage Railroad, the canal boats were lifted from the water in Hollidaysburg and placed on rail cars. Soon, specially designed boats had been devised that could be separated into two or three sections, without disturbing the load. Hemp lines reached down the Alleghenies and connected the loaded cars with a large steam engine at the mountain's summit. Once engaged, the steam engine hoisted the loaded car up the slope.

Regardless, the amount of cargo carried in the early 1800s made an important contribution to trade between Philadelphia and the Ohio River basin. This created a vital link between the interior and the eastern seaboard. Subsidiary canals were subsequently constructed in Pennsylvania as local communities pressured the state legislature for extension of canals as a path to local prosperity. In eastern Pennsylvania, the Schuylkill Canal and others were built specifically to carry anthracite coal from the mines to Philadelphia. To the south, the system involved the Chesapeake region.

At Little Falls, Maryland, on July 4, 1828, President John Quincy Adams began construction of the Chesapeake and Ohio (C & O) Canal. It was officially the work of the C & O Canal Company, with approximately $3.6 million of private and public investment, including the federal government, the states of Maryland and Virginia, and the cities of Washington, Georgetown, and Alexandria. Each of these investors hoped that the canal would link with the Baltimore and Ohio Railroad (B & O), America's first, which also began construction.

Conclusion: Little Future in a Limited Technology

From the beginning, however, canal construction was slower than expected. Legal problems slowed construction until 1832. Even with only small sections completed, boats began to appear as early as 1831. Trade increased as other segments opened in western Maryland, with boats carrying cargoes of flour, grain, building stone, and whiskey to Georgetown. When the canal reached Cumberland, cargo of coal raised the tonnage to more than 850,000 tons. After spending $11 million, the canal would finally open in 1850; however, developers had given up on their original idea to extend through the Allegheny Mountains.

Other areas took up canal building to provide alternative modes of shipping as well as to lower transportation costs. New England had several canals, including the Farmington, which was built north from New Haven toward Canada. This canal, however, had only reached Massachusetts before being supplanted by a railroad line in the 1840s. Farther south, the James River and Kanawha Company extended the canal begun in the 1780s in Richmond, Virginia, with the support of George Washington.

In 1840, Americans had more than 3,000 miles of canals in operation, mostly in the northeast. For a variety of reasons, including shortage of funds, many of these canal efforts were not successful in creating efficient modes of travel, and yet they did contribute to financial speculation, particularly in state bond sales. This canal-building frenzy was part of a more general speculative boom that helped to expand the scale of industrial development through the 1850s (Shaw 1990, 24–39).

More than any other contribution to the United States, however, canals ushered an era of centralized development to boost trade networks. The infrastructure for establishing these networks shifted to the more stable technology of railroading, which had much more possibility for growth.

Sources and Further Reading: Gordon and Malone, *The Texture of Industry*; Opie, *Nature's Nation*; Rice, *The Allegheny Frontier*; Shaw, *Canals for a Nation: The Canal Era in the United States, 1790–1860*; Sheriff, *The Artificial River: The Erie Canal and the Paradox of Progress, 1817–1862*; Wertime, *Citadel on the Mountain*.

AMERICAN LEADERS CREATE A CULTURE AND SYSTEM OF EXPANSION

Time Period: 1800s
In This Corner: Federal government, development interests, settlers
In the Other Corner: Native American tribes
Other Interested Parties: Traders
General Environmental Issue(s): West, expansion, aridity

There were many reasons for Americans *not* to expand westward. Little was known about the vast area west of the Appalachians in the early 1800s, and the area was already populated by thousands of people who had been living in the region for generations. The infrastructure of trade and transportation that made life in the east profitable was largely nonexistent in the American West at this early juncture. Once the Louisiana Purchase had expanded the surface area of the United States, there could no longer be any debate. By the early 1800s, it was clear to most Americans that every effort needed to be made to create the infrastructure for settlement in the West.

For many years, historians viewed the evolution of the infrastructure for settlement as a portion of the march of American "progress." Over the past few decades, however, some historians have offered a "new" or "revisionist" interpretation of settlement. Historian Patricia Nelson Limerick refers to this movement westward as a "Legacy of Conquest."

An important portion of settling on an interpretation of the westward movement is to establish intent. Did this process unfold "naturally" or "organically"? The infrastructure offers an important tool for effectively understanding the conquest of the West, because, in short, it makes obvious that the settlement of white Americans in the western United States was a methodical, systematic process organized around cultural, political, and economic ideas and tools.

The preeminent of these tools was, of course, the iron horse. Ultimately, this conquest of the West, its space and its people, would succeed because of the technological innovation of the railroad. However, in this earlier era, the culture of conquest overcame each of the challenges presented by the West and the difficulties cited by critics. The elements of this infrastructure included physical elements, such as trading posts and trails, as well as entities such as the U.S. military. In addition, however, the effort to settle the West included a cultural spirit that became known as "Manifest Destiny" that rationalized this "conquest."

Modes of Transport Westward

Throughout all of American history, ways of travel and shipping have played a major role in shaping people's lives and their use of nature. From the earliest American Indian inhabitants up to today's modern world, paths or roads have served as an essential link between groups of people. The earliest roads were the paths and trails made by the American Indians. This intricate, interwoven series of trails and hunting paths would eventually lead to the modern road system we have today.

When the first European settlers came to this continent, they traveled over the ocean to what was then a new land, a land not connected by roads. Small paths and trails were used by people on foot, not on horses or in wagons. To overcome this obstacle, the colonists did what was logical to them: they built simple dirt roads. The majority of these roads were constructed by hand using private funds and were often toll roads.

Even with the development of roads and use of rivers, the Appalachian Mountains remained a major obstacle for settlers interested in expanding inland. There were two solutions for the colonists and the British to consider: build roads or attack the French colonies to the north and seize their access to the interior waterways such as the Mississippi. The British eventually did go to war with France, for many reasons, leading the colonies into the French and Indian (or Seven Years) War.

One major objective of the British was to take Fort Duquesne, at the forks of the Ohio River, the modern-day site of Pittsburgh, Pennsylvania. To do this, the British had to get troops and supplies over the Appalachian Mountains, and this meant building a road, the first major road to the West. Eventually this road would be called Braddock's Road in honor of British general Edward Braddock, whose orders were carried out by colonial soldiers including George Washington. The road reached the present-day site of Braddock, near Pittsburgh, and then stopped abruptly because of a major defeat at the site. Several years later, Washington again worked on a road westward, although this time the orders came from General Forbes. After the French and Indian War, these roads served as vital links to the West and also served the young United States after 1776. Over time, the federal government realized that these two old military roads were inadequate for the number of people and goods that traversed them. The plan was set for the National Road.

The story of the National Road is the story of the federal government's first major civilian construction effort. The nation's first government-funded road became the first government toll road, which marked the first step toward our modern highway system. Begun with approval from President Thomas Jefferson in 1806, the National Road would eventually run from Will's Creek in modern-day Cumberland, Maryland, to Vandnella, Illinois, then the state's capital. This road saw traffic ranging from Conestoga wagons to herds of animals. There would be routine stops on the road to rest, eat, and get fresh teams of horses. These structures were normally known as "taverns" or "wagon stands."

Although the National Road was an important beginning, road travel could not compete with water travel. Most often, roads were used to get people and goods between navigable rivers. By using each of these modes of travel, the northeastern and Mid-Atlantic United States reached inland to develop and settle lands toward the Appalachians. Important trade networks extended deep inland from the eastern coasts. The South, however, did not completely follow this model of development, creating enduring examples of regional difference in the United States.

Emanating from the Mid-Atlantic

Although Spanish societies had already settled portions of the West from the Pacific inward and native peoples called nearly every inch of the region home, Americans were steadily following their government's design and leaving homes in the eastern United States to settle new plots cordoned off by the surveyors who carried out the nation's new land policies. Using trails such as the National Road as a jumping-off point for movement westward, significant numbers of American citizens left settled areas of the Mid-Atlantic and eastern United States to seek new opportunities in the west from 1810 onward. Much of what is known about this first wave of settlers comes from cultural geographers who have traced these patterns from the landscapes and artifacts that settlers created and then left behind.

Along what geographer Terry Jordan refers to as the "backwoods frontier," Scotch-Irish settlers left homes in the Mid-Atlantic region to cross the Appalachians and form the first wave of American settlers. Typically, these settlers constructed log cabins in a form that Jordan traced back to Swedish roots (Jordan and Kaups 1986, 91). Remnants of these cabins offer geographers a traceable pattern of western settlement. These settlers typically converted a forested plot to agricultural farm land and then rather quickly moved farther to the west, leaving behind their farm land for the next wave of settlement.

By using the cabin structures as well as the barn designs found throughout the midwestern United States, geographers have constructed a model of the population's shift westward that is referred to as cultural hearths. Organized around the idea of cultural diffusion, geographers have traced westward movement in North America from three primary coastal hearths: New England, Mid-Atlantic, and South (Jordan and Kaups 1986, 20–23). Similar to the hearth in front of a fireplace, these settlement clusters served as incubators for cultural exchange. Settlers then took these ideas and traditions with them when they moved to new regions. At each stop, the settlers' primary task was to transfer a land normally found in forest to the Jeffersonian model of a productive, agricultural landscape. The conversion of land for agriculture marked the most significant change in American geography during the first half of the nineteenth century. Although this change had been observable throughout American history, during these years, it became a national compulsion.

Farmers most often felled forest to clear the way for their fields. The majority of settlers filed claims in untouched forest lands, convinced that their energy and enthusiasm to create a holding would compensate for their lack of capital and help them through the difficulty of building and clearing. In the budget of the settler, clearing land was the most expensive item. Historian Michael Williams estimates the cost of clearing as follows. In 1821, a New York farmer made a plan for making a small farm of fifty forested acres. First, the settler purchased thirty acres for approximately $1 to $6 per acre. Clearing this land became the priority. Typically, five acres included the house, barns, orchards, and gardens, ten acres were meadow and pasture, and fifteen acres were reserved for crops.

Clearly, a farmer needed funds to help him and his family initiate the farm and establish its future productivity. Most accounts estimated that one settler might hope to clear ten acres and erect a cabin during the first year. In the second year, he might have enough crops to enable him to buy supplies, including a cow, oxen, sheep, hogs, and a plow and harrow. During each of the following years, the farmer might hope to clear fifteen acres. At this pace, the entire one-hundred-acre farm could be cleared within a decade.

Some fortunate settlers possessed enough capital to hire setup men, itinerant laborers who assisted in chopping down trees and clearing the land. Equipped with axes and hoes, these laborers often charged 50¢ for a full day's labor. Some farmers who could not clear their own land fast enough to reap a profit might spend the year clearing others' land.

The season in which the clearing was performed also influenced the cost. Although a spring chopping was quickest, the green wood made firing difficult. In summer, chopping was a bit quicker because the wood was drier and took less effort to burn. This made it cheaper than a spring clearing; however, a winter chopping was cheaper still. The great advantage of winter, of course, was that clearing took no time away from the major tasks of farming that took place during other seasons.

Whether the land was cleared through clear-cutting or girdling, the process left behind many stumps. Farmers hoped that, after five to ten years, the stumps would rot sufficiently to be pulled out of the ground. Hardwood stumps, including hickory and oak, could rot in five years. However, many types of pines did not rot for twenty years. All totaled, the best estimates place the cost of clearing land at between $10 and $12 per acre. Given all of the demands on a homesteader's time, clearing the land took five to ten years (Williams 1992, 126–27). Farmers would clear the land until they had sufficient acreage to grow their crops. At this point, remaining forest became little more than a source of fuel. This process was the norm for settling any region. However, once settlers established their farms, regional distinctions became clear.

Steadily, throughout the 1800s, settlers passed through the Midwest on their way to other areas of settlement. Although the process of making the frontier primarily extended west through the Mid-Atlantic, the settlement effort also got a boost from the opposite direction. The Pacific coast of North America had opened as the 1700s closed. Various European powers still had settlements on the coast, but Americans began to arrive after the 1790s. The most famous outpost was Astoria, which as a hub for the fur trade placed a pocket of white settlers in the distant West by early 1811.

Astoria Establishes an Anchor in the West

When Robert Gray guided his ship the *Columbia* between Cape Adams, Oregon, and Cape Disappointment, Washington, in 1792, he placed a goal before the American spirit of expansion. For the first time, Gray bounded the far end of the continent. On landing, the American trader placed a U.S. flag in the land, now known as Washington. There was much discussion over who rightfully owned the land. The British, Russians, Spanish, and Americans all feuded over ownership. Thomas Jefferson put an end to the discussion when he commissioned his personal secretary, army captain Meriwether Lewis, to explore the Northwest Territory. For others, however, Gray's landing presented a goal entirely separate from its relationship to the interior United States. The first form that this place took was a trading post known as Astoria.

Astoria was primarily the dream of one man: New York financier John Jacob Astor, who sent traders in 1811 to the Pacific coast to establish a trade route. They established a fort to help get them through the harsh winters. In 1813, British armies took over the fort, but it was retaken by U.S. forces in 1818. The site of that fort is now named Astoria, after the man who funded the expedition to establish a trade route that would provide for the future.

Pioneers would soon follow the Oregon Trail, laid down by the brave explorers who had come before them. They settled the lands of Clatsop County, and, by 1860, the population was 498. Only twenty years later, the population had boomed to 7,222. Some settlers farmed, others involved themselves in trade, but almost all hunted.

Astoria functioned as an outpost for hunters from a variety of cultures. Native peoples of tribes including the Chinook and the Yakima used their age-old hunting skills to trade with white merchants recently arrived to the region. After venturing into the unsettled surrounding regions, trappers returned to trade and sell in Astoria. Their wares pooled together to form gigantic piles of pelts awaiting shipment by boat to ports of call throughout the world.

Astor's vision of this place began with economics, specifically, the profits to be made by organizing and concentrating the pursuit of furs that could then be sold throughout the world. The fur frontier had begun for European settlers in the northeastern and then Great Lakes regions of the United States. Now Astor and others sought to develop new frontiers in the distant West.

Astor had little knowledge of the western United States. He emigrated to New York in 1783, where he learned about prosperous fur-trading opportunities in America. By 1786, he had opened his own shop to sell furs and other materials. He soon met Alexander Henry, a businessman in Montreal, who further educated Astor concerning western trading posts. Astor began to formulate his plan to make a fortune; however, in addition to gaining capital and technical knowledge, his plan also required Astor to become aware of the politics of westward expansion. Astor began courting political influence with U.S. officials as well as those representing foreign governments in the region. As he became more knowledgeable of the fur companies expanding to exploit the American Northwest and Canada, he devised his own vision for a fortune in fur.

In 1809, Astor chartered the Pacific Fur Company, which he hoped would create a chain of fur-trading posts in the Pacific Northwest. To compete with companies and traders from all over the world, Astor decided to send a ship to the mouth of the Columbia River. This would be partly a symbolic gesture to claim the area for himself, but he would also order the ship's crew to construct a fort. This fort, then, would be ready for the arrival of a second party belonging to his company that would make the long overland trip from St. Louis. In early 1810, the final piece of the Astoria plan fell into place when Astor enlisted the cooperation of the Russians who traded along the Pacific coast. He received assurances from the

Russians that no weapon brought to the region via Astoria would be sold to the region's native population. In this arrangement, Russian traders would purchase all of their supplies from Astor. Astor, as the middle man in the trade network, took responsibility for carrying Russian furs to Canton, China. Now Astor needed to make sure the Astoria settlement took proper form (White 1991, 4–11).

Leaving in March 1811, the Astor's land-based team followed Lewis and Clark's route before shifting to a more southerly route. The seagoing team traveled on the *Tonquin*, which Astor had purchased in 1810 for $37,860. The *Tonquin* departed New York Harbor in September 1810 and arrived at the mouth of the Columbia River on March 22. On their arrival, the Astorians met the Chinook Indians and began the regional trade networks that would define the region's early history. The fort was built, and, almost a year later, the overland crew arrived. Despite the hardships that both the overland and seaward expeditions endured, both arrived at Astoria in good enough shape to carry on Astor's grand business enterprise (White 1991, 70).

Enterprises such as Astoria used trade opportunities to create pockets of population in the American West. By finding value in the everyday nature of this region, American traders and settlers began to remake the region. Initially, some of these undertakings were quite small affairs. For instance, Astorians worked closely with the Chinook people and other native groups to harvest the Columbia River's seasonal supply of salmon (White 1996, 14–20). By the mid-1800s, however, the movement of Americans was escalating. In addition, trade networks made the movement westward easier than ever before.

In general, the fur trade surged westward from 1700 to 1850. Beaver colonies suffered severe population decline, which also affected stream ecology in the region. However, no animal targeted by trappers for its fur became extinct. Beaver trapping continued until about 1840, when the demand for fur declined. Although beaver skin had fetched $6 per pound in the 1820s, the price had dropped to 50¢ per pound by 1850.

Trails West Fuel Western Development

The primary flow of settlers continued to flow from the east, however. As the cultural support of this effort became more widespread, the trails became more formal. However, the majority of settlers before 1850 traveled on trails that extended the paths and roads of the Mid-Atlantic over the Ohio River and into the West.

The major overland trails included the Oregon, Sante Fe, California, and Mormon Trek. Each allowed wagon, horse, and foot access to different portions of the American West. Along each trail, settlement grew, and, ultimately, trade began to flow (White 1996, 44–46). By the mid-1800s, surveying for wagon roads had also contributed to the pace of settlement. Between 1850 and 1860, engineers constructed thirty-four separate roads throughout the western territories (White 1996, 127).

As settlers moved farther west, they made the territorial claims by other nations increasingly untenable, not to mention opening consistent and persistent conflicts with native peoples. In the 1844 presidential election, James Polk brought the nation a federal policy of unapologetic expansionist zeal. At this time, about 5,000 settlers occupied Oregon, which obliged the British government to cede all parts of the territory south of the 49th Parallel to the United States in 1846. In Texas, some of the 20,000 settlers protested Mexican rule in

1835, which ultimately resulted in the annexation of Texas in 1845. The Mexican War, then, added territories from Texas to the Pacific.

The Indian Problem

Throughout the settlement process, one of the primary difficulties was contending with the population already inhabiting the land. In fact, it is, in particular, the federal treatment of native peoples that takes away any honest debate on whether or not the movement westward was, in fact, a legacy of conquest or not. Whether in the South or Mid-Atlantic, each extension of the young nation came at a cost, because each region of North America had been occupied previously for hundreds of years. What historian William Cronon described as a "widowed land"—areas left uninhabited by shifting population and the impact of European-borne disease—eventually gave way to a prolonged period of contact between native populations and European settlers. At times, this interaction resulted in mutually beneficial trade. The general trend, however, saw native peoples on increasingly constricted areas of land with increasingly limited possibilities for economic development.

After the Revolutionary War, Congress treated Native Americans as conquered peoples. The national government made treaties with various tribes, arbitrarily drew boundaries, and as a general rule refused to offer compensation for lands. Although native peoples were generally forced to cede their lands, in the frontier areas in which natives interacted with European settlers, a vibrant economic relationship evolved. However, American leaders knew that the nation could not develop until it had confronted what was referred to nationally as "the Indian problem."

By 1786, it had become clear to the federal government that the treaties were not effective. A war against the dispersed native groups would be too costly for the U.S. government. Therefore, in 1787, Congress turned from coercion to a policy of conciliation and purchase. One of the first expressions of this policy was the Northwest Ordinance of 1787. This policy, of course, assumed that Native Americans would willingly negotiate and sell their lands. This assumption resulted in warfare during the 1790s and the Greenville Treaty of 1795, which resulted in the further loss of native lands north of the Ohio River.

Initially, reservation policies allowed native traditions to endure, albeit in difficult circumstances. As settlement pressures increased, the U.S. Army formally and informally made independent life impossible for native communities.

In reaction to the fighting at Little Bighorn and other forms of resistance, the federal government abandoned the reservation policy and adopted a new approach. Federal Indian policy during the period from 1870 to 1900 marked a departure from earlier policies, which were dominated by removal, treaties, reservations, and even war. The new policy focused specifically on breaking up reservations by granting land allotments to individual Native Americans. Very sincere individuals reasoned that they were helping native people by assisting them in adopting the culture of Americans. Many truly believed that they were helping the native peoples by making them responsible for their own farm and assimilating them into American life. Today, many critics argue that such paternalism destroyed native culture and life.

Creating policies that would at once break down native culture and free up western lands for American settlement, the U.S. government made these ideas official on February 8,

1887, when Congress passed the Dawes Act. Named for its author, Senator Henry Dawes of Massachusetts, the law is also referred to as the General Allotment Act. Essentially, it gave the U.S. president the right to break up reservation land, which was held in common by the members of a tribe. The act allowed the president to break the common ownership into small allotments that were then parceled out to individuals. Primarily, however, this act made it easier for Indian lands to be sold off to American settlers. The basic structure of the allotments included the following: each head of family would receive one-quarter of a section (120 acres); each single person over eighteen or orphan child under eighteen would receive one-eighth of a section (sixty acres); and other single persons under eighteen would receive one-sixteenth of a section (thirty acres).

Initially, the act exempted the Cherokees, Creeks, Choctaws, Chickasaws, Seminoles, and Osage, Miamies and Peorias, and Sacs and Foxes, who were located in the Indian Territory as well as lands belonging to the Seneca Nation of New York and the Sioux Nation in Nebraska. In later negotiations, however, most of these groups agreed to allotment if they would dissolve their tribal governments. To receive the allotted land, tribal members needed to enroll with the Bureau of Indian Affairs, which placed the individual's name on the "Dawes rolls." At this point, the individual (who now had no tribal affiliation) was eligible to receive land.

At the time, the federal government argued that the Dawes Act protected Indian property rights, particularly during the land rushes of the 1890s. Most often, however, the new land owners received land of little worth: desert or near-desert lands unsuitable for farming. Even if the land proved capable of being farmed, agricultural techniques used by settlers were often quite foreign to Indian understanding. Often, the landowners could not afford the tools, animals, seed, and other supplies necessary to get started.

Gadsden Purchase Claims the West for America

Under the pressure of war in 1836, Mexican President Antonio López de Santa Anna, who was a prisoner of war with Texas, negotiated a settlement that recognized the Rio Grande River as the boundary between Texas and Mexico. However, once the Texans freed Santa Anna, he renounced the Treaty of Velasco. The status of this boundary remained the same for the next few decades. Seventeen years later, Santa Anna was still leading the Mexican government, but now his nation was going bankrupt. He opened negotiations not with Texas but with the United States, which now controlled Texas.

From the American perspective, the area known as the Mesilla Strip had been creating difficulties for years. Negotiations had broken down but President Franklin Pierce was adamant that a purchase was needed that would clarify the boundary disputes left open by the errors and miscalculations within the Treaty of Guadalupe Hidalgo. Pierce entrusted this task to his Ambassador to Mexico, James Gadsden, who was president of the South Carolina, Louisville, Charleston, and Cincinnati railroads. One of the primary reasons for the purchase and the need to settle the boundary disputes was the interest in connecting the West via railroads; so Gadsden was the perfect man for the job. Pierce gave Gadsden authority to pay up to $50 million for large portions of the northern states of Coahuila, Chihuahua, Sonora, and all of Baja California.

Ultimately, Gadsden and Santa Anna negotiated the treaty many times, changing boundaries and prices. Finally, Gadsden and Santa Anna agreed on a narrow belt of land comprising today's southern Arizona and New Mexico. The area was bordered by three rivers: on the east by the Rio Grande, on the north by the Gila River, and on the west by the Colorado River.

Santa Anna's sale of this valuable section of northern Mexico to the United States took place with the Treaty of La Mesilla, also known as the Gadsden Purchase, which added 29,649 square miles to the United States, including what is now southern Arizona and New Mexico. Signed at the end of 1853, the Gadsden Purchase was the last major American land acquisition on the North American continent. Additionally, the treaty bound the United States to prevent incursions of Indians from the United States into Mexico and to restore Mexican prisoners captured by such Indians. For this responsibility and the vast acreage, the United States paid Mexico $10 million. The text included the following language:

> IN THE NAME OF ALMIGHTY GOD: The Republic of Mexico and the United States of America desiring to remove every cause of disagreement which might interfere in any manner with the better friendship and intercourse between the two countries, and especially in respect to the true limits which should be established, when, notwithstanding what was covenanted in the treaty of Guadalupe Hidalgo in the year 1848, opposite interpretations have been urged, which might give occasion to questions of serious moment: to avoid these, and to strengthen and more firmly maintain the peace which happily prevails between the two republics, the President of the United States has, for this purpose, appointed James Gadsden, Envoy Extraordinary and Minister Plenipotentiary of the same, near the Mexican government, and the President of Mexico has appointed as Plenipotentiary "ad hoc" his excellency Don Manuel Diez de Bonilla, cavalier grand cross of the national and distinguished order of Guadalupe, and Secretary of State, and of the office of Foreign Relations, and Don Jose Salazar Ylarregui and General Mariano Monterde as scientific commissioners, invested with full powers for this negotiation, who, having communicated their respective full powers, and finding them in due and proper form, have agreed upon the articles…. (Gadsden Purchase)

Conclusion: Catlin and Others Rationalize the Move West

When President Thomas Jefferson argued for western expansion, he likely could not have imagined the elements of culture and politics that would be essential to its success: the available land combined with the grid system to form the infrastructure for American expansion. However, a unique culture was needed to grease the wheels of this important change in American history. In the spirit of John Winthrop and others, Americans began applying their "choseness" to the western United States. An attitude of American "exceptionalism" took root during the 1830s. In the 1840s, this spirit found expression in the policies of Franklin Pierce and James K. Polk as they struggled to find logic for the American right to take the western United States from native peoples and European control.

The belief that American expansion westward and southward was inevitable, just, and divinely ordained was first labeled "manifest destiny" by John L. O'Sullivan, editor of the *United States Magazine and Domestic Review*. This logic was quickly attached to expansionist rhetoric during the 1840s. Among the long-standing objectives of expansionists was the Republic of Texas, which in addition to Texas included parts of present-day Oklahoma,

Kansas, Colorado, Wyoming, and New Mexico. After winning its independence from Spain in 1821, Mexico encouraged the development of its rich but remote northern province, offering large tracts of land virtually free to settlers. The settlers in turn agreed to become citizens of Mexico, and, by 1835, 35,000 Americans lived in Texas. Expansionists began clamoring for the nation to accept the inhabitants' desire for statehood. The annexation of Texas, O'Sullivan wrote in 1845, was "the fulfillment of our manifest destiny to overspread the continent allotted by Providence for the free development of our yearly multiplying millions."

It was exactly this spirit that Polk would parlay into a rationale for the contrived 1848 war with Mexico. Thanks to this expansionist war and other negotiations, by 1850, the United States stretched from sea to sea. This expression of manifest destiny forced the nation to confront territorial issues such as slavery, ultimately culminating in the Civil War.

In addition, the economic power of the railroad allowed it to take a dominance over Western areas that was quite peculiar from that seen in other regions. When Norris used the railroad as a symbol of destruction in *The Octopus* in the late 1890s, he wrote from a perspective of reform and change over the great control exerted by the railroad over western shipping. Although the iron horse made trade possible in even the most remote stretches of the West, railroad companies fostered the reliance of western farmers and then rather ruthlessly managed shipping rates for their own gain.

In the end, trails, policies, and, even new technology were not the primary elements of American settlement. The politics of "manifest destiny" bred a permanent portion of American culture that identified with stereotypes, imagined and real, of the settlement process that became part of a unique "frontier culture." The popular interest in the culture of the frontier West helped to propel the conquest of the West into permanent settlement. It began, however, in piecemeal fashion.

One of the most important elements of this "cultural" expansion was the unique nature of the American West. Very quickly during the mid-1800s, the odd nature of the western United States captured the imagination of many Americans. George Catlin was one of the first artists to try to capture the everyday nature of this unknown region. In addition to the landscape and plants of the region, however, Catlin went to great lengths to include details of the West's threatened human population. Native peoples, in this perception, were a part of this unique place and its natural environment.

In the spring of 1830, Catlin left Pennsylvania for St. Louis, where he met General William Clark, who by this point was serving as superintendent of Indian Affairs for the western tribes. After seeing Catlin's portfolio of paintings of eastern tribes, Clark assisted Catlin in forming the contacts that would allow him to paint portraits of some of the West's great native leaders as well as the landscapes that they inhabited. Catlin conducted approximately forty-eight different trips and painted more than 300 portraits. In addition, Catlin approached his task much like an anthropologist by painting 200 more images of Indian religious ceremonies, games, buffalo hunts, and of the peoples' living environment. He opened the North American Indian Gallery in 1837 and published his findings in a book in 1841. Catlin observed that to reach the true frontier,

One is obliged to descend from the light and glow of civilized atmosphere, through the different grades of civilization, which gradually sink to the most deplorable conditions along the extreme frontier.... Through the dark and sunken vale of wretchedness one hurries, as through a pestilence, until he gradually rises again into the proud the chivalrous pale of savage society in its state of original nature beyond reach of civilized contamination. Here he findes much to fix his enthusiasm upon, and much to admire. (Catlin, *Letters and Notes*)

Catlin offered a sensibility that was largely sensitive to the massive changes being wrought on the American West by American settlement. Most Americans, however, were buying into a national penchant for perceiving the vast amount of unused space and resources seemingly available in the western lands, and this perception was about to get even more intense.

Although art continued to celebrate America's natural beauty, it was no longer for religious or spiritual sake. By the 1860s, art was often accompanied by expectations for a change in the treatment or administration of nature. Increasingly, these expectations involved the federal government. This was particularly true regarding the American West, with which federal policy was just being created during the late 1800s. The "settlement" process became a symbolic one for many romantics. As the western United States became more and more civilized, Americans began discussing what was being lost.

Historian Roderick Nash wrote, "With a considerable sense of shock, Americans of the late nineteenth century realized that many of the forces which had shaped their national character were disappearing. Primary among these were the frontier and the frontier way of life" (Nash 1982, 145). This perception stimulated a romantic view of elements of the raw western United States, including native peoples, cowboys, open land, and wild animals.

In painting, Romantic art returned to the idealized landscape but not the landscapes of classical civilizations. Instead, painters such as Albert Bierstadt, Frederic Church, and Thomas Moran used their keen observations of the West to transform it into the promised land of America. Bierstadt's paintings of the Rockies or Moran's portrayals of the geological wonders of the West depict the American landscape with a primeval majesty intended to carry the viewer to the land's prehistoric era, often before human intrusion.

Artists who explored the West, whether or not they were members of the Hudson River School, painted with a sensibility that magnified the beauty of the American West. Some artists, including Bierstadt and Thomas Moran, wanted to convey grandiose impressions of the wilderness they saw. This subjectivity, however, was unknown to the audiences in eastern cities who viewed the magnificent panoramas of western landscapes.

So, although Americans may have never seen the western United States, by the late 1800s, they identified settling the West as a unique American project, one that defined their nation from all others.

Sources and Further Reading: Catlin, *Letters and Notes*; Conzen, *The Making of the American Landscape*; *The Gadsden Purchase Treaty*, http://www.yale.edu/lawweb/avalon/diplomacy/mexico/mx1853.htm; White, *It's Your Misfortune and None of My Own*.

THE WAR OF 1812 STEERS AMERICANS' ENERGY FUTURE TOWARD COAL

Time Period: 1812–1818
In This Corner: Industrial interests, coal owners, transportation developers

In the Other Corner: Industrial interests with vested interest in lumber, foreign nations
Other Interested Parties: Federal leaders, citizens
General Environmental Issue(s): Economic development, energy, coal, Appalachia

Many factors can dictate a shift in technology that is known as an energy transition. Often, such shifts do not occur at once and, therefore, factors can accumulate over a period of time that, ultimately, provide a tipping point that causes revolutionary changes in the way that people live. Often, these factors will include discovery of a new supply or depletion of a previous one. At other junctures, tips may come from new cultural understandings or expectations. Although these factors may not deal specifically with issues related to supply and demand, they may, by implication, influence choice and affordability. At the nation's founding, the primary energy source for Americans was wood. When the young United States swayed from this course and began deriving more energy from coal, the tipping point was war.

Similar to other moments of energy transition, new ideas were not welcomed with unanimity. Industrialists powered by water power and those involved in the charcoal and wood industry undoubtedly viewed a transition to coal with great trepidation. In addition, many critics felt that coal was unclean. One of the standard critiques of Britain, for instance, was of its abundant soot and smoke, yet in the spirit of expansion that swept Americans in the early 1800s, it was a national good to locate and make use of one of the nation's most ubiquitous resources.

Soft Coal Ushers in the Industrial Era

Although British settlers to North America were well aware of the capabilities of coal, settlers had little use for North America's massive reserves of coal before American independence. With abundant supplies of wood, water, and animal fuel, there was little need to use mineral fuel in seventeenth-and eighteenth-century America. The growth of cities in the northeastern United States, however, offered new opportunities for coal use, and American forges and furnaces began to appear in areas outside urban concentrations.

The earliest coal that was used in American industry was very likely imported from Great Britain. By the 1700s, however, a small bituminous trade developed in the fields outside of Richmond, Virginia, and near Pittsburgh, Pennsylvania. In terms of anthracite, Native Americans near Nazareth, Pennsylvania, are reported to have begun making use of it by 1750. Nearby settlers had begun using the anthracite for commercial and industrial purposes by the end of the eighteenth century.

The demands of the American Revolution spurred colonists to put anthracite to work in forges making weapons. After the war, manufacturing developed further and with it brought additional markets for coal. Imported coal, however, became much less common. Voices that called for tariffs and trade limitations also suggested that the nation's coal trade, at that time centered in the Richmond coal basin of eastern Virginia, would serve as a strategic resource for the nation's growth and independence. Although the Richmond Basin was the nation's first major coalfield, it had grown on the back of slave labor and the southern plantation economy. Miners found growth potential in the region to be limited.

New interest, however, grew in a related mineral, anthracite coal. Nicknamed "stone coal," the harder anthracite also possessed a higher carbon content. Early on, the coal proved difficult to burn; however, as furnace temperatures grew, the coal proved to last longer as an energy supplier.

However, before either type of coal could really develop as an energy source, developers needed to construct the infrastructure to move coal where it was needed. Although the pace and intensity of American life had changed by the early 1800s, nature remained at the core of its existence. For this reason, it makes sense that the new century would be ushered in at a seaport, which harkened back to the earliest days of European settlement. The conflict that had begun with England had not been settled with the Revolution. Especially in matters relating to the natural resources of North America and their shipment to Europe, great animosity still swirled between the nations during the early 1800s.

The War of 1812, which pitted the United States against Great Britain, occurred when these disagreements escalated. The war started in 1812 and ended in stalemate in 1815. The root of the conflict concerned the rights of American sailors who were being pressed to serve in the British Navy. The major military initiative of Britain during the war, however, was more related to trade: the British blockade of ports such as Philadelphia nearly crumbled the economy of the young republic. Although the British fleet carried out attacks on Yorktown and along the Great Lakes, they clearly set their sites on the port of Baltimore.

To capture Baltimore, the British planned a combined land and naval attack. Neither proved successful. On land, Baltimore troops fought for two hours on September 12, 1814 and delayed the British from reaching the city. At Fort McHenry (which had been built to protect Baltimore's harbor at a cost of $35 million!), the troops sustained a twenty-five-hour naval bombardment on September 12 and 13, 1814. This bombardment, of course, inspired Maryland lawyer Francis Scott Key to write the "Star Spangled Banner" when he spied the huge flag still flying over the fort by dawn's early light.

The attacks on Baltimore, however, derived from its economic significance in the global age of sail. The port benefited from the construction of internal trade routes, including roadways and two canals: the Chesapeake and Delaware Canal, which opened in 1829 to link the Bay with the Delaware River, and the Susquehanna and Tidewater Canal along the lower Susquehanna, which diverted Pennsylvania produce away from Philadelphia.

The blockades of the War of 1812, however, became instrumental in moving the United States more swiftly toward its industrial future. Depleting fuelwood supplies combined with the British blockade created domestic interest in using anthracite or hard coal, particularly around Philadelphia. Historian Martin Melosi wrote, "When war broke out … [Philadelphia] faced a critical fuel shortage. Residents in the anthracite region of northeastern Pennsylvania had used local hard coal before the war, but Philadelphia depended on bituminous coal from Virginia and Great Britain." Coal prices soared by more than 200 percent by April 1813. Philadelphia's artisans and craftsmen responded by establishing the Mutual Assistance Coal Company to seek other sources. Anthracite soon arrived from the Wilkes Barre, Pennsylvania area.

After the war of 1812, anthracite coal gained much larger acceptance in urban markets that were near coal fields. Shortages of soft coal, whether from Britain or Virginia, created new opportunities for anthracite, which some sellers called "stone coal." Scientists, primarily those from Philadelphia's American Philosophical Society and Franklin Institute, tested the new energy source and widely reported their positive findings. In addition, new transportation networks opened in the 1820s that could more effectively bring northeastern

Pennsylvania's anthracite to markets. The industrial use of hard coal continued to increase slowly until 1830. Between 1830 and 1850, the use of anthracite coal increased by 1,000 percent. The energy contained in this black rock would alter the nature of American industry during the nineteenth century (Melosi 1985, 26–27).

Coal mining, both hard and soft, expanded into states beyond Virginia and Pennsylvania by the mid-1800s. For instance, by 1850, Ohio's bituminous fields employed 7,000 workers and produced more than 320,000 tons of coal. Maryland followed, and then the expanding railroad system made coal industries profitable in Illinois and Missouri. By the advent of the Civil War, coal industries appeared in at least twenty states. In addition to the influence of the War of 1812, new industries stimulated the viability of coal as an energy source.

Small-scale Railroad Increases Reliance on Coal

It was a cruel irony that the industrial era that evolved in the late 1800s relied intrinsically on transportation. Long, slender mountains stretched diagonally across Pennsylvania, creating an extremely inhospitable terrain for transporting raw materials. Opening up the isolated and mountainous region required the efforts of a generation of capitalists and politicians, who used their resources and influence to create a transportation network that made the coal revolution possible. Canals were the first step in unlocking the great potential of the coal fields. Soon, however, industrialists focused on a more flexible transportation system that could be placed almost anywhere. Railroads quickly became the infrastructure of the industrial era. Knitting together the raw materials for making iron, steel, and other commodities, railroads were both the process and product of industrialization.

The iron rails produced in anthracite-fueled furnaces extended transportation routes throughout the nation. This revolution in transportation led to corresponding revolutions in the fueling of industries and the heating of urban residences, which in turn required more and more miners and laborers.

Where coal occurred, new possibilities now unfolded for community development. In the case of eastern Pennsylvania, Mauch Chunk is a town that demonstrates how the acceleration of industrial development was nearly as much a cultural development as technological. Although harvesting, processing, and shipping industrial materials involved hundreds of different communities, one was specifically designed to serve as a showcase of industrial development. The town of Mauch Chunk, which is now known as Jim Thorpe, was designed to demonstrate industrial processes to others involved in industry as well as to the general public.

Mauch Chunk was owned and developed by the Lehigh Coal and Navigation Company from 1818 to 1831. If anthracite coal had not been found, the town of Mauch Chunk would have been much different, but in 1791, Philip Ginter discovered coal on Sharp Mountain. Then, in 1818, the Lehigh River became the center of anthracite coal transportation, and Mauch Chunk became an integral link along this shipping corridor. Roads were made to connect the primary coal mine to the river. In 1827, a new era in technology was ushered in when the wagon road was transformed into a gravity railroad. One of the nation's first small-scale rail lines, Mauch Chunk's railroad became known as the Switchback Gravity Railroad. From the mine's location at higher elevation, the coal could then travel down the mountain by means of this railroad system using gravity. The empty cars were then hauled up the mountain by mules.

All the coal from Summit Hill and Panther Valley arrived at the canal landing in this fashion. Then, in 1872, a large tunnel opened that ended the need for the gravity railroad. At approximately the same time, the gravity railroad was replaced for coal hauling by the steam-locomotive-powered Panther Creek Railroad. This, however, did not end the gravity line's history. In 1874, developers purchased the switchback gravity railroad and made it the centerpiece attraction in Mauch Chunk, the tourist attraction of the industrial era. Eventually, the switchback railroad inspired roller coasters.

Historian Stephan Sears contrasts Mauch Chunk with Niagara Falls, the other great tourist attraction of the northeast, while also making an important statement about nineteenth-century tourists' interest in visiting industrial sites. He wrote, "At Niagara Falls industry intruded on visitors' expectations of what they had come to see, but at Mauch Chunk it was an integral part of the scene" (Sears 1989, 191–92). This is the appeal that made Leo Marx investigate technology's role in nineteenth-century literature. Historian David Nye refers to this attraction to industrial blight as the industrial or technological sublime (Nye 1996, 25–27).

Industrial Expansion through Coal during the Civil War Era

Throughout the antebellum period, the seams that were first worked in the anthracite fields of eastern Pennsylvania or the bituminous fields in Virginia, western Pennsylvania, and Ohio tended to lie close to the surface. A skilled miner and a handful of laborers could raise several tons of coal per day by using a "drift" or "slope" mine to access a vein of coal running along the subsurface of a hillside.

Most often, a few skilled miners would be aided by a larger cast of less skilled laborers. Strong labor unions did not develop in the early years of mining for a variety of reasons. One was actually the type of mining that went on. All the miners worked relatively close to the surface but spread widely over a far-flung area. In this manner, the nature of the extractive process during these early years also helped to minimize worker organization.

As the pursuit of coal changed, workers began to feel that collective action was necessary. Unions became a significant portion of coal's story in the late 1800s. The rate of coal production also grew significantly during these years. In 1840, American miners raised 2.5 million tons of coal to serve the needs of industrialization. As these needs only grew by 1850, coal production grew to 8.4 million tons.

By the late 1850s, coal development expanded to influence nearly the entire nation. Short-line railroads allowed the access and development of new bituminous coalfields in new producers, including Maryland, Ohio, and Illinois. In the established anthracite coal regions of Pennsylvania, the scale and scope grew as railroad companies reaped great profit from the increased traffic of the Civil War years. West of the Mississippi River, railroads also helped open up massive coal reserves. Although small coal mines grew in Missouri and Illinois in the 1850s, all their production traveled by steamboat on the river. As railroads expanded through the Great Plains, this region could use the same infrastructure being used in the Mid-Atlantic. Large-scale coal fields grew in distant locations, including Colorado, New Mexico, and Wyoming. Therefore, by the late 1800s, coal had become a dominating and unifying endeavor in the United States.

New mining methods also assisted this expansion. Early slope or drift mines intersected coal seams relatively close to the surface and needed only small capital investments to prepare. Most miners still accessed the coal seam by using picks and shovels; however, others had begun to expand their capabilities by using black powder to blast holes in the seams. After detonation, the miners broke up larger chunks and loaded the broken coal onto wagons by hand. These methods worked as long as the mines did not extend too deep into the earth.

As the desire for additional amounts of coal drove miners deeper, mines extended below the water line. New technological innovations provided the pumping, ventilation, and extractive implements required by the deeper locations. Most important, steam power was brought into most mines. By the 1890s, electric cutting machines replaced the blasting method of loosening the coal in some mines. By 1900, these new methods were used to mine approximately one-quarter of America's supply of coal. As the century progressed, miners raised more and more coal by using new technology. Along with this productivity came the erosion of many traditional skills cherished by experienced miners, and, of course, these new technologies meant that coal mining required larger amounts of capital. By necessitating larger corporate involvement and control, new technologies also stimulated the need for workers to join together and unionize to protect their interests as much as possible.

These changes in the mining industry allowed for a huge increase in the coal that was available for various industrial pursuits. As a prime mover energy source, coal fueled the growth of many other sectors. The next section will list a series of these pursuits.

Iron Industry Demands Energy

Iron making was the preeminent coal-burning industry of the antebellum years. It was also the first and primary industry to shift from soft coal to anthracite.

As early as the 1780s, British iron makers had been using bituminous coal or coke, which is bituminous coal with the impurities burned off. Anthracite, which was more prevalent in the United States, was successfully used by American iron manufacturers who had modeled their original industry on that of Britain. In either nation, highly concentrated heat was the priority for making iron. The level of heat required to smelt iron ore required a blast of excess air to aid the combustion of the fuel, whether it was coal, wood, or charcoal. Whereas British iron makers in the 1820s attempted to increase the efficiency of the process by using superheated air, known commonly as a "hot blast," American iron makers continued to use a "cold blast" to stoke their furnaces. The density of anthracite coal resisted attempts to ignite it through the cold blast, and, therefore, although it was plentiful, hard coal initially appeared to be an inappropriate fuel for most American iron furnaces.

Using hot blast technology that he imported from Wales, David Thomas is reportedly the first American to smelt iron with anthracite in Pennsylvania in 1840. His Allentown-based Crane Iron Company had been chartered in 1839. His innovation radically altered American iron making, and similar furnaces began to appear across the Pennsylvania anthracite-producing region. The influence of this innovation grew in 1841 when industry backers cited that significant savings had been acquired by furnace owners using anthracite. In Pennsylvania, anthracite investment was paying great dividends for the state. This success also indicated that anthracite might be used in other industrial purposes. By 1854, 46 percent of all American pig iron had been smelted with anthracite coal as a fuel, and, by 1860, anthracite's share of pig iron was more than 56 percent.

The emergence of coking techniques that baked out the impurities of bituminous coal at high temperatures also fed growth for iron manufacture using coal. It did not, however, stop with its use for iron. The discovery of excellent coking coal in southwestern Pennsylvania spurred the aggressive growth of coke furnaces that were simply designed to bake the bituminous coal and produce coke that could be used to fire furnaces for other purposes elsewhere. By 1880, the Connellsville region contained more than 4,200 coke ovens, and the national production of coke in the United States stood at three million tons. Two decades later, the United States consumed more than twenty million tons of coke fuel. This fuel source could be used anywhere. Most often, it was used in iron and steel manufacture.

Although iron manufacture continued, new technologies allowed for steel manufacturing to become popular by the end of the 1800s. Combining a variety of materials and alloy, steel provided new flexibility in construction.

Conclusion: Coal for a Growing Nation Fuels Pollution Problems

The expansion of industry and coal usage largely went hand in hand. New technology made it possible to increase the supply. The increased supply allowed coal to remain the most inexpensive source of energy throughout the twentieth century. Eventually, of course, coal would additionally serve as the primary supplier of electrical power for use in homes and businesses. Regardless of how and from where it was gotten, the United States became increasingly dependent on coal during the nineteenth century, consuming seventeen million tons in 1861 to nearly seventy-two million tons in 1880.

Historians such as John McNeill and Alfred Crosby argue that inexpensive energy represents a primary unifying characteristic of human life in the late twentieth century. It may be the lack of such supplies that then define the twenty-first century. As the beat of the steam engine came to define the pace of many Americans' work day, lighting such as that made by coal gas allowed work to extend into the night. This also altered the type of workers who were needed. Wrote one observer, "With coal power to substitute for adult muscle, and machinery to substitute for adult skill, factory owners found that children were not only adequate for many jobs, but cheaper and far easier to discipline" (Freese 2004, 77).

With the increase in productivity came clear pollution and health impacts. In the twenty-first century, changes in temperature have even been attributed to coal-burning industry during the last two centuries.

Sources and Further Reading: Adams, *The US Coal Industry in the Nineteenth Century*, http://eh.net/encyclopedia/article/adams.industry.coal.us; Freese, *Coal*; Gordon and Malone, *The Texture of Industry*; Melosi, *Coping with Abundance*; Opie, *Nature's Nation*; Steinberg, *Down to Earth*.

WORKING IN A COAL MINE

Time Period: 1800s
In This Corner: Laborers, reformers
In the Other Corner: Coal company owners, industrialists
Other Interested Parties: Political leaders
General Environmental Issue(s): Workers' rights, coal, energy, mining

Mining coal has always required a serious impact on the land. When job creation was the nation's top priority, few people complained. In fact, entire generations performed harrowing work for little pay—labor that ultimately contributed to serious health problems. Mining methods changed during the twentieth century; however, some critics continue to question whether or not our reliance on this dirty fuel is worthwhile.

Coal Mining as a Growth Business

Although petroleum would become a vital cog in portions of the industrial era, coal was the prime mover that achieved most of the work. Of course, coal deposits are scattered throughout the globe; however, northeastern Pennsylvania holds a 500-square-mile region that is unique from any other. When coal was formed more than a million years ago, northeastern Pennsylvania accelerated the process with a violent upheaval known as the Appalachian Revolution. Geologists speculated that the mountains literally folded over and exerted extra pressure on the subterranean resources. In northeastern Pennsylvania, this process created a supply of coal that was purer, harder, and of higher carbon content than any other variety. Named first with the adjective hard, this coal eventually became known as anthracite. Geologists estimate that 95 percent of the supply of this hard coal in the western hemisphere comes from this portion of northeastern Pennsylvania.

William Henry Jackson's 1892 view of the Elk Mountain coal mine shows the denuded surroundings of the coal landscape. Short-line railroads and hopper cars carried loads of coal out of the mines. Library of Congress.

This supply defined life in the state during the late 1800s. Thousands of families of all different ethnic backgrounds moved to mining towns to support themselves by laboring after coal. In other areas, mills and factories were built that relied on the coal as a power source. In between, the railroad employed thousands of workers to carry coal and raw materials to the mills and finished products away from them.

Coal would alter every American's life through the work it made possible, but Pennsylvania was "ground zero" for the ways that coal culture would influence the nature of work and workers' lives in the United States. The rough-hewn coal communities that sprouted up during the anthracite era reflected the severe organization that defined labor in the coalfields. An elite class of coal owners and operators often lived in magnificent Victorian mansions while their immigrant laborers lived in overcrowded, company-owned "patch towns." The class disparity was perpetuated by a steady change in ethnic laboring groups. Waves of European families arrived to live and work in the company towns found throughout the Appalachian Mountains. The original miners from Germany and Wales were soon followed by the Irish, and later, the Italians, Polish, and Lithuanians.

Despite difficult living conditions and ethnic discrimination from more established groups, these diverse ethnic groups ultimately created vibrant enclaves. In each patch town, each ethnic group built churches, formed clubs, and helped others from their nation of origin to get a start in the fields.

Working a Coal Mine

With its high carbon content, anthracite quickly became the predominant prime mover for American industry. Each step in its production (extracting it from the ground, getting it to market, and preparing it for commercial and household use) helped to develop related industries and to devise new technological advances. Mining in the 1860s sent men, usually small in size, into the earth to chip away at underground seams of coal. Innovations enabled miners to stay in the ground longer and to reach greater depths. By the 1860s, some anthracite miners in northeastern Pennsylvania reached mines at depths of 1,500 feet. Deaths in the mines were an accepted part of the industry. Owners paid little attention to rectifying problems or improving safety. Typically, mine owners simply replaced fallen miners, often with less expensive laborers.

Mining at such depths demanded new technology, which also helped to develop the occupation of mining engineer. Most of the trained mine engineers in the world came from the mines of Wales, England, and Scotland. Their jobs were to build the deep mines and keep them operable. To do so, they needed to wrestle with two primary problems: first, the mines produced water, which had to be drained with a constant system of pumps; second, methane and other poisonous gases that seeped from the ground required a reliable ventilation system. Without ventilation, miners could be poisoned or explosions could start mine fires. The necessary pumps used electricity, which generally required steam power. Power shafts were run to powerful ventilation fans that could pull air through the shafts.

If either the drainage or ventilation systems failed, the mines instantly became pits of death. The greatest danger, however, was igniting mine timbers deep in the earth that created fires that would consume the mine's entire oxygen supply and suffocate the miners.

Case Study: The Avondale Disaster

Coal mining has had long-term health effects on miners and the land on which the mining took place. There were also some very clear impacts that occurred almost immediately. This account tells about one such event, know as the Avondale Disaster.

The following is an excerpt from *The Coal Mines*, by Andrew Roy, State Inspector of mines of Ohio, 1876 (134–37):

> The great calamity of the Avondale shaft, furnishes another case where the whole population of the mine perished for want of means of escape, in time of accident, to the only opening of the mine. This shaft … was divided into two compartments for ventilation, by means of a wooden partition, and had but one outlet.
>
> The Avondale shaft is situated on the right bank of the Susquehanna river, four miles from Plymouth, in Luzerne county, in the heart of the anthracite coal regions of Pennsylvania. The catastrophe occurred on the morning of September 6th, 1869…. One hundred and ten men and boys were in the shaft….
>
> The catastrophe was caused by the ventilating furnace setting fire to the woodwork in the shaft. The fire was discovered about nine o'clock in the morning, by the stable boss of the mine, who had just gone down the pit with a load of hay for the hauling mules. On reaching the bottom, and discovering the fire, he immediately gave the alarm, and in a few minutes afterwards, a cloud of smoke, followed by a mass of living flame, rose through the upcast compartment of the mine. The flames set fire to the breaker, and spreading to the engine house, drove the engineer from his post. The people on top of the shaft became paralyzed with terror, knowing the fate of the miners in the distant chambers of the mine. Dispatches were sent to all the neighboring cities, and in a short time the fire departments of Wilkes-Barre, Scranton, Kingston, and adjoining towns, were on their way to the scene of the conflagration. The news of the accident spread like wild-fire, and people rushed to the burning mine in thousands, to assist in rescuing the imperiled miners; but they were powerless before the burning elements….
>
> On the arrival of the fire engines, streams of water were turned into the burning mine; but the monster volume of lurid flame appeared to bid defiance to the water, and for several hours the fire raged with unabated fury. When at length it had become subdued, a band of volunteers, fifty in number, composed of miners, mine superintendents and colliery proprietors, offered to go down into the shaft to rescue the imprisoned men, or perish in the attempt. The shaft was choked up for nearly forty feet with fallen debris, and it was half past five in the evening, before any attempt at descent could be made.
>
> A dog and lamp were first let down as far as possible, and on being withdrawn, the dog was still alive, and the lamp still burning. An hour later a miner was lowered, who returned in a few minutes, nearly exhausted. Soon after, a shift of men went down with tools to clean out the rubbish. Having effected a landing on the bottom, they advanced for sixty yards along the main gallery of the mine, and came upon three dead mules in the stables. The main door, for directing forward the ventilating current of air, was found closed; they rapped on it with a club, and shouted with all their might,

but on receiving no response, they returned to the bottom of the shaft, and were drawn up to day.

An exploring party was lowered, but were unable to withstand the influence of the deadly gases, and they soon returned. On being raised to the surface, they were nearly overcome. A ventilating fan with canvas hose leading into the shaft, was then erected and fresh air blown into the mine. The next explorers found the ventilating furnace still burning, and also a heap of loose coal lying near the fire. The gases from these fires had been driven forward into the interior of the mine by the ventilator, and it was found necessary to extinguish them before any further attempt was made to penetrate the interior of the mine. All night efforts were made to extinguish the furnace fires, but without success, as it was found impossible to get the water hose to play upon them. The miners, however, reported that the fires were dying out of their own accord.

During the second day several attempts were made to reach the entombed men, but the accumulated gases prevented any extended search. At midnight the air had become greatly improved, and at two o'clock in the morning an exploring party came upon two dead bodies, but they could not recognize their features, owing to their blackened and distorted appearance.

The explorers returned to communicate the fact to the people above-ground. Preparations were at once made for the descent of several bands of explorers, to be divided into groups of four each. At half-past six, as an exploring party was traversing the east side of the plane, they discovered the whole force of the mine lying behind an embankment which they had erected to shut off the deadly gases. Fathers and sons were found clasped in each other's arms. Some of the dead were kneeling, as if in the attitude of prayer; some lay on the ground with their faces downward, as if trying to extract a mouthful of fresh air from the floor of the mine; some were sitting with clasped hands, as if they had vowed to die with each other; and some appeared to have fallen while walking. In two hours, sixty dead bodies were sent to the surface, and by noon the last of the unfortunate one hundred and ten men who had gone down to work three days before, full of health and vigor, were sent up to find their last resting place in the tomb. (Roy, 1876)

Case Study: Breaker Boys and the Call for Mine Safety

Coal mining was dangerous for any miner. However, small spaces underground and the ability to work for low wages made young boys one of coal's most exploited work forces. Although industrialization altered the nature of youth in America in positive and negative ways, the health hazards to young miners mark an unconscionable example of the ethics of extraction.

Often, young boys worked on the long trough known as the coal breaker, on which larger chunks of coal were shaken and jostled to break them apart into smaller pieces. Writers and investigators began to alert the public to the life faced by so-called "breaker boys" in accounts such as this one from the *Child Labor Bulletin*:

The next trapper boy we passed was John. John wanted to go to school but his parents made him work. They didn't know that he could earn better wages later, if he went to school now. The trap door was the nearest thing to a blackboard he had, so he drew

Breaker boys in Woodward Coal Mines, Kingston, Pennsylvania, in the late 1800s demonstrated a willingness to put young people in unhealthy and unsafe working conditions. Library of Congress.

pictures on that. John liked birds, and couldn't see any out of doors, because it was after dark evenings when he left the mine. So he drew them on the trap door, and played they were alive and he wrote on the door, "Don't scare the birds!" and this was all the fun he had.

When we passed a place where the roof had caved in, old Mr. Wise Coal shuddered. "I hope no boys and men are buried there," he said, "they often get killed in that way."

As we came out of the mine we met James. They call him "a greaser" because he has to keep the axles of the car greased so that they run smoothly. He had grease all over himself and his clothes.

Next we met Harry. He does odd jobs about the mine. When he first started at work, he wanted to go to school, but now he does not care. He is too tired to think about it, even.

At last our car full of coal came to a building, called a "coal breaker." Here the coal was put into great machines, and broken into pieces the right size for burning.

Then the pieces rattled down through long chutes, at which the breaker boys sat. These boys picked out the pieces of slate and stone that cannot burn. It's like sitting in a coal bin all day long, except that the coal is always moving and clattering and cuts their fingers. Sometimes the boys wear lamps in their caps to help them see through the thick dust. They bend over the chutes until their backs ache, and they get tired and sick because they have to breathe coal dust instead of good, pure air. (Child Labor Bulletin)

Nationwide, it was estimated that 18,000 persons were employed as slate pickers by the close of the nineteenth century. The majority of these were boys aged ten to fourteen years. Historian Peter Roberts reported at the turn of the century that his representative investigation of 4,131 persons dependent on the mines found that sixty-four workers were under fourteen years of age. There were twenty-four boys employed in the breakers who were not yet twelve years of age.

Such figures and the descriptions of everyday mining life spurred a national call for change in these ethics. Roberts wrote in 1904:

No industry demands the service of boys whose bone and muscle are not hardened and whose brain has not been developed for continuous and effective thinking. Muscle without intelligence is annually depreciating, being displaced by machinery which does nearly all the rough work. To stunt the body and dull the brains of boys in breakers is to rob them of the mental equipment which is essential to enhance their social worth and enable them to adjust themselves to the requirements of modern life. (Roberts 1904, 174–81)

Conclusion: Fighting for Workplace Changes in Coalmines

When change finally came for coal, it brought serious alterations to techniques and methods used for mining. During the twentieth century, high labor costs contributed to companies' interest in adopting methods such as strip mining that required more use of earth-moving devices and less use of laborers underground.

Sources and Further Reading: Freese, *Coal*; Gordon and Malone, *The Texture of Industry*; Montrie, *To Save the Land and People: A History of Opposition to Surface Coal Mining in Appalachia*; Nye, *Technological Sublime*.

GEORGE PERKINS MARSH SPURS CONSIDERATION OF INDUSTRIALIZATION

Time Period: 1860s
In This Corner: Marsh, reformers, conservationists, romantics
In the Other Corner: Most other Americans
Other Interested Parties: Political leaders
General Environmental Issue(s): Industry, conservation, environmental ethics

Through his writings, George Perkins Marsh introduced a new way of approaching land use. Library of Congress.

In each romanticized approach to nature, little criticism was leveled at nineteenth-century life. Primarily, romantics celebrated nature while not detracting from wealthy consumers. However, as nature's beauty was seen worthy of celebration and reverence, some observers were growing increasingly unwilling to overlook the abuses wrought on it by the insensitive. Leading this vocal criticism, George Perkins Marsh emerged in the 1860s as one of the only spokesmen for a scientific orientation who was willing to take on the culture of industrialization.

With the growth of cities in the United States during the nineteenth century, there was a dramatic increase in industry, and, as industry grew, the natural environment was adversely impacted in immediately visible ways. For example, the machinery of many factories was fueled by coal that caused smokestacks to belch black smoke into the air, and industrial byproducts flowed into the waterways, leaving them polluted. The impact of these industries did not go unnoticed to young Marsh, who wrote the following in 1849:

I spent my early life almost literally in the woods; a large portion of the territory of Vermont was, within my recollection, covered with natural forests; and having been personally engaged to a considerable extent in clearing lands, and manufacturing, and dealing in lumber, I have had occasion both to observe and to feel the effects resulting from an injudicious system of managing woodlands and the products of the forest (letter to the botanist Asa Gray, 1849). (Lowenthal 2000)

A trained geographer, George Marsh still had influences that could instill in him a romantic view of the natural world. For instance, his cousin, the philosopher and University of Vermont president, James Marsh, helped to redefine transcendentalism. Instead of the idealism of much of New England transcendentalism, with its interest in conservation or primitivism, Marsh's view of transcendentalism advocated taming wilderness. He advocated for practical informed decisions and increased command over nature (Lowenthal 2000, 252). Concerning human use of natural resources, he felt that it was important to weigh the results and act accordingly.

Seeing the damage to the natural environment occur right before their eyes, some people became alarmed and began to search for ways to create a balance between industrial progress and the preservation of natural resources. These very early "conservationists" included George Perkins Marsh, who wrote *Man and Nature*. Marsh argued that the growth of industry was upsetting the natural balance of nature (Nash 1982, 104–5). The scale and scope of this action overwhelmed knowledgeable observers such as Vermont statesman Marsh. While acknowledging the need for human use of the natural environment, Marsh used his 1864 book *Man and Nature* to take Americans to task for their misuse and mismanagement of their national bounty. Marsh wrote the following:

Nature, left undisturbed, so fashions her territory as to give it almost unchanging permanence of form, outline, and proportion, except when shattered by geologic convulsions.... In countries untrodden by man, the proportions and relative positions of land and water ... are subject to change only from geological influences so slow in their operation that the geographical conditions may be regarded as constant and immutable. Man has too long forgotten that the earth was given to him for usufruct alone, not for consumption, still less for profligate waste.... But she has left it within the power of man irreparably to derange the combinations of inorganic matter and of organic life ... man is everywhere a disturbing agent. Wherever he plants his foot, the harmonies of nature are turned to discords ... of all organic beings, man alone is to be regarded as essentially a destructive power. (Marsh 1865, 29–37)

To reach his conclusions in *Man and Nature*, Marsh drew from his observations as a youth in Vermont, as well as those from travels in the Middle East. The philosophies expressed above, such as referring to humans as "disturbing agents," contradicted the conventional ideas of the time.

In geography, for instance, the work of scholars including Arnold Guyot and Carl Ritter argued that the physical aspects of the earth were entirely the result of natural phenomena, such as mountains, rivers, and oceans. To suggest that humans could disrupt and, ultimately, manipulate these forms and patterns was profound. Marsh was the first to describe the interdependence of environmental and social relationships. Lowenthal wrote, "Like Darwin's *Origin of Species*, Marsh's *Man and Nature* marked the inception of a truly modern way of looking at the world. Marsh's ominous warnings inspired reforestation, watershed management, soil conservation, and nature protection in his day and ours" (Lowenthal 2000).

In addition to constructing this intellectual framework for future generations, Marsh used various occupations to influence approaches to land use, including lawyer, newspaper editor, sheep farmer, mill owner, lecturer, politician and diplomat. As a congressman in Washington (1843–1849), Marsh helped to found and guide the Smithsonian Institution. He served as U.S. Minister to Turkey for five years, where he aided revolutionary refugees and advocated for religious freedom, and spent the last twenty-one years of his life (1861–1882) as U.S. Minister to the newly formed United Kingdom of Italy.

One of the lasting influences of his thought was to celebrate and, eventually, to preserve the remaining unspoiled places. Years of living in the Middle East afforded him time to travel throughout Egypt and part of Arabia. On one of these journeys, he developed an obsession for the camel and was convinced that the animal might thrive in the American deserts. In addition to transportation, Marsh thought that the camel could prove useful in wars in the Southwest. Inspired by a lecture Marsh delivered at the Smithsonian on his return to the States, Congress ordered seventy-four camels from the Middle East to be shipped to Texas in 1856. The experiment failed, mostly because of the onset of the Civil War and the unfamiliarity with the ways of the camel on the part of the army's equestrian division.

Regardless, however, Marsh brought an alternative paradigm to ideas of development and land use.

Sources and Further Reading: Lowenthal, *George Perkins Marsh: Prophet of Conservation*; Marsh, *The Camel—His Organization, Habits and Uses*; Marsh, *Man and Nature*.

MAKING A FUTURE FOR FREED SLAVES

Time Period: 1860–1876
In This Corner: Freed slaves, abolitionists
In the Other Corner: Southern property owners, foreign interests
Other Interested Parties: Politicians, American legal authorities
General Environmental Issue(s): Slavery, the South, agriculture

As the southern United States endured invasion, military defeat, and, ultimately, military occupation, life in the region changed forever. Was the conflict about slavery? Historians would continue to debate this point for generations. Regardless of the rationale for war, however, the fighting and the southern defeat demanded institutional change from those owning and using slaves. During the war, the status of slaves as contraband of war was awkward for all concerned. These complications, however, became even greater when the end of the Civil War also brought a permanent end to the social, economic, and cultural system of slavery. Forced on most southern slaveholders, this transition required the authority and capabilities of the federal government.

Ending Southern Traditions

Some traditional elements of the South were more easily eradicated than others after the war. For instance, worm fences were quickly dismantled and used as firewood by the Union soldiers. Consequently, livestock, particularly hogs, suddenly roamed the countryside freely. With property and ownership in disarray, soldiers from both armies appropriated and consumed millions of southern meat animals, boiling or roasting beef and pork over fires fueled by torn-down fence wood. As a whole, the impact of the war marked what historian Jack Templeton Kirby referred to as the "... beginning of the end of southern rural life as it had been known for at least two centuries" (Kirby, *The American Civil War: An Environmental View*).

Thus, the war destroyed not only thousands of miles of fences but consumed range cattle and hogs. In one area of southeastern Virginia (below the James River) and adjacent northeastern North Carolina, there were nearly 360,000 hogs in 1860 (Steinberg 2002, 106–9). According to the next federal census, in 1870, after five years of peace, there were still less than half that number. By 1880, the swine population had grown to approximately 60,000, but during the depressed decade that followed, the population fell again.

When war had broken out, the South was already far behind the North in terms of industry. At the time of the Civil War, the North was responsible for more than 92 percent of the nation's manufacturing. The North's superior industrial power not only helped it win the war, it also strengthened the North's position during Reconstruction. The growth of northern railroads, steel and iron industries, canning factories, farm mechanization, and the rise of mass production contributed to the North's economic growth. With its lack of industry and largely agricultural economy, the South lagged behind the North.

Creating a Policy Framework for Freedom

Of course, one of the most valuable resources that many southerners owned was suddenly stripped away. Southern blacks who lived as slaves were freed by the Emancipation Proclamation as well as the Union victory in the war. However, for most freed slaves, their life became, at least temporarily, more confused than ever. For southern whites, of course, their labor infrastructure was gone. Spending thousands of dollars, each landowner staffed his home or plantation based on a system of servitude that commodified the labor of black workers. The entire system was a shambles, and any capital that they had invested in slavery was suddenly stricken away. Although the nature of property needed to be rethought in the South, a more important intellectual shift was demanded: the alteration of the citizen's very idea of the nature of humans. Structured around an assumption of the basic inequity of black southerners, every member of American society had to rethink the nature of the human species and the role of black citizens in the United States.

One of the Union occupiers' early attempts to unravel the future status of former slaves is referred to as the Freedmen's Bureau, which was created in 1865 by the Freedmen's Act. The text of the policy reads as follows:

CHAP. XC.—An Act to establish a Bureau for the Relief of Freedmen and Refugees.

Be it enacted by the Senate and House of Representatives of the United States of America in Congress assembled, That there is hereby established in the War

Department, to continue during the present war of rebellion, and for one year thereafter, a bureau of refugees, freedmen, and abandoned lands, to which shall be committed, as hereinafter provided, the supervision and management of all abandoned lands, and the control of all subjects relating to refugees and freedmen from rebel states, or from any district of country within the territory embraced in the operations of the army, under such rules and regulations as may be prescribed by the head of the bureau and approved by the President. The said bureau shall be under the management and control of a commissioner to be appointed by the President …

SEC. 4. And be it further enacted, That the commissioner, under the direction of the President, shall have authority to set apart, for the use of loyal refugees and freedmen, such tracts of land within the insurrectionary states as shall have been abandoned, or to which the United States shall have acquired title by confiscation or sale, or otherwise, and to every male citizen, whether refugee or freedman, as aforesaid, there shall be assigned not more than forty acres of such land, and the person to whom it was so assigned shall be protected in the use and enjoyment of the land for the term of three years…. At the end of said term, or at any time during said term, the occupants of any parcels so assigned may purchase the land and receive such title thereto as the United States can convey, upon paying therefore the value of the land, as ascertained and fixed for the purpose of determining the annual rent aforesaid. (Freedmen's Act)

Most ex-slaves, as well as poor whites, had very few skills to offer in hopes of earning a decent living. Some blacks became tenant farmers who owned their own tools and rented land from plantation owners. Most, however, became sharecroppers. Sharecroppers bought tools on credit from plantation owners and paid for the land by giving the owners part of their crops. Neither the tenant nor sharecropping systems were profitable for either the farmers or the landowners. In general, southern farmers never owned the land they worked and rarely got out of debt. Meanwhile, the landowners watched their large farms get smaller and smaller (Cowdrey 1983, 100).

To make matters worse, landowners usually demanded that sharecroppers plant cotton, which was the traditional crop of the South. During the war, the North had cut off the South's seaports. The Union's disruption of southern trade during the war had created a glut of cotton. When sharecroppers continued to plant large amounts of cotton after the war, they found that demand had decreased significantly. By not allowing sharecroppers to plant food crops, the landowners worsened the situation. From 1859 to 1866, Southern farming declined 70 percent (Steinberg 2002, 106–9).

Federal efforts such as the Freedmen's Bureau accomplished little that could genuinely help the freed slaves. Neither federal nor state governments offered much material support to a people who began to negotiate free life and labor without property or education. There was no "Forty Acres and a Mule"—black folks' modest dream of reward during the war— nor were there to be sufficient feral cattle and hogs in the woods and swamps, which might have provided the most basic sustenance and independence.

If the southern range had actually functioned freely, many black peasants might have used the economies that poor white men had enjoyed already for two centuries by claiming loose pigs and ranging them in the woods. Then they might have fed themselves their own meat, traded for other food and necessities, and, ultimately, been able to expand hold-ings sufficiently to market surpluses for cash and purchased land. Instead, most ex-slaves fell into a dependency of an especially onerous sort, sharecropping, taking rations from landlords and merchants who bought Ohio meat by the barrel, falling into near-perpetual debt.

Creating the Exoduster Tradition

Through the work of settler abolitionists, Kansas became a symbol of tolerance for many African Americans. When the Civil War ended, many freed slaves considered Kansas a uto-pia of racial tolerance in the United States. This was part of the consideration of freed slaves in 1877 when a group of them sought their own opportunity in the American West. With Reconstruction a confusing morass of unfulfilled promises, poverty, customs, and laws pre-vented the newly freed slaves from buying land after the Civil War. Many returned to work for their old masters as sharecroppers, but dishonest treatment by many landlords and lack of education kept African Americans in economic slavery (Painter 1992, 5–8).

When southern blacks considered leaving their homeland, one destination stood out above all others: Kansas. Well known for its heroic effort to enter the Union as a free state, Kansas seemed a beacon of fairness. When it welcomed thousands of refugees in 1879, the Kansas Freedman's Association proclaimed publicly that the state had "shed too much blood for this cause now to turn back from her soil these defenseless people fleeing from the land of oppression" (Painter 1992, 199–201). A black Louisianian wrote to the Kansas governor,

An early homestead of freed slaves in Nicodemus, Kansas, in the late 1800s. Known as "exodusters," freed slaves came here for many of the same reasons that attracted settlers throughout the West. In Nicodemus, though, they hoped they would control their own future. Library of Congress.

"I am anxious to reach your state … because of the sacredness of her soil washed by the blood of humanitarians for the cause of freedom" (Painter 1992, 194–95).

This group of westward-moving, former slaves became known as "exodusters." Their interest in opportunities in the West spurred the formation of the Nicodemus Township Company by Reverend W. H. Smith and a white land promoter named W. R. Hill. They recruited former slaves in the South with fliers that promised great opportunity in Kansas. Very often, however, these "Ho For Kansas!" fliers overstated the certainty of life on the Kansas prairie.

Nicodemus became the first all-black town in the West in 1877; however, by 1886, its population still hovered only around 200. The town boasted a bank, four general stores, three groceries, four hotels, three pharmacies, two millineries, two liveries, and two barber shops along with other businesses (Painter 1992, 149–53). Most important, its African Americans controlled their own destiny, but the town still needed to succeed within an American society rife with bias.

Although the early years of farming went well for Nicodemus settlers, Nicodemus's future, similar to each western community, could turn abruptly when it came to access to the railroad. When the Union Pacific's construction team bypassed Nicodemus, the black town's future was sealed. On September 7, 1888, the *Nicodemus Cyclone* wrote the following:

> "We are sorry to see several of our business men making preparations to move to the proposed new town. We consider this a very unwise move and one they will regret. With a thickly settled surrounding, already established in business and as reliably informed in the extension of the Stockton road in the near future, Nicodemus and her business men have nothing to cause them alarm. For every one that goes now we will get ten wide awake men next spring. Don't get frightened hold on to your property and be ready to enjoy the real boom that will surely come."

African Americans felt the same call westward that many Americans followed after the Civil War. Few blacks, however, set out on the exodusters' trek. Nicodemus's experience with the railroad proved to many freed slaves that the West would not be color-blind, and yet the opportunity for advancement still enticed many blacks to the West. Nicodemus ultimately survived, although it lacked the vital connection to the railroad that would have allowed it to prosper.

Conclusion

The efforts by the federal government to force the end of slavery essentially concluded with the end of federal Reconstruction after the 1876 presidential election. The color line that took shape (primarily in the South) was that of a divided society. Slavery had been ended, but the curse of racism fed segregation and sharecropping for another century. During this era, the federal government that had initiated changes during Reconstruction backed off and allowed southerners to determine their own rules and ideas about racial distinctions. Only when the federal government intervened again during the Civil Rights movement of the 1950s and 1960s did the color line begin to fade.

Sources and Further Reading: Cowdrey, *This Land, This South*; Freedmen's Act; Kirby, *The American Civil War: An Environmental View*; Painter, *Exodusters*.

HITCHING THE NATION'S FUTURE TO THE RAILROAD

Time Period: 1800s
In This Corner: Expansionists, railroad and industrial interests
In the Other Corner: Protectors of the status quo
Other Interested Parties: Politicians, American citizens
General Environmental Issue(s): Development, the West, railroads

When new innovations require a complete shift in the way that very basic things are done, the costs might seem too great to justify the benefits. Significant changes might be spaced out over a generation or pursued gradually to spread out expenses. The United States has proven itself to be one of the greatest societies in history at accepting such change and adjusting very basic living patterns. Culture often provides the salve to help make the changes seem less radical.

The evolution of railroads in the nineteenth century provides a fascinating example. Critics argued that, although the upside potential was promising, the initial investment to shift American society from a horse-and-wagon basis for trade made any such switch untenable. For railroads, the method for making this infrastructural shift possible grew from minimizing the initial scale, in other words, first emphasizing short lines, and from creating an unmistakable cultural niche to increase the new technology's appeal.

Particularly in the Northeast and the Mid-Atlantic United States, railroads were used in many small-scale instances during the early 1800s, and their growing popularity and intricacy would remake an industrial place such as Pennsylvania. At a basic level, however, railroads remade the nature of human travel. The inroads to everyday human life began in 1826 when John Stevens demonstrated the feasibility of steam locomotion on a circular experimental track constructed on his estate in Hoboken, New Jersey. Stevens received the first railroad charter in North America in 1815. The experiments in 1826, however, indicated the potential scale of the steam engine's application.

Initially, the expansion of railroads was slowed by the difficulty in preparing and organizing their construction. Surveying, mapping, and construction started on the B & O in 1830, and fourteen miles of track were opened before the year ended. This roadbed was extended in 1831 to Frederick, Maryland, and, in 1832, to Point of Rocks. Until 1831, when a locomotive of American manufacture was placed in service, the B & O relied on horsepower. Soon other operating lines joined the B & O, the Mohawk and Hudson, and the Saratoga Railroads.

Expanding Scale: Connecting the Dots

The planning and construction of railroads in the United States progressed rapidly from this point. Some historians say it occurred too rapidly. With little direction and supervision from the state governments that were granting charters for construction, railroad companies constructed lines where they were able to take possession of land or on ground that required the least amount of alteration. The first step to any such development was to complete a survey of possible passages.

Before 1840, most surveys were made for short passenger lines that proved to be financially unprofitable. Under stiff competition from canal companies, many lines were begun only to be abandoned when they were partially completed. The first real success came when the Boston and Lowell Railroad diverted traffic from the Middlesex Canal in the 1830s. After the first few successful companies demonstrated the economic feasibility of transporting commodities via rail, others followed throughout the northeastern United States.

The Railroaders' New View of the Landscape

The process of constructing railroads began with reconstructing humans' view of the landscape. Issues such as grade, elevation, and passages between mountains became part of a new way of mapping the United States. Typically, early railroad surveys and their subsequent construction were financed by private investors. When shorter lines proved successful, investors began talking about grander schemes (Stilgoe 1983, 3–8).

The possibility of railroads connecting the Atlantic and Pacific coasts was soon discussed by Congress, and this initiated federal efforts to map and survey the western United States. A series of surveys showed that a railroad could follow any one of a number of different routes. The least expensive, however, appeared to be the 32nd Parallel route. The Southern Pacific Railroad was subsequently built along this parallel. Of course, this decision was highly political, and southern routes were objectionable to northern politicians and the northern routes were objectionable to the southern politicians.

Although the issue remained politically charged, the Railroad Act of 1862 put the support of the federal government behind the transcontinental railroad. This act helped to create the Union Pacific Railroad, which subsequently joined with the Central Pacific at Promontory, Utah, on May 10, 1869 and signaled the linking of the continent.

The "wedding of the rails," May 10, 1869, at Promontory Point, Utah, connected the broad United States with the technological innovation of an intercontinental railroad. Library of Congress.

Making Railroaders of All Americans

Railroading became a dominant force in American life in the late nineteenth century, and one of its strongest images was its ability to remake the landscape of the entire country. After 1880, the railroad industry reshaped the American-built environment and reoriented American thinking away from a horse-drawn past and toward a future with the iron horse. The luxury passenger expresses boomed over grade crossings and passed small-town depots. In industrial zones, the slow freight chugged slowly but undeniably toward its next stop. Finally, morning and evening commuters shuttled back and forth between suburban stations and underground urban terminals, creating new types of communities as well as the landscapes to support them.

For this system to function, each American needed to learn to live with the needs of the railroad. This ubiquitous infrastructure included the following: the actual railroad right-of-way of roadbed and tracks; signals and depots; bridges and junctions; and the omnipresent wail of the passing locomotive. Trains and the "right-of-way" that cleared a path for them transformed adjacent built environments as well, modifying them in novel, sometimes startling, ways. The railroad also influenced many related patterns of land use. For instance, it nurtured factory complexes and electricity generating stations in accessible locations, and it contributed to the development of commuter suburbs while initiating the decline of "Main Streets" and many urban centers. It should be pointed out, however, that, although trains initiated suburbanization, their transportation centers in the hearts of major cities made cities vital and dynamic in a new manner.

The suburban trend, however, was the railroad's effect on the living patterns of most Americans. In the years between 1880 and 1930, hundreds of thousands of Americans declared their dependence on railroads and moved to suburbs. Magazines presented the railroad suburbs as miniature edens that freed wives and children from the unhealthful aspects of urban life. Most towns created railroad stations, and railroad suburbs entirely laid themselves out around these. Terminals and rail yards became sprawling centers of commercial activity, inspiring awe in their scale and scope.

With the railroad at their center, it is not surprising that the rhythm of the railroads defined many communities in this era. Many stations included clock towers on their stations, which publicly displayed time for the first time in many American communities. The typical small-town depot existed to serve several purposes, each clearly defined by its builders. It provided accommodation for passengers waiting to board, sheltered people waiting for arrivals, and received people having business in the telegraph office. With the imposition of standard time in 1886, the railroad made itself timekeeper for the nation. Although the public enjoyed chiming church-steeple clocks, the station clocks provided perfect accuracy. Once a week, the telegraph flashed the exact time, and clocks would be adjusted.

Finally, the American passion for trains is seen in the toys with which children chose to play. In fact, as railroads became less popular for passenger travel in the early 1900s, the passion of nostalgia seemed to increase. The Lionel Company began making toy trains in 1900, but, by 1930, the toy train was firmly ensconced as the most-desired toy by boys across the nation. Powered by electricity, toy trains remain a rallying point for hobbyists. As American transportation shifted to other forms in the twentieth century, these toy trains became one of the best reminders of the bygone era of railroading.

Sources and Further Reading: Nye, *Technological Sublime*; Stilgoe, *Metropolitan Corridor: Railroads and the American Scene.*

INDUSTRIAL ETHICS AND THE LESSONS OF THE DISASTER IN JOHNSTOWN

Time Period: 1889
In This Corner: Laboring classes, reformers
In the Other Corner: Industrial elite, protectors of the status quo, political leaders
Other Interested Parties: Political leaders
General Environmental Issue(s): Dams, water management, industrial ethics

When is the cost of job production too severe? During the nineteenth century, extractive industries tore through areas of the United States, such as Pennsylvania. The goal, of course, was to harvest the raw materials that would help to make the United States the world's industrial leader. On the ground level, these extractive enterprises created a generation of employment and development, resulting in factories, towns, and regions growing where there had previously been none. Often, however, this development brought with it a price. In some cases, this price was not obvious until a century later; in other cases, it became immediately obvious as a result of tragic failings that came directly from the ethics of industrial extraction.

The ethics of extraction drove business owners to see human workers as an expendable resource. In certain industrial areas, extraction and the profits that it could generate far outweighed any ideas of community or town development. The single best example of this might have occurred in Pennsylvania's Allegheny Mountains, in Johnstown, a town that likely would not have existed without the iron and steel industry. In 1889, Johnstown was a steel town filled with German and Welsh workers. With a population of 30,000, it was a growing and industrious community known for the quality of its steel.

When it was established, Johnstown had been built on a flood plain at the fork of the Little Conemaugh and Stony Creek Rivers. The flooding was exacerbated by the growing city's need to narrow the river banks to clear building space. In addition, fourteen miles up the Little Conemaugh River was the three-mile-long Lake Conemaugh. This lake was held 450 feet above Johnstown by the old South Fork Dam.

The dam had been poorly maintained, and every spring residents claimed that *this* was the moment when the dam might give way. An earthen dam, the South Fork Dam was made of piled dirt that required annual replenishment, but, in the 1880s, supervision of the dam had fallen to the South Fork Fishing and Hunting Club, which claimed as its members some of the wealthiest industrialists in the nation. Carnegie, Frick, and other members overlooked the need to repair erosion that had compromised the earthen dam.

The inevitable break occurred at approximately 4:07 P.M. on the chilly, wet afternoon of May 31, 1889. The inhabitants of Johnstown heard a low rumble that grew to a roar. When the South Fork Dam finally broke, it sent twenty million tons of water crashing down the narrow valley. Filled with huge chunks of debris, the floodwater is reported to have reached sixty feet in height and moved at forty miles per hour. In its path, the water flattened everything.

The Johnstown Flood, as seen along the town's Main Street, demonstrated some of the prevailing ethics of the industrial era. Library of Congress.

Thousands of people desperately tried to escape the flood. Over in only ten minutes, the flood left debris piled up against the stone bridge below Johnstown. The tragedy was compounded when oil ignited on the water and burned more than eighty people who survived the initial flood. Many bodies were never identified, and hundreds of others were never found. Emergency morgues and hospitals were set up, and commissaries distributed food and clothing. It was the worst flooding disaster in American history, killing more than 2,200 residents.

Near the dam, workers made heroic efforts to stabilize the collapsing structure. The account of William T. Showers, who was an employee of the Unger farm in South Fork, attests to these efforts in his recollection:

> All the men at hand were engaged continuously at the work about the dam until about 12.45, when we went to dinner. At that time the water was up to within about two feet of the top of the new bank which we had thrown up across the top of the dam. This bank, I think, was about two feet high, so that when we went to dinner, the water was up to about level with what had been the top of the dam before we threw up the additional bank. We got back from dinner about two o'clock and the water was then running over the top of the dam; that is, over the top of the new two foot bank which we had thrown up. It had broken through it by that time at one place. We had exhausted all our efforts to stay the water, and were powerless to do anything further. I did not watch any marks, and am unable to say positively whether the water continued to rise after two o'clock, or whether the running of the water over the dam after that time was due entirely to the fact of the stream wearing away the dam. I know that the action of the water did out away the face of the dam, that is, the side down stream, and that the break worked back towards the main body of

water, so that the quantity of water running through increased, and finally about 2.45 the wearing away of the dam became more rapid, so that it was cut out quite fast and the water began to go through in great and still increasing quantities, so that by about four o'clock the main body of water had gone out.

Except as I heard the matter discussed by others, I was not sufficiently familiar with the country to know to what extent the breaking of the dam and the water therefrom [sic] would cause disaster, but I heard that matter spoken of by others who were there, and in any event, I knew that the destruction of the dam would of course cause injury to property, both of the South Fork Club and of others down the stream, and all of us who were there were willing and anxious to do anything and everything that we could do to preserve the dam and hold back the water. Every one who was there worked earnestly and continuously to this end, but I do not think it was in the power of those who were on the ground to have done better than they did. (National Park Services, *Flood Witnesses*)

In the aftermath, most survivors blamed the dam's failure on the South Fork Fishing and Hunting Club. Later investigations showed that the wealthy club members had raised the lake level and built cottages without following up on engineers' urging to repair the dam. Their wealth and power, however, ensured that, in the Gilded Age, they would be found guilty of no wrongdoing. As a historical story, however, the Johnstown flood serves as a type of parable for the exploitation of workers in the age of heavy industry. The opulent club of the robber barons, after all, cared little for the safety of its members' own workers who lived in the towns below. In the end, their negligence and the priorities of the industrial age washed away thousands of workers' families (McCullough 1968, 244–50).

Sources and Further Reading: McCullough. *The Johnstown Flood*; National Park Service, *Flood Witnesses*, http://www.nps.gov/archive/jofl/witness.htm.

MORMONS CREATE A MODEL FOR INTERPRETING THE ARIDITY OF THE WEST

Time Period: Mid-1800s
In This Corner: Mormon leaders and followers
In the Other Corner: Most Americans, expansionist interests
Other Interested Parties: Political leaders
General Environmental Issue(s): The West, aridity, hydraulic societies, religion

The open spaces of the American West represented a tremendous resource for settlers interested in ranching and for the railroad. For exodusters and other ethnic or religious groups, the West also offered a sanctuary from a lack of tolerance found elsewhere in American society. Any long-term settlement, however, required that groups come to terms with the ecological idiosyncrasies of the West. Possibly no ethnic group of settlers did this better than the Mormons.

The Church of Latter Day Saints, or "Mormons," which was founded by Joseph Smith, who was born in Sharon, Vermont, on December 23, 1805, is the prime example of a group of settlers using the West to achieve separateness while remaining within the territorial limits of the United States (Stegner 2003). Smith moved to western New York State in 1816 and became a devout believer in Jesus Christ but remained confused about what he called conflicting doctrines. In 1830, Smith and five other men organized themselves under the name of the Church of Christ. The church officially took its present name eight years later.

Although many Americans were interested in the church's ideas, others were vehemently opposed to them. In 1837, Smith and Brigham Young, one of the most prominent of the twelve leaders of the church, fled for their lives to Missouri. Immediately a dominant force in politics and economic development, Mormons made immediate enemies in Jackson County. In 1838–1839, the state militia drove the Mormons from Missouri.

On the banks of the Mississippi River, the Mormons next set up the city of Nauvoo in Illinois. Smith became mayor of Nauvoo, newspaper editor, and lieutenant general of the Nauvoo Legion. In these posts, Smith promoted economic development and eventually dabbled in national politics, running for the presidency of the United States in 1844. It was also in Nauvoo that Smith introduced the belief in plural marriage and other controversial

In the distant stretches of Salt Lake City, Utah, the Mormons found their safe haven. They worked with the limited natural environment and created a successful settlement. By 1880, they had added their first Mormon Temple, in center background. Library of Congress.

tenets of the church. As the Mormons' economic strength grew, rumors of their ideas and practices led to eventual alienation. In the midst of such animosity, Joseph Smith and his brother Hyrum were arrested in June 1844. While under arrest, the men were murdered by an angry mob. In the wake of these events, Brigham Young led the Mormons in 1846 on another exodus to locate a permanent and safe colony.

Initially, the Mormon settlers followed the Oregon Trail that had carried many others westward. Soon they met a number of people who directed them toward the Great Basin as a promising site for settlement. Eventually, Young settled in the Salt Lake Valley; however, the Deseret Kingdom, which was intended to protect the settlers from further persecution, included parts of what is now California, Oregon, Arizona, New Mexico, Colorado, and Wyoming and all of Nevada and Utah. Of course, one of the reasons that the Mormon settlers could expect to be left alone in the Great Basin was its relative inhospitableness to settlement. Annual rainfall was well below that needed for prosperous agriculture. Thus, irrigation would be necessary.

The Mormon migration to Salt Lake City continued over a number of years. Some converted members of the church arrived from Europe and took wagons or handcarts across the country from New York. Later, these immigrants would take the train to Missouri and walk with handcarts from there. Although they desired separation, the Mormons quickly seized the trail and travel industry as a route to financial success. Working with handcarts that could be used by settlers or pulling them themselves, Mormons created trading posts that attracted many settlers who were passing through.

The irrigation canals in the Uinta Basin represent turn-of-the-century water development efforts of the Bureau of Indian Affairs and Mormon settlers. Originally intended to help the reservation Indians develop small self-sufficient farms, the majority of the irrigation water was eventually appropriated by Mormon settlers who homesteaded on the ceded lands of the Uinta Indian Reservation. Library of Congress.

For long-term development, however, Mormon elders studied natives. Learning from the irrigation methods of the Pueblo and other peoples who survived centuries in the American Southwest, the Mormons used their centralized authority to administer the construction and maintenance of dikes and irrigation pools. This same central authority, of course, also enforced serious restrictions on the amount of land turned over to agriculture and its placement (Stegner 2003).

Sources and Further Reading: Stegner, *Mormon Country*; White, *It's Your Misfortune and None of My Own*.

SETTLEMENT SYSTEMATIZES AND SIMPLIFIES THE ECOLOGY OF THE WEST

Time Period: 1800s
In This Corner: Settlers, political leaders
In the Other Corner: Native Americans, foreign nations
Other Interested Parties: Citizens of other regions within the United States
General Environmental Issue(s): The West, development, aridity, technology

The railroad grew incrementally in many areas of the United States throughout the 1800s; however, in the Western territories, many Americans viewed the railroad as more than infrastructure for trade. Instead, it was a civilizing force that might make an inhospitable, distant place habitable. Few people debated the potential impact of the railroad on western settlement. The undertaking, however, was very different from that in any other region. In the West, the railroad needed to grow first in its largest scale and only afterward could it fill in the gaps.

This technology proved to be the tool for which European settlers had been searching for centuries. With its wide-open expanses of land, the West offered the greatest resource of all to Americans: space. The purposes of trade and settlement, however, demanded that even the most distant reaches needed to be linked to larger systems of exchange.

Using the Horse to Tame the West

The effort to close the great expanse of land began with horses. Horses had emerged in the Old World and came to the New World when the Spanish used them to conquer Mexico. It was likely that Coronado introduced them to the Plains in the 1540s. Indians across the West acquired horses after the Pueblo drove out Spanish settlement in 1680. High plains people such as the Comanche transformed their lives by exploiting the transportation use of the horse for hunting, trade, and war. Their primary change grew from the animal's need for grass. Historian Elliot West wrote, "The plains horse culture was complete by about 1780, as that other revolution was taking its course in the Atlantic colonies."

Most early settlers traveled over land trails in covered wagons that served as equivalent to modern-day U-Haul trucks. These trails would quickly become institutionalized, with trading posts and forts along the way to aid settlers in their movement westward. The Oregon and Santa Fe Trails are the most well known. The slow wagon travel along these trails would continue for settlers, but Asian, Irish, and other immigrants would work through the 1860s to complete the Transcontinental Railroad at Promontory Point, Utah, in 1869. From this spine, the rest of the West would rapidly be entwined in a national system of commerce. Seen as a civilizing force, the railroad also brokered political and economic power in the West.

Pony Express and Efforts to Systematize the West's Space

The horse provided individuals the opportunity for individual travel throughout the West. If organized by a consistent system, horse travel also offered a much needed conduit for communication in the West. By the first half of the nineteenth century, the population of the United States began to flow steadily into the newly acquired territories of Louisiana, Oregon, and California.

The discovery of gold in California in 1848 spurred the Post Office Department to award a contract to the Pacific Mail Steamship Company to carry mail to California. In this remarkable trek, mail traveled by ship from New York to Panama, then by rail across Panama, and then finally by ship to San Francisco. The goal was for each letter to arrive in three to four weeks.

Overland mail also began to be carried throughout the West in 1848. Moving through military outposts such as Fort Leavenworth and Santa Fe, mail was carried by stage coaches. This system grew more efficient when, in 1858, the Post Office issued a contract to the Overland Mail Company stage line of John Butterfield. His company's stages traveled the 2,800-mile southern route between Tipton, Missouri, and San Francisco. His goal was to deliver mail within approximately four weeks.

In March 1860, William H. Russell advertised in newspapers for employees: "Young, skinny, wiry fellows not over 18. Must be expert riders willing to risk death daily. Orphans preferred." Although he repeatedly failed to get federal support for an express mail route of horseback riders, Russell set out to create a 2,000-mile central route to the West from St. Joseph, Missouri. Without this support, his first step was to form the Central Overland California and Pike's Peak Express Company. The company built new relay stations in necessary locations and prepared existing stations for expanded use. The raw material for this communication network were horses and talented riders. The company gathered good horses from throughout the West that could stand up against the challenges of the region's deserts and mountains. Before being hired, riders were required to swear on a Bible that they would not cuss, fight, or abuse their animals.

The new undertaking, which became known as the Pony Express, started on April 3, 1860 and ran through portions of Missouri, Kansas, Nebraska, Colorado, Wyoming, Utah, Nevada, and California. The first mail by Pony Express via the central route from St. Joseph to Sacramento took ten and a half days, cutting the Overland Stage time via the southern route by more than half. A typical rider was capable of covering seventy-five to one hundred miles per day. To keep the time to a minimum, a rider transferred himself and his mochila (a saddle cover with four pockets or cantinas for mail) to a fresh horse at relay stations, which had been constructed approximately every ten or fifteen miles.

The Pony Express was only efficient compared with its alternatives. In the case of this communication route, its moment lasted for approximately one year. From April 1860 through June 1861, the Pony Express operated as a private enterprise. From July 1, 1861, it operated under contract as a mail route until October 24, 1861. The completion of the transcontinental telegraph line that ran with the new railroad made the remarkable ten-day trip across the region on horseback seem outmoded and inefficient.

Synching the Nation Together by Rail

Although each of these efforts to shrink time and space acted as an important step in the settlement of the western United States, it was the railroad that finally provided the system to support widespread settlement. The idea of a continent-spanning railroad had been broached in the late 1830s and grew larger with the acquisition of California from Mexico and the creation of Oregon Territory in the late 1840s. Approximately five routes were given serious consideration in 1853 as the railroad's construction became a national political issue. The debate continued until the urgency of the Civil War brought the topic to resolution. At this crucial historic moment, stronger access to the West Coast became an urgent concern for the Union. Of course, the lack of representation from the southern United States also simplified the argument.

Construction on the aptly named Union Pacific Railroad was authorized by the Pacific Railroad Act of 1862. No longer debating the role of the federal government in such internal improvements, politicians and President Abraham Lincoln provided land grants and government-backed financing for both the Union Pacific and the Central Pacific rail lines. Omaha, Nebraska, served as the eastern terminus, and the primary trunk line followed a central route. From this point, however, survey teams needed to establish the most sensible route.

Engineering crews such as this one for the Deadwood Central Railroad in 1888 moved ahead of the crews creating railroads in order to assess the landscape and to find the best passages through it. Library of Congress.

These famous survey parties took over where Lewis and Clark and others left off. However, they analyzed the landscape with an eye to the needs of railroad technology. Their work began in 1863, and, by 1866, the majority of the line had been established: from Omaha, it ran west along the Platte River and then followed the South Platte River and Lodgepole Creek to the base of the Rocky Mountains. Approximately five hundred miles to the West, the line arrived at the Wyoming Territory and what would eventually become Cheyenne. After crossing the Black Hills, the line covered four hundred miles of the Wyoming Basin before reaching the Wasatch Mountains. At approximately the 1,000-mile marker from Omaha, the line passed the Great Salt Lake before reaching Promontory. Before the railroad, there were no towns in the area and only a few forts and trading posts (Haycox 2001).

From the other direction, the Central Pacific had nothing but long, flat expanses through which to add rail. The Central Pacific had to run track through the Sierra Nevada Mountains. This difficult work was largely performed by Chinese immigrants. It is estimated that up to 8,000 of the 10,000 Chinese men worked for the Central Pacific at one time.

The progress through the mountains occurred because of the great risks taken by Chinese workers. Often, Chinese workers were lowered in hand-woven reed baskets to drill blasting holes in the rock. They placed explosives in each hole, lit the fuses, and were, hopefully, pulled up before the powder was detonated. In these difficult circumstances, handcarts moved the drift from cuts to fills. Bridges, including one 700 feet long and 126 feet in the air, had to be constructed to ford streams.

On May 10, 1869, at Promontory Summit, Utah, a golden spike was hammered into the final tie as the two construction projects met. Construction of the transcontinental railroad took six years. The undertaking had also helped to make the West one of the most culturally diverse regions of the United States, including Irish and German immigrants, former Union and Confederate soldiers, freed slaves, and Chinese immigrants who elected to remain in the United States.

Unlocking the West's Tourist Potential

The railroad also functioned to unlock one of the West's most productive and enduring resources: tourism. Advertising and camping lodges constructed by railroad companies brought tourists to oddities of the western regions as early as the 1870s. This contributed to the first efforts at preservation of natural resources, focusing on Yosemite and Yellowstone National Parks.

In 1865, after thirty-five years of steady, cautious development in the northeast and midwest, the American railroad system was far from complete. From 35,000 miles in 1865, the network grew to embrace nearly 200,000 miles by 1897. As a unit, the trunk line railroads of the northeast may rank as the most impressive concentrations of economic capital in human history. This infrastructure, then, allowed late-nineteenth-century American society to achieve unrivaled industrial and commercial development. From the start, the New York Central and Pennsylvania Railroad were successes; the B & O and the Erie Railroad played significant roles in this development, but they were secondary to the two former rail lines. All lines pointed westward for expansion.

The growth westward moved from trunk lines and through Chicago and secondarily St. Louis. These lines included the following: the Chicago, Burlington and Quincy; Chicago and

North Western; and the Chicago, Rock Island, and Pacific. The Chicago, Rock Island, and Pacific was the first line to cross the Mississippi, which was a watershed achievement for the development of railroads. Then, the Burlington set out to control the route to the Pacific coast. This same strategy inspired James J. Hill as well, as he created a new rail empire out of small but strategically important lines in Minnesota. Soon he dominated the region, which meant that he was a significant player in any railroad's effort to reach the Pacific.

The completion of the transcontinental railroad in 1869 at Promontory Point, Utah, was a great psychological boost for the nation, but the nation's rail network would not function well for years. More important, the connection of the transcontinental track marked a moment of pause for the nation when it might contemplate the future of a region and a nation tied directly to the iron horse. The continued development of the network was largely carried out by Hill's Great Northern, and, by 1901, he controlled the Northern Pacific, the Burlington, and the Great Northern, constituting one of the best-conceived regional consolidations in the nation. It was through these efforts that America used the railroad better than any other nation to minimize the constraints of space and time.

Conclusion: Oversimplifying a Complex Environment

Other essays will investigate some of the difficulties that arose from these patterns on the West. In short, however, the nineteenth century clearly marked a consolidated effort by capital investors, political leaders, and American settlers to organize and carry out American population of the region. The twentieth century would require westerners to confront some of the difficulties created by this rapid alteration.

Sources and Further Reading: Haycox, "Building the Transcontinental Railroad, 1864–1869"; West, *A New Look at the Great Plains*, http://www.historynow.org/09_2006/historian2.html; White, *It's Your Misfortune and None of My Own.*

BORDER DISPUTES AND THE SETTLEMENT OF TEXAS

Time Period: Late 1700s to 1850
In This Corner: Anglo settlers
In the Other Corner: Native peoples, Mexico
Other Interested Parties: U.S. government
General Environmental Issue(s): Territorial expansion, native rights, land use, ranching, agriculture

Throughout the West, efforts at Anglo settlement proceeded differently. However, if there was one similarity throughout the region, it would likely be described with words such as contact, conflict, and border. Whereas states in the interior West needed to be politically defined and its native population contended with, Texas represented contested land between nations for an extended period of time. Although the West represents one cast border of sorts, no place represents the borderland idea better than Texas.

Throughout its history, Texas has been defined by the idea of the borderland: a place contested by various cultures and defined by the flux of each group's effort to control it. Out of necessity, a borderland becomes a site in which radically different cultures come together to form a new, entirely unique culture. Although this culture is a hybrid of many others, one group can still remain dominant. The key element, however, is conflict and contest. Whereas today's borderland culture is a product of a certain degree of reconciliation between differences, the Texas territory spent many years with an uncertain future and with very severe divisions.

The efforts of each ethnic interest group to gain and wield political power defined this contested area in more pronounced ways than most western territories. In many cases, Texas became a military front, and its future was defined by weapons.

Remember the Alamo

The Alamo stands as a monument to the fortitude of one portion of Texas culture, the settlers from the midwest and eastern United States. These settlers moved along the backwoods frontier, ahead of the government, military, and any assistance from other Americans. Most early Texans came from Scots-Irish ancestry. Independence drove them westward and made them willing to operate outside of the protection of federal authority. Although Texas eventually became part of the United States, Texans fought for independence from any authority.

The San Antonio River originally brought settling people toward the interior of Texas. Spain constructed the Alamo as part of the Mission San Antonio de Valero, one of five missions near the center of Texas in San Antonio. By 1793, Spain had abandoned the complex, which was surrounded by thick, twelve-foot walls. In 1803, a company of Spanish soldiers from Alamo de Parras, Coahuila, Mexico, reoccupied it. The Alamo gained its name during this era, for either a nearby stand of cotton wood trees (*alamo* in Spanish) or the soldiers' hometown. After liberation from Spain, Mexican forces occupied the Alamo until 1835.

In the mid-1830s, American settlers in Texas organized to gain independence and to liberate their land from Mexican rule. The movement became known as the Texas Revolution, and it resulted in a war with Mexico. Against Mexico's trained army, Texan squatters and Anglo-American volunteers offered mainly their commitment to control the destiny of this region. Combat began with the battle of Gonzales in October 1835.

After various skirmishes, the Texas army took San Antonio from Mexican troops in the late fall of 1835. After this great victory, the majority of the volunteers in the "Army of the People" left service and returned to their families, many of whom needed them desperately to survive the trials of frontier life. Texan leaders, however, needed to establish a free-standing nation while also finding some way to prepare for the inevitable border offensive from Mexico.

The Mexican forces could approach the Texas interior by only two roads: the Old Santonio Road and the Atascosito Road. The remaining Texas forces fortified forts along each road, the Alamo and Presidio la Bahia. Under the command of James Clinton Neill, the Alamo possessed ample room for supplies of food and plenty of water, an oddity in Texas. To serve effectively as a fort, however, the mission complex was overly large, and Texas leaders considered the Alamo too remote to protect if it were attacked first. In fact, General Samuel Houston openly considered pulling troops from the vulnerable fort. However, the troops remained. Soon, volunteers of the highest caliber arrived from elsewhere, including Colonel Jim Bowie and frontiersman David Crockett.

Family matters called Neill from the compound, and command of the Alamo fell to Bowie and Lt. Colonel William B. Travis. Soon, the two commanders learned that General Antonio Lopez de Santa Anna and his Mexican forces had reached the Rio Grande, which marked the southern border with Texas. By February 23, the Mexican forces were within sight. Santa Anna sent a courier to demand that the Alamo surrender. Travis replied with a cannonball.

While the Alamo awaited a full Mexican assault, Texas's leaders made sure that there would be a nation to defend. On March 1, 1836, Texans voted to replace their leaders with a group of delegates to attend a convention and establish leaders for a new, independent nation. On March 16, the convention adopted a document—mainly composed of excerpts from constitutions of the United States and a few other states—as its national constitution and established the nation of Texas. No word of nationhood reached the Alamo.

The Texans held out for thirteen days, with Travis's daily pleas for more volunteers going largely unheeded. It seemed as if Santa Anna planned to starve the rebels out. Then, on March 6, he surprised his own soldiers by abruptly ordering a full assault on the fort. The meager troops in the Alamo lasted only ninety minutes. Santa Anna ordered the execution of the few survivors. Between 189 and 257 defenders were killed that day, including Daniel Boone and Crockett. The Mexican dead numbered 600–1,600. After the Alamo, Santa Anna delayed his offensive for three weeks (a move that historians argue cost him additional victories in the war). During this pause, Texas declared itself a nation, which added new meaning to the sacrifice of those killed at the Alamo.

By April 1836, the Texans, with cries of "Remember the Alamo!", had Santa Anna on the run. The Republic of Texas lasted from 1836 to 1845, with Houston elected its first president. By 1845, politicians such as James K. Polk publicly called for the United States to annex Texas. After Polk was elected president, annexation took place on February 28, 1845. Americans now would use the existing methods to move westward with greater numbers than ever before.

The Railroad and Texas's Cattle Future

Similar to the rest of the West, the railroad was the key to linking the area into the new western region. The main rail line to the West offered the access that was essential to further development. A system of smaller rail lines and trails took shape to connect towns with the main rail line. These connections allowed remote areas to access the markets and factories of the midwest and northeast. In Texas, the main thoroughfare was a grassy strip that led to the north. Eventually, this swath would provide a great resource to help make Texas one of the nation's capitals of cattle ranching.

The main rail line to the West offered the access that was essential to further development of the region. With the main trunk line in place, however, different regions chose a variety of methods for connecting areas of the hinterland to the main conduit. A system of smaller rail lines and trails took shape to connect towns with the main rail line. These connections allowed remote areas to access the markets and factories of the Midwest and Northeast. In Texas, the main thoroughfare was a grassy strip that led to the north. Thanks to the

Cowboys, such as this one shown on a Texas knoll looking down at a herd of cattle at the LS Ranch in 1907, managed as they grazed on the range. Then, they drove the herds northward to the railroad for butchering. Library of Congress.

appreciation of this swathe by early settlers, Texas became the nation's first prime area for growing cattle. By the end of the nineteenth century, this would lead to difficulties for the state.

Similar to filling stations along the interstate highway, the swathe of prairie in the American interior created a transportation corridor for cattle. In this humid grassland, alkaline soils prohibited extensive tree growth. Big and little bluestem, wire grass, Texas winter grass, and buffalo grass rose toward the wide-open sky, whereas the grass's real life, its root structure, extended as much as seven feet into the earth. The prairie impressed everyone who saw it; however, only a few observers appreciated it as the raw material for Texas's future as a cattle empire.

The railroad was the most important factor in the settlement of the western United States, but not every region immediately received rail stops. Cattle proved portable enough to close this divide. However, long trail rides meant that the cattle would lose weight and be worthless by the time they arrived at cattle yards in Omaha and Chicago. Cattlemen searched for grassy corridors that would allow the cattle to munch as they walked to market hundreds of miles away.

Throughout the world, humans have long raised cattle for meat. Americans, however, tied cattle raising into larger economic markets. This required transportation. Cattle were moved by trail in the eastern United States as early as the late 1700s. The scale of trailing in Texas and other western states expanded immensely by the 1860s. Plus, Texas ranchers moved wilder beasts: longhorn cattle and mavericks.

After the Civil War, the longhorns were Texas's primary asset (estimated to number from three to six million, which was more than six times the human population in the state). However, in 1868, Texas cattle were credited with starting the deadly "Texas Fever" that threatened all of American cattle. Although the Texas longhorns remained healthy, they carried the bacteria. Therefore, Missouri and Kansas banned all cattle from Texas.

With access to outside markets cut, postwar trailing in Texas might have ended completely, but Joseph G. McCoy of Illinois actively sought a new model for getting Texas beef to consumers. In the spring of 1867, he convinced officials of the Kansas Pacific Railroad to lay a siding at the hamlet of Abilene, Kansas, on the edge of the quarantine area. Then he persuaded Kansas not to enforce the quarantine in this area of the state. Kansas offered access to the railroad and a grassy passage to central Texas. McCoy began building pens and loading facilities and sent word to Texas cowmen that a cattle market was available. That year, he shipped 35,000 head, and the number doubled each year until 1871, when 600,000 head glutted the market.

The first herd to follow the future Chisholm Trail to Abilene belonged to O. W. Wheeler. In 1867, Wheeler had purchased 2,400 steers in San Antonio. He planned to winter the cattle on the Plains and then move them by trail to California, where a $14 steer could garner more than $100 in the gold fields. They followed wagon tracks and wound up on a trail made by Scot-Cherokee Jesse Chisholm, who, in 1864, began hauling trade goods to Indian camps from his camp in Wichita, Kansas. Normally, Texas herds followed the old Shawnee Trail from San Antonio, Austin, and Waco to the North. The Chisholm Trail continued to Fort Worth and then passed east to cross the Red River to Abilene.

Thanks to the rich prairie grasses, when conditions were favorable, cattle actually gained weight on the trail. A trail boss, ten cowboys, a cook, and a horse wrangler could move 2,500 cattle for three months at a cost of 60–75¢ a head. This was the era of the cowboy, although not the romanticized experiences seen in film and fiction.

Most often, the drudgery of moving cattle was punctuated with only rare violent weather or stampedes. Cattle drovers protected herds from rustling, which was performed by Indian and Mexican groups but most often by disgruntled trail hands or cowboys. To make rustling more difficult, many ranchers began using brands, which were used first by Spanish mission ranches. Unique to each ranch, brands were burned into the flesh of each steer and could not be removed. Branding cattle brought some order to cattle on the open range.

Texas Cattle Fever grew worse in the late 1800s. In 1885, Kansas completely outlawed cattle driving from Texas over its border. By then, the Chisholm Trail had moved 1.5 million head of cattle. By the 1890s, bacteriologists traced the disease to ticks and devised an immunization regimen for Texas cattle. Although the future of Texas's cattle industry brightened, the long trail rides were now unnecessary because of new railroad expansion.

Conclusion

Once Texas was connected to the cattle yards in Omaha and Chicago, its future was sealed. Across the border came some of the most experienced cowboys in the world. Using the vast open land for grazing, Texas settlers would then gather the herd and drive them toward railroad hubs to the North. The strips of grass allowed cattle, whether on the Chisholm Trail or elsewhere, to graze all the way to the slaughtering house. Therefore, their weight and health remained at its highest level.

Eventually, of course, other resources, such as petroleum, would help to define Texas's future. Coming through the mid-1800s, however, Texas grew out of various ways that agriculture could profit from its vast open spaces. By moving in these economic directions, the region also moved from its ethnic divisions to form a unique, blended culture.

Sources and Further Reading: Opie, *Nature's Nation*; White, *It's Your Misfortune and None of My Own*; Worster, *Dust Bowl: The Southern Plains in the 1930s*.

MANAGING WETLANDS AND THE SWAMPLAND ACTS OF THE MID-1800S

Time Period: 1800s

In This Corner: Corps of Engineers, conservationists, water resource managers, stream conservationists, environmentalists

In the Other Corner: Property-rights advocates, Corps of Engineers

Other Interested Parties: American public, politicians

General Environmental Issue(s): Rivers and streams, wetlands, water management

Watersheds have been an important part of environmental planning and management for more than a century. Although the importance of water resources was always appreciated, it was not initially clear that restraint was required of land users and owners over such delicate sites. Often, wetlands were viewed as a nuisance that prevented fertile land from being cultivated and bred mosquitoes and disease. Throughout most of our history, federal policies reflected this perspective by encouraging the conversion and/or drainage of wetlands.

Swampland Acts Stimulate Land Conversion

For Congress in the nineteenth century, the priority was settling the public domain. Wetlands were a menace and slowed land development. To assuage this limitation to settlement, Congress passed the first Swampland Act in 1849, which granted to Louisiana all swamp and overflow lands then unfit for cultivation. The goal in doing so was to help control floods in the Mississippi River Valley. In 1850, the act was made applicable to the other twelve public-domain states, and, in 1860, its provisions were extended to Minnesota and Oregon.

Originally, the Swampland Act enabled the States to reclaim their wetlands by the construction of levees and drains. The states were supposed to carry out a program of reclamation that not only would lessen destruction caused by extensive inundations but also would eliminate mosquito-breeding swamps. As of June 30, 1954, a total of 64,895,415 acres of wetlands had been patented to the fifteen states included in the program. Minor adjustments are still going on, although it is unlikely that the figure will ever reach sixty-five million acres.

In many cases, the land that was turned over to the states was not dealt with as instructed. Once it was turned over to the counties for dispersal, it was bartered for all sorts of considerations, such as public buildings, bridges, and like purposes foreign to the intent of the acts granting the land. In other cases, the land was sold to themselves by the county commissioners for a minimal payment. Other counties gave their wetlands to railroad companies. Of approximately sixty-five million acres of wetlands given to the states, nearly all are now in private ownership. The landowners can do with them as they want.

Table 1.
Acreage Granted to States for Swamp Reclamation

State	Acres
Alabama	441,289
Arkansas	7,686,575
California	2,192,875
Florida	20,325,013
Illinois	1,460,164
Indiana	1,259,231
Iowa	1,196,392
Louisiana	9,493,456
Michigan	5,680,310
Minnesota	4,706,503
Mississippi	3,347,860
Missouri	3,432,481
Ohio	26,372
Oregon	286,108
Wisconsin	3,360,786
Total	64,895,415

Action authorized by Swamp Land Acts of 1849, 1850, and 1860. From the USGS.

Inventorying Wetlands

In 1906, Congress began the next phase in watershed management by requesting that the U.S. Department of Agriculture (USDA) conduct a national inventory of remaining wetlands. The agency was urged to supplement and verify existing data on the subject as well. To this end, the agency sent questionnaires to a sampling of people in each county in states east of the 115th meridian. In his letter requesting the information, the Chief of Irrigation and Drainage Investigations of the Office of Experiment Stations stated the following:

> This office is being called upon by Members of Congress and others interested in the matter for information as to the amount and location of swamp and overflowed lands in the United States that can be reclaimed for agriculture. These frequent inquiries, together with the fact that numerous bills were introduced in both Houses of the last Congress for the drainage of swamp lands, show that the reclamation of these lands is fast becoming a matter of national importance. (U.S. Geological Survey)

For the inventory, eight of the public-land states in the arid West and all coastal lands overflowed daily by tidewater were excluded.

The second inventory of wetlands, conducted in 1922, is the most complete nationwide survey of wetlands ever conducted. Today, it remains the basis of most reclaimable wetland estimates. The 1922 inventory reported the following:

> 91,543,000 acres, of which 7,363,000 acres were listed as tidal marsh and the remainder as inland marsh, swamp, and overflow land. After subtracting sixteen million acres of very deep peat and some coastal-marsh areas, the investigators believed that

seventy-five million acres of wetlands would be suitable for crops after drainage. Of this amount, about two-thirds would have to be both drained and cleared of trees or brush (swamps and timbered overflow lands), and one-third required only drainage (herbaceous marshes). (U.S. Geological Survey)

To clarify treatment and regulation of America's wetlands, the Soil Conservation Service (SCS) was established during the 1930s New Deal. This agency estimated the original, natural wetlands of this country at 127 million acres. Its statisticians assumed a loss of approximately forty-five million acres and, therefore, established that the United States now has about eighty-two million acres of land that, by the agency's standard, was too wet for crop or pasture use, lands on which drainage or flood-control operations so far have had little effect on their original wet condition.

To remedy this situation, the service began an active policy of draining wetlands once the inventories and reports had established their location and status. In connection with the 1930 census of drainage, the service estimated that more than fifty million of the eighty-four million acres (approximately 60 percent of the land then in organized drainage-management enterprises) could be classed as "fair" to "good" for agriculture before any drainage improvements were undertaken. A major initiative to help stabilize agriculture in the western United States became wetland conversion.

Today, more than one-fifth of this country's cropland is in drainage enterprises, meaning that the land has had to or must continuously be drained. Farmers in the humid parts, and in some of the semihumid parts, of the United States (including the two Dakotas) drain to take surplus rainfall off some of their lands and to be used for other purposes. Most of this is gravity drainage, although pumps are sometimes used. In the western states, where irrigation is practiced, drainage is mainly for the purpose of taking seepage water off irrigated lands and carrying away alkali salts.

Conclusion: Accounting for the Necessity of Wetlands

The Swampland Acts of 1849, 1850, and 1860 transferred sixty-five million acres of wetlands to fifteen states on the condition that they be reclaimed for a productive use such as farming. Various other policies encouraged or funded the damming of rivers and streams to provide flood control, irrigation, and hydroelectric power. The Army Corps of Engineers played a pivotal role in these activities in straightening and channelizing rivers for navigation. The conversion of wetlands in some areas has reduced the ability of some river systems to buffer floods, because wetlands often provide temporary storage of runoff water.

Overall, most experts agree that agricultural drainage and flood control have been the greatest destroyers of wetland habitat in the country. Other land-use practices, however, also deserve some of the blame.

Development, of course, is a primary culprit. Both coastal marshes and interior marshes and swamps are being dissected by more and more roads that drain or fill wetlands and induce additional exploitation of adjacent areas. Throughout the nation, the expansion of cities, industrial sites, and resorts is often accomplished at the expense of good wetland-wildlife habitat. Wetlands are often filled in to allow development of airports and beach properties. Since the end of World War II, such initiatives have been clearly construed as a common good. At times, the federal government has sought legal mechanisms to foster such development.

Some officials, however, are hopeful that this attitude might be changing. Hydrologists and planners concur that the impact of Hurricane Katrina in 2004 was made worse by the shrinking amount of wetlands on the Gulf Coast. For all of the reasons listed above, wetlands have disappeared from this region for the last two decades. Now, officials argue for the secession of such policies and even the initiation of wetland restoration efforts in delicate ecosystems such as that found along the Gulf Coast.

Sources and Further Reading: Opie, *Nature's Nation*; Reuss, *Water Resources Administration in the United States: Policy, Practice, and Emerging Issues*; U.S. Army Corps of Engineers, *Navigational Improvements before the Civil War*, http://www.usace.army.mil/inet/usace-docs/eng-pamphlets/ep870-1-13/c-1.pdf; U.S. Army Corps of Engineers, *Hydrogeomorphic Approach to Assessing Wetland Functions: Guidelines for Developing Regional Guidebooks*, http://el.erdc.usace.army.mil/wetlands/pdfs/trel02-3.pdf; U.S. Geological Survey, *Wetlands of the United States*, http://www.npwrc.usgs.gov/resource/wetlands/uswetlan/century.htm.

GOLD OPENS UP THE WEST

Time Period: 1840s to 1860s
In This Corner: California developers, settlers
In the Other Corner: Landowners, native people
Other Interested Parties: Asian immigrants, U.S. government
General Environmental Issue(s): Resource/regional development, the West, minerals, mining

Many different people saw different things in the American West. Debate raged over which development strategy should guide the effort to populate the region with white Americans. At times, however, resources known elsewhere were found unused and undeveloped. By the late 1800s, the settlement form that came to follow such discoveries occurred often enough that it was referred to as a type of settlement strategy: boom and bust. In truth, however, there was little that was strategic or systematic about it other than raw greed. Although many resource frontiers emerged in the nineteenth-and twentieth-century West, the form was largely set by the region's most famous rush: the rush that occurred when gold was found in California in the late 1840s.

In the case of the California gold rush, the population and economic shifts also significantly altered the settlement in surrounding territories and, ultimately, in the entire American West. Rapid settlement brought significant changes to micro-settlements. Such rapid shifts in economic production and population leak into surrounding areas can completely alter entire regions. In the case of gold's discovery in the California territory, an entirely new scale to resource development was established for the American West. A variety of people converged on this resource to make a living from its harvest and, possibly, to achieve a fortune impossible anywhere else in the world.

Although the paths westward were well trodden by the mid-nineteenth century, word of the discovery of gold stimulated more human movement than ever before. The stunning example of gold demonstrated the great value that could be attached to natural resources. The profits to be made in such enterprises could cause shifts in human history and even influence the history of entire regions. In the early 1840s, California was a distant outpost that only a handful of Americans had seen. The sleepy port that would become San Francisco had just a few hundred residents.

John Sutter, an immigrant from Switzerland, was one of the wealthiest residents of the region. Intent on establishing a private empire, Sutter arrived in 1839, constructed a fort, and amassed 12,000 head of cattle and hundreds of workers. In 1847, Sutter sent James Marshall and about twenty men to an area along the American River, just fifty miles from the fort. Their assignment was to build a sawmill that would provide lumber for Sutter's ranch. When they had almost completed the mill, Marshall noticed a reflection in the water. Later, he described his discovery on January 24, 1848 as follows: "I reached my hand down and picked it up; it made my heart thump, for I was certain it was gold. The piece was about half the size and shape of a pea. Then I saw another."

Although Sutter and Marshall concluded that the sample was gold, they tried to keep the discovery secret. They were interested less in preserving secrecy than in not disrupting their agricultural development. Soon, however, rumor of the discovery filtered out to the public. The first booster was a San Francisco merchant named Sam Brannan. Although he would never mine gold, the rush would make Brannan rich. He ran through the streets of San Francisco shouting about Marshall's discovery and showing everyone a bottle of gold dust. As the rush of speculators began, Brannan sold them supplies. In just nine weeks, he made $36,000.

In December 1848, President Polk fanned the flames of gold fever when he specifically discussed the discovery: "The accounts of the abundance of gold in that territory are of such extraordinary character as would scarcely command belief were they not corroborated by authentic reports of officers in the public service." Americans rushed to figure out how they could get their own chance at fortune. Writer Horace Greeley left little doubt about how he viewed the potential in gold: "Fortune lies upon the surface of the earth as plentiful as the mud in our streets. We look for an addition within the next four years equal to at least One Thousand Million of Dollars to the gold in circulation."

These calls helped to provide the existence of gold in California with a mythical grandeur even before it had truly been developed. The attraction of gold in the American West helped to enlarge trails into roads and to establish new cities and towns. In 1849, 80,000 new arrivals joined the 20,000 residents already in California. Many immigrants came by ship: either the 17,000 miles around Cape Horn or across the Isthmus of Panama. By 1852, the state's population had catapulted to 250,000 (Brands 2002).

The inefficiency of the early search demanded improvement to become more reliable and profitable. At first, prospectors for gold used pans to separate the small specks of fortune from the sand and gravel. Because gold is a heavy metal, it settled to the bottom as prospectors swirled water around in the pan. Time consuming and with scant payoffs, panning soon gave way to the use of rockers. Although rockers washed gravel more quickly, soon they were replaced by the sluice box. In this inclined trough, riffles or bars trapped the gold as the water swept the gravel down the sluice. With these devices, miners found an estimated 3,950,000 troy ounces of gold in all of California by 1852, a modern-day value of approximately $1.5 billion.

By the mid-1850s, the scale and scope of mining reflected the industrial age. Hydraulic mining replaced the previous mining processes and very quickly made them seem primitive

Hydraulic mining for gold became popular in California throughout the late 1800s. Washing away the topsoil, though, created serious ecological problems for the region. Library of Congress.

and inefficient. In hydraulic mining, hoses shot water under tremendous pressure against hillsides that were thought to contain gold. This technology allowed miners to break up huge amounts of earth in a matter of hours that would have previously taken weeks working with shovels and picks. The profits were significant: in 1867, a mining engineer estimated that, if wages were $4 per day, the cost of washing one cubic yard of gold-bearing rock with a pan would be $20; with a rocker, $5; with a long tom, $1; and with a hydraulic operation, $0.20.

With financial backing from San Francisco investors, mining soon grew to include the steam-powered Blatchley diamond rock drill and the "Little Giant" hydraulic monitor. The latter, which began use in 1870, was similar to a piece of artillery that forced a jet of water into the hillside. From its beginning with a solitary miner panning the sand of stream beds, mining had been overtaken very quickly by companies that represented powerful economic and political interests within California. Often, the powerful corporate interests in the water and railroad industries sold stock in potential mines to wealthy investors. These individuals or companies then hired low-paid miners to do the hard work of finding gold. Because there was limited labor immediately available, mining opened new opportunities for Native and African Americans as well as Asian immigrants.

The high-pressure jets tore away at the soil, consumed enormous amounts of water, and intensified erosion, but the gold that they loosened fed the rush westward and changed California forever. The silt and gravel debris from hydraulic mining left the Sacramento River and others polluted. After many farmers complained of hydraulic mining's impact on their water supply, the U.S. Circuit Court in 1884 permanently prohibited mining companies from using this method.

Conclusion: The Legacy of California's Gold Rush

The pattern in the California gold fields would be seen with many mineral developments in the western United States. Eventually, many critics would decry the environmental impacts of such rapid development; however, throughout the twentieth century, the West remained a remote area in which such resource development could occur well ahead of most criticism. The product, of course, is that many areas of the West were left polluted for future generations.

Sources and Further Reading: Brands, *The Age of Gold*; Opie, *Nature's Nation*; White, *It's Your Misfortune and None of My Own*.

POSITIONING FOR BATTLE: GETTYSBURG, JULY 1863

Time Period: 1860s
In This Corner: Confederates
In the Other Corner: Union
Other Interested Parties: slaves, Native Americans, western settlers, trading nations
General Environmental Issue(s): geographical landscape, military positioning, Civil War

In each theater of battle, the natural topography was altered from being the stage for agriculture or everyday life to being a site of battle. Such a process requires that battlefield commanders view the resources before them very differently. For instance, in the American Civil War as Union and Confederate officers studied the land around Gettysburg, Pennsylvania, in late June 1863, they each interpreted a language of strengths and weaknesses. The natural environment in this place, and in others that became battlefields, could turn the entire war in his favor. In the case of Gettysburg, the battle's location grew first from a macroscopic analysis of the environment of the Mid-Atlantic United States and, later, from a more microscopic interpretation of the hills and general topography of the town and its surroundings.

When Confederate President Jefferson Davis ordered General Robert E. Lee to take the war to the Union, he did not point specifically at any single site. He and Lee agreed that too much of the suffering and fighting had taken place in the South. The fight must be taken to the northern states in an effort to further diminish Lincoln's tenuous hold on public support for the war. Once the Army of the Potomac caught up with Lee's soldiers near Gettysburg, topography played a significant role in situating the action (Pfanz 1993, 1–6).

About four miles north of the town, Gen. Richard Ewell, who traveled with Robert E. Rodes' Division, heard cannon fire ahead. As the Confederates cautiously neared, they encountered the 17th Pennsylvania Cavalry. After the morning fight, Ewell and Rodes stood on Oak Hill, which is just north of Mummasburg Road. The early fight had settled on the small hills, almost bluffs, on this side of the town. The Confederate artillery could be seen along Chambersburg Pike, on Herr Ridge. The Federal First Corps was nearer on McPherson and Seminary Ridges. The exchange of cannon fire was the only fighting that remained by midday. Ewell and Rodes brought their troops around to prepare a strike against the Union artillery and sent orders to Major Gen. Jubal Early that he should join in the battle. The Confederate push was nearing its goal.

One by one, each group of blue-clad soldiers gave up their forest hideaways. The Confederate forces pressed Union soldiers off of each hill in sequence. As they took positions along the high points on the western side of town, it became obvious that the town lay in the center of the conflict. Union fighters, who became known as the Iron Brigade, slowed the Confederate push toward town from McPherson's Ridge beyond the Seminary Ridge. When they dropped back, they saw 3,500 Confederates approaching. A Union colonel observed, "There was not a shadow of a chance of our holding this ridge" (Pfanz 1993, 20–22).

General John Buford watched the action from the cupola at the Seminary. By noon, the Confederate forces were nearly on the ridge. The Union forces set themselves along the railroad bed that ran along Chambersburg Pike. The ridge was cut away for the railroad to pass through it. The crevasse had steep sides. Buford could tell that the Confederate forces were growing in number, although the Union artillery had inflicted casualties. He fell back to town after conferring with John F. Reynolds.

Soon, Henry Heth realized how narrow the Union line must be along the ridge. He sent two additional brigades forward, commanded by Brigadier Generals James Archer and Joseph Davis. The Confederates dropped back for protection. Rufus Dawes incorrectly assumed that they were in retreat. Union forces converged on Davis's Confederate regiment in a slit in the ridge known as the Railroad Cut. Dawes realized the cut could be used as a trap to slow the Confederate withdrawal. Two hundred of Davis's regiment were taken prisoner. More importantly, Union forces could now hold the defensive position for four more hours on this side of town (Pfanz 1993, 39–44).

Reynolds passed the Union orders on to Oliver O. Howard, James S. Wadsworth, and Abner Doubleday that they were to converge on Gettysburg. The marches began between 6:00 and 8:00 A.M. Howard arrived first and became the first Union officer to digest the features of Gettysburg for battle. He watched as the Union line crumbled and individual soldiers fled through town for safety.

Confederate artillery under the command of Early used the heavy forest north of Barlow's knoll to the north of town and lobbed flank fire at the Union soldiers. Additionally, a Georgia brigade advanced across Rock Creek under the cover of the forest.

Ewell watched the Union retreat across town. Early and Rodes urged him to pursue the opponent; however, Ewell believed that he must wait for Lee before initiating a full-blown confrontation. The terrain also beckoned Confederate pursuit. As the three Confederate commanders looked across town and studied Cemetery Hill, they could see that it could prove a formidable position for the Union. In addition, the Confederates could help to undermine it by promptly taking the high ground on the north end of Cemetery Ridge, known as Culp's Hill, which was unoccupied by Union forces. From this spot, Cemetery Hill would be at their mercy, but Ewell resisted. In response, Hill sent Lt. James P. Smith to find Lee and find out for certain what his wishes were. Smith found Lee with Longstreet in the field off of Seminary Ridge. From this vantage point, a curve from where Smith had begun, the men could use Lee's binoculars to study Cemetery Ridge from a different angle. Union forces moved throughout the ridge (Pfanz 1993, 56–60).

Although Cemetery Ridge lay occupied, Lee relayed through Smith his permission for Ewell to proceed if he thought it prescient. Ewell had little support for any attack. In addition, he had come to believe very strongly that attacking Cemetery Hill from town would unduly endanger his troops. Any staging area, it appeared, would be susceptible to artillery fire from the hill. General Trimble, who was familiar with the area, urged Ewell to take possession of the unoccupied Culp's Hill, which he later described as "the key to the whole position about here" (Pfanz 1993, 81). Standing pat to reorganize his army, Ewell would only later send Johnson's division to take Culp's Hill if it seemed possible. The Confederates proceeded by the light of the moon. Unbeknownst to the Confederate leaders, Johnson's party would turn back and give up their assault after encountering Union forces.

It was the lay of the land that led the officers to create the famed "fishhook formations" along Seminary Ridge and the Roundtops. The town of Gettysburg, of course, wound up in no-man's land between the enemy lines. The battle of Gettysburg contained urban warfare and close, hand-to-hand combat in forests. The decisive maneuvers of the battle, however, came when the Confederate army ventured from its defensive, ridge position and pressed across the seemingly flat land between the two lines. The folds of the Union's ridges misled Lee: he saw the forward Union troops and assumed that was the extent of the enemy forces when he ordered the bulk of his forces to charge on July 3, 1861. "Pickett's Charge," as the maneuver is called, proved catastrophic for the Confederate army. Union forces that were concealed by the additional ridges forward from Little Round Top opened fire, and the bulk of the 15,000 charging Confederates were stopped even before reaching Emmittsburg Road. A despairing Lee then ordered his Confederate survivors to flee back across the Potomac.

In the case of Gettysburg, the efforts of early preservationists have assisted later generations in appreciating the importance of the terrain to the course of the battle. After the establishment of the Soldier's National Cemetery in November 1863, private preservationists worked to establish the sequence of the battle. Then, preservationists set out to acquire the

View of Little Round Top, Gettysburg, Pennsylvania. In the fighting of July 1–3, 1863, this ground proved to be the center of the Union's strong defensive line. With its clear sight lines, it offered them a commanding artillery line of fire to the Confederates. Library of Congress.

land and to set it aside from development. This historic endeavor created the sacred land-scape that most Americans know so well.

Sources and Further Reading: Black, *Contesting Gettysburg: Preserving an American Shrine*; Pfanz, *Gettysburg.*

MANAGING RESOURCES IN THE CIVIL WAR

Time Period: 1860s
In This Corner: Confederates
In the Other Corner: Union
Other Interested Parties: slaves, Native Americans, Western settlers, trading nations
General Environmental Issue(s): Regional ecology, agriculture, energy resources, Civil War

The Civil War was a complicated conflict for the United States on many levels. Throughout the conflict, strategic and economic warfare demanded that Americans reevaluate resources of many different types. Debate ensued at every turn, but the changes wrought by the war would have lasting impacts on specific resources and the pace of approaches to resource use, including industrialization.

Slaves

There is no resource of this era that is more troubling than that of human labor, particularly that performed by African-American slaves. Although many Americans argue that the Civil War grew out of slavery, this is not entirely accurate. Slavery was the most obvious product of differences in sectional approaches to economic development. Throughout the war years, great care and attention was used by the Union to address the concept of slavery while attempting to reconcile the monetary value that many Americans had invested in this horri-ble resource.

The enslavement of African Americans in what became the United States formally began during the 1630s and l640s. At that time, colonial courts and legislatures made clear that Africans, unlike white indentured servants, served their masters for life and that their slave status would be inherited by their children. By the time of the Civil War, scholars estimate that fewer than 10,000 families owned more than fifty slaves. Fewer than 3,000 families owned more than one hundred slaves, almost all of which were located in the South. The typical Southern slave owner fell into a much smaller category, typically owning one or two slaves. Most often, however, southern males owned no slaves. Although this is a crucial real-ity in understanding slavery in the South, slaves presented a serious political and logistical challenge during the Civil War. In the South, many people were convinced that northern threats to undermine slavery would unleash the pent-up hostilities of four million African-American slaves who had been subjugated for centuries.

With the confusion and flux of federal policies toward slavery, a significant number of slaves took matters into their own hands and attempted to escape from servitude. Beginning in 1861 and continuing throughout the war, whenever the proximity of Union troops made successful escape likely, slaves abandoned their plantations by the hundreds, even the

THE (FORT) **MONROE DOCTRINE.**

On May 27, 1861, Benjamin Butler, commander of the Union army in Virginia and North Carolina, decreed that slaves who fled to Union lines were legitimate "contraband of war" and were not subject to return to their Confederate owners. The declaration precipitated scores of escapes to Union lines around Fortress Monroe, Butler's headquarters in Virginia. In this crudely drawn caricature, a slave stands before the Union fort taunting his plantation master. The planter (right) waves his whip and cries, "Come back you black rascal." The slave replies, "Can't come back nohow massa. Dis chile's contraban." Library of Congress.

thousands. Escaped slaves, however, presented problems of legal classification. Declared "contraband of war"—enemy property that could be used against the Union—the federal government achieved a designation that avoided the question of property against Southerners.

In border areas, this designation fueled slaves to seriously weigh escaping. In Virginia, for instance, within a month of this new designation, approximately 900 slaves had fled to Fort Monroe (Kirby). By war's end, there were more than 25,000 escaped slaves in and around Fort Monroe. Ultimately, some of these escaped slaves served as members of the Union army. On an even larger scale, thousands of slaves in the Sea Islands off South Carolina fled into the woods when the Union navy landed nearby. In the New Orleans area, emancipation became widespread.

Most often, the first slaves to escape were the most educated: typically house servants and skilled craftsmen. In addition to education, these slaves had access to changes in the policies and troop movements of the Union. If successful, these escapees often sent for family or friends to follow them.

The spike in the number of escapees produced significant logistical challenges for the Union. There was no plan in place for how to care for these black refugees. For this reason, escapees often found themselves at least temporarily in worse conditions than those of slavery from which they were fleeing. Most often, escaped slaves were herded into camps and given tents and rations in exchange for work. Eventually, some northern benevolent organizations, such as the Freedmen's Aid Societies, and religious groups, such as the American

Missionary Association, sent missionaries and teachers to the South to aid the blacks. They provided much of the food and clothing that enabled the refugees to survive this difficult transitional period. Often, the schools and churches created by these agencies were the freed slaves' first.

To complicate the situation further, Lincoln went against the requests of his cabinet in January 1863 and made the war about slavery by issuing the Emancipation Proclamation. In actuality, it proclaimed free only those slaves outside the control of the federal government, that is, only those in areas still controlled by the Confederacy. However, the text squarely made rejecting slavery federal policy. It reads as follows:

> That the Executive will, on the first day of January aforesaid, by proclamation, designate the States and parts of States, if any, in which the people thereof, respectively, shall then be in rebellion against the United States; and the fact that any State, or the people thereof, shall on that day be, in good faith, represented in the Congress of the United States by members chosen thereto at elections wherein a majority of the qualified voters of such State shall have participated, shall, in the absence of strong countervailing testimony, be deemed conclusive evidence that such State, and the people thereof, are not then in rebellion against the United States.
>
> Now, therefore I, Abraham Lincoln, President of the United States, by virtue of the power in me vested as Commander-in-Chief, of the Army and Navy of the United States in time of actual armed rebellion against the authority and government of the United States, and as a fit and necessary war measure for suppressing said rebellion, do, on this first day of January, in the year of our Lord one thousand eight hundred and sixty-three, and in accordance with my purpose so to do publicly proclaimed for the full period of one hundred days, from the day first above mentioned, order and designate as the States and parts of States wherein the people thereof respectively, are this day in rebellion against the United States, the following, to wit:
>
> Arkansas, Texas, Louisiana, (except the Parishes of St. Bernard, Plaquemines, Jefferson, St. John, St. Charles, St. James Ascension, Assumption, Terrebonne, Lafourche, St. Mary, St. Martin, and Orleans, including the City of New Orleans) Mississippi, Alabama, Florida, Georgia, South Carolina, North Carolina, and Virginia, (except the forty-eight counties designated as West Virginia, and also the counties of Berkley, Accomac, Northampton, Elizabeth City, York, Princess Ann, and Norfolk, including the cities of Norfolk and Portsmouth), and which excepted parts, are for the present, left precisely as if this proclamation were not issued.
>
> And by virtue of the power, and for the purpose aforesaid, I do order and declare that all persons held as slaves within said designated States, and parts of States, are, and henceforward shall be free; and that the Executive government of the United States, including the military and naval authorities thereof, will recognize and maintain the freedom of said persons.
>
> And I hereby enjoin upon the people so declared to be free to abstain from all violence, unless in necessary self-defence; and I recommend to them that, in all cases when allowed, they labor faithfully for reasonable wages. And I further declare and

make known, that such persons of suitable condition, will be received into the armed service of the United States to garrison forts, positions, stations, and other places, and to man vessels of all sorts in said service.

Officially, the legal end to slavery in the United States came in December 1865 when the Thirteenth Amendment was ratified. It declared, "Neither slavery nor involuntary servitude, except as a punishment for crime whereof the party shall have been duly convicted, shall exist within the United States, or any place subject to their jurisdiction." Although this amendment would be resisted by southerners, slavery could no longer take place on American soil. Was the Civil War about slavery? Of course, the proper answer is that the war was about many things. However, the Union victory brought about the official outlawing of slavery in the United States.

Horses

Although slaves were the most complicated property related to the Civil War, other important resources included animals. Animals, especially horses and mules, were essential participants in nineteenth-century warfare, and they, too, suffered and died in appalling numbers. Marshalled (like humans) in cities, camps, and fortifications, they also exchanged pathogens and died by the thousands before a single cavalry charge, artillery caisson pull, or wagon haul could take place. Disease deaths necessitated resupplies from farther and farther afield, so the war's equine impact was continental in scope. Many of the horses and mules that survived epidemic disease were maimed and killed by the thousands in battle. Because their carcasses were so much larger than dead men, horses and mules presented daunting sanitary challenges on battlefields. Onsite burial was usually hasty and incomplete.

Edmund Ruffin, the Virginia secessionist who reputedly fired the first shot against Fort Sumter and then was an eyewitness to eastern fighting in the first half of the war, reported that nature was little help in cleaning up decaying dead animals and men. Vultures, common and plentiful throughout the regions where the war was fought, stayed away, apparently discouraged by the noise of artillery. The only benefit that may be wrangled from this particular carnage is that modern equine medical science began, arguably, during the great disease kill-offs early in the war.

Approximately 3.5 million horses and mules were killed in battle during the Civil War. Scholars estimate that the animals died at a rate six times higher than humans in the war. Union forces are estimated to have had 825,000 horses, which cost them approximately $124 million. In the battle of Gettysburg, more than 72,000 horses were used, with between 7,000 and 10,000 killed in action.

Illuminating Oil: Petroleum

A tumultuous time of war helped to define the nature of the oil industry that was taking shape on land as well. Although petroleum had little impact on the war's outcome, the war significantly influenced the development and production of the newest energy commodity. When General Robert E. Lee drove the Army of the Potomac into Pennsylvania in 1863, oilmen suspended business and the price shot upward. As the "Great Stone Fleet" fell to the bottom of Charleston harbor, 700 soldiers were mustered from Pennsylvania's oil regions. Stories circulated of a few wealthy oilmen who paid $300 in exemption fees to assign their roles to other Pennsylvanians.

By far the most dramatic impact of war on one oilman's holdings occurred outside of Pennsylvania along the Ohio River, where it forms the border of western Virginia and Ohio. In 1859, the Rathbone brothers drilled for salt in the region after they heard of Drake's oil well. Late in 1859, they started drilling for oil at Burning Springs on the Little Kanawha River, Virginia. Their well began to produce in 1860 at a rate of one hundred barrels per day. A second successful well brought thousands of oil hunters to the region. The area, known as Burning Springs, became contested terrain when Virginia seceded. Although it lay on Confederate soil, Burning Springs was in a portion of the state with strong Union senti-ments. Union forces policed the area to defend the B & O, and therefore Confederate gueril-las consistently harassed them. Cal Rathbone used his oil profits to muster a cavalry detachment that he called the "Home Guard." This militia of forty horses and two compa-nies of men had instructions to protect Rathbone's oil well at all costs.

Harassment continued but did not escalate for two years. In May 1863, Confederate leaders decided to attempt to disrupt the supply of kerosene and lubricating oils that aided the Union cause. Ironically, those supplies on Confederate soil were the most susceptible to attack. Under orders, General William E. "Grumble" Jones led 3,000 Confederate soldiers into the Virginia oil country. Was Burning Springs part of the Confederacy, the Union, or neutral? Jones' Confederate troops did away with the confusion of ownership by simply destroying the oil works. As Confederate troops lit all portions of the industrial site aflame, Burning Springs became a very apt name. General Jones reported that "a burning river carry-ing destruction to our ... enemy, was a scene of magnificence.... It will be many months before a large supply can be had from this source." Losses from the destruction of "oiltown," as Jones called it, were estimated at 150,000 barrels of crude, valued around $40 million. The wells would not be resurrected for nearly two decades. (Black, 2000)

Burning Springs marked the only attack on a petroleum installation during the Civil War. On another front, however, the war made an obvious impact. Innovators discovered new applications for petroleum that aided the Union in a number of ways. For instance, pe-troleum could act as a satisfactory substitute for turpentine, a product of the southern pine forests that was lost by the announcement of war. By substituting petroleum, the production of varnish, paint, and other items could continue during the war.

Illuminating Oil: Whale Oil

As early as 1861, Confederate leaders identified whaling as one of the Union's most impor-tant industries. Confederate efforts to thwart Union commerce were not based on future promise but on current reliance, and so commenced the whalers' problems of the 1860s. In summary, during the space of five years, the fleet was cut in half. Between January 1, 1861 and January 1, 1866, whaling tonnage fell from 158,746 to 68,536, and the number of ves-sels fell from 514 to 263 (Hohman 1928, 296–300).

The initial blow came from the whalers' own sacrifice for the Union cause. In 1861, the federal government purchased forty older whaling ships, laden with stones, and deliberately sunk them off Charleston and Savannah harbors in an effort to make the navigable channels unsafe for blockade runners. The cargoes of stones were intended to prevent the hulks from being raised or washed away after settling into place in the channels. As an experiment in blockading tactics, the enterprise was dubious; but as a drain on the resources of the whaling industry (of course, not the intention of Union leaders), the "Great Stone Fleet," as it was called, proved eminently successful. Although the government purchased the vessels outright, their destruction left an immediate gap in the ranks of the whaling fleet that was never filled.

The "Great Stone Fleet" was roughly the extent of the direct impact of the war while the conflict raged. Ironically, the next front of the whalers' Civil War arrived after the actual fighting had ceased.

On June 28, 1865, more than a month after the surrender by General Robert E. Lee at Appomattox, Virginia, the final front of the Civil War opened off of the Bering Strait. The whaling ship *Brunswick* had experienced an unexpected wind surge and crashed onto the ice. Other whaling ships had congregated to divide up the ship's cargo, when one observer noted another steamship approaching on the horizon. A few whalers thought the ship might be a Confederate raider that had been seen rigging out in Australia in February. However, the approaching vessel flew the U.S. flag and gave every indication of being a supply vessel for the Western Union Telegraph Expedition, which was attempting to lay a cable across the Bering Strait to link Paris and New York. Despite their vulnerable location in the remote Arctic, the whalers continued their work in the open sea.

In minutes, the shocked whalers watched as five armed boats carrying uniformed Confederate soldiers put off from the steamer. The soldiers announced to the crews of the becalmed ships that they were the prizes of the CSS *Shenandoah*. The men were ordered to abandon their vessels and come aboard the steamer as prisoners of war or be blown out of the water. The panicked whalers moved to the steamer, and the soldiers axed open casks of oil and set each whaling ship afire. Two ships were allowed to leave to transport the crews. Eight whaling ships, however, burned amid the Arctic ice as the prizes of *Shenandoah* and the defunct Confederacy (Hohman, 290).

For those whalers who wanted to take no chances confronting a possible Confederate raider, the only protection available lay in patches of loose ice. If given enough of a head start, the whalers could navigate themselves into areas where *Shenandoah*'s commander, Thomas Waddell, dared not follow. Their experience in the region saved some in this manner, but the losses were still great. Repeatedly, Waddell's captives attempted to convince him of Lee's surrender. Waddell, however, refused to trust such claims. Finally, on June 23, Waddell's log verifies that one captured vessel used newspaper accounts to convince the Confederate raider that Lee had surrendered at Appomattox. Despite the danger of being declared a pirate, Waddell chose to press on. In his log, Waddell emphasized that, without orders to discontinue his efforts, he felt that he must continue as an officer of the Confederate Navy. With the number of prisoners rapidly growing, *Shenandoah* led an odd ocean-going train of whaleboats (twelve was the largest number).

Despite his success, the pressure on Waddell's conscience soon grew too great. He commanded his crew to abandon their raid and flee to safe harbor in Liverpool, and the reign of terror on the Arctic whalers was over. Waddell had taken more than 1,000 prisoners, destroyed twenty-five ships, converted four others for prisoner transport, and seized cargo valued at about $1.5 million. Added to other ships' losses in the war, forty-six whaling ships were taken from service. From a New Bedford fleet of 186 whaling ships in 1861, these losses added to the forty in the Great Stone Fleet to seriously diminish the industry. The final chapter in this episode played out when the United States claimed that the total cost of the ships destroyed by *Shenandoah* was $14 million and Britain should be held responsible because of its involvement in the construction and launch of the Confederate vessels. An international tribunal agreed and awarded recompense of $15.5 million in gold (Hohman 1928, 293–94).

Conclusion

Many other examples of commercial impacts of the Civil War can be found. These examples, however, demonstrate that the American Civil War was a complex, multifaceted undertaking. Its implications would play out in many different aspects of American development after 1865.

Sources and Further Reading: Black, *Petrolia: The Landscape of America's First Oil Boom*; Black, *Contesting Gettysburg: Preserving an American Shrine*; Kirby, *The American Civil War: An Environmental View*; Steinberg, *Down to Earth*.

SETTING THE STRATEGY FOR THE CIVIL WAR

Time Period: 1860s
In This Corner: Confederates
In the Other Corner: Union
Other Interested Parties: Slaves, Native Americans, Western settlers, trading nations
General Environmental Issue(s): Territorial expansion

Civil War buffs, of course, love to debate the strategic choices made by leaders of the Civil War. Often, these considerations possessed specific connection with the natural environment or resource use.

Should the War Reach the West?

One of the basic strategic debates of the American Civil War concerned to what extent the unique American environment should influence the pace and organization of the fighting. How would the great expanse of the American territory influence the war? Obviously, the need for new western states to enter the United States as either a slave or free state had created one of the flashpoints that led to actual fighting over the conflict between the northern and southern states. Although most of the fighting in the Civil War occurred in the eastern portions of the United States, the western territories and states saw fighting that played an important strategic role in the outcome of the Civil War.

Focusing the effort on controlling or limiting trade, Union forces concentrated on disrupting river traffic in the three or four primary areas of western trade. In February 1862, for instance, Union gunboats forced Confederates back from their Tennessee River stronghold at Fort Henry to Fort Donelson. On February 16, 1862, General Ulysses S. Grant won the

Union its first great victory of the war far from the primary fields. His victory came when his Army of the Tennessee captured Nashville, Tennessee, without a struggle.

Attacks coordinated around regional topography helped to make this campaign successful. Union gunboats gained control of the upper Mississippi, which led to the fall of Memphis on June 6, 1862. This coordination of the navy also aided Grant's Vicksburg campaign. With control of the Tennessee River, however, the Union controlled the whole Mississippi River, and the trans-Mississippi West was severed from the rest of the Confederacy. The fighting in that area held Missouri for the Union and led to the partial conquest of Arkansas. However, after the fall of Vicksburg, the war in that area was primarily confined to guerrilla activity.

As the Civil War dragged through 1863, Abraham Lincoln urged his military advisers to broaden the conflict and its impact on Confederate life. Union leaders looked toward the South's western frontier and particularly at the volatile coast near Texas and Louisiana. By a coincidence of nature and politics, a significant Union attack occurred near the Sabine Pass, where ships normally took refuge.

Sabine Pass

The southeastern coast of Texas was particularly active during the mid-1800s. Cotton grown in Texas would often be shipped out the Sabine River to textile mills in the northeast. A pass or gap is the term given to a break in a string of mountains. Typically, the term implies passage by land, meaning that one can get through the mountains easiest at this point. The Sabine Pass offered ships laden with cotton an avenue through the steep cliffs into the Gulf of Mexico, which made it one of Texas's most important corridors for trade.

The new state's economy relied on cotton, a product that depended on slave labor. When voters elected Abraham Lincoln president in 1860, Texas put aside its strong support for the Union and voted to secede from the United States to maintain its cotton production. From 1821 to 1865, cotton production and slavery spread rapidly throughout eastern Texas.

When it was written in 1836, the Constitution of the Republic of Texas demonstrated the state's feelings about slavery, including stipulations that slaves would remain the property of their owners, the Texas Congress would not prohibit the immigration of slave holders bringing their property from other areas, and slaves could be imported from within the United States to Texas. In such an environment, the number of slaves in Texas increased rapidly: in 1845, there were 30,000 slaves, and, by 1860, there were 182,000, 30 percent of the state's population. The great majority came with their owners from older slave states. Some, however, were traded through New Orleans, Galveston, and Houston. Slavery significantly influenced the economic development of Texas, including a 600 percent growth in cotton production in the 1850s.

So, Texans supported the Confederate cause out of economic necessity, albeit with some reluctance. During the war, 25,000 Texans served in the Confederate Army, with an overwhelming majority serving in the cavalry, But political support for the Confederate cause remained lukewarm. Events at Sabine Pass changed all that.

In an effort to stall the economy of the South, Union ships blockaded the Texas ports in the summer of 1861. Most important, Union leaders wanted to ensure that valuable cotton would continue to be shipped to northern textile mills. They also wanted to prevent shipments from going to France or other European nations that might then be moved to support the Confederate cause. The blockade led to further Union action in Texas, including the capture and occupation of Galveston.

In January 1862, Confederate forces retook Galveston and solidified forts near the Sabine Pass. Two forts, Sabine and Griffin, were constructed to defend against further Union attacks at this important opening in the land. These forts were temporarily evacuated during an outbreak of yellow fever later in 1862. The epidemic, of course, also temporarily dissuaded Union invaders, although the blockade continued offshore. When it was reestablished at Sabine Pass, Fort Griffin, an earthwork fortress, included six large cannons, with forty-six troops.

Heeding Lincoln's call to increase pressure on the Confederacy, Union leaders David G. Farragut and N. P. Banks planned a major offensive against Texas in 1863. They believed that landing Union troops deep within Texas would allow them to maintain better control of trade and potentially lead to a southwestern front in the war. To begin such a maneuver, there was really only one place to concentrate their forces.

The Union launched its attack on Sabine Pass on September 8, 1863. For this task, the Union leaders sent four gunboats and twenty-three transports, bearing about 5,000 troops for the initial landing. First, Major General William B. Franklin planned to use his gunboats to take out the guns set along the channel at Sabine Pass. Then, safe landings could be made into Sabine Lake, which would open up the Texas interior.

The planning was perfect, but events immediately did not cohere with the plan. Early in the day, one of the Union ships missed its appointed meeting and thereby tipped off the Confederates about the invasion plan. Confederate Lieutenant Dick Dowling, who commanded Fort Griffin, began to furiously gather forces. The Texas Heavy Artillery Regiment, which was led by Captain Frederick Odlum, set up along both channels along the pass. The Texans had no doubt where the attacking ships would concentrate. The Sabine Pass Light marked the only route inland.

As four Union gunboats neared the pass around 3:00 P.M., the Confederate cannons opened fire. The cannoneers averaged a shot every two minutes for the next half-hour, which ranked far better than the average for heavy artillery during this era. The battle ended quickly, with the Confederates capturing 300 Union prisoners and two gunboats. More importantly, they had successfully defended the Sabine Pass and helped to dissuade any further attacks there.

Navies Bring the Civil War to the Sea

Many of the western battles involved naval warfare. However, the navies also reached beyond land and into the sea. In the most famous naval battle of the war, the USS *Monitor* battled the CSS *Virginia* (more commonly known by its former designation, the *Merrimack*) near Hampton Roads, Virginia, on March 9, 1862. Although the battle itself was inconclusive, the *Monitor* trapped the *Virginia* in the James River. Neither ship played much of a subsequent part in the war.

Such armored vessels were a revolution to naval warfare. The ship consisted of a heavy, round iron turret on the deck, which housed two large cannons. The armored deck was barely above the water line. Aside from a smoke stack and a few fittings, the bulk of the ship was below the water line to prevent damage from cannon fire (torpedoes, that is, anchored naval mines, were a concern, but self-propelled torpedoes would not be a worry for another fifty years).

Confederate forces did not have a true navy in their arsenal. Instead, specific vessels were developed with very specific duties or missions. One of the most interesting of these missions actually occurred after Lee's surrender in April 1865, which was discussed in an earlier essay.

Sherman's March to the Sea

Although many moments are credited with being the turning point of the Civil War, the essential turn came when Union commanders altered the nature of the campaign. The March to the Sea, the Civil War's most destructive campaign against a civilian population, began in Atlanta on November 15, 1864, and concluded in Savannah on December 21, 1864. General William T. Sherman abandoned his supply line and marched across Georgia to the Atlantic Ocean. His goal was later called "annihilation." Primarily, he attacked the South's will to fight. He hoped that his campaign might prove to the Confederate population that its government could not protect the people from invaders. Partly, Sherman practiced psychological warfare: he truly believed that, by marching an army across the state, he would demonstrate to the world that it was hopeless for the Confederacy to resist the Union's war machine.

On November 12, 1864, Sherman marched out of Atlanta toward the Atlantic coast. He had his troops march a line between Macon and Augusta, through which they carved a sixty-mile-wide swath of destruction in the Confederacy's heartland. For defense, the Confederacy could only bring Wheeler's cavalry and a tattered collection of militia. It was no match for the 62,000 Union veterans with whom Sherman left Atlanta. Before the army left Atlanta, the general issued an order outlining the rules of the march, but soldiers often ignored the restrictions on foraging.

Sherman's March to the Sea, destroying tracks on its way, is credited with breaking the Confederate will to fight in the Civil War. Library of Congress.

Sherman next prepared to place Savannah under siege to limit shipments in or out and to initiate a constant barrage of artillery. Refusing to allow their city to be leveled like Atlanta, the small army of Confederate soldiers retreated to South Carolina. In celebration on December 22, Sherman wrote and offered President Lincoln a Christmas present of 25,000 bales of cotton.

The march had the effect for which Sherman had hoped. And, finally, Lincoln had a commander who directed the war where southerners would feel it most: Sherman ordered his men to destroy all sources of food and agricultural goods. His ruthlessness left southerners appalled, frightened, and unable to support themselves or their war cause. The destruction and violence destroyed any morale and confidence and made every southerner feel threatened. In retaliation, Confederate president Jefferson Davis ordered Georgia residents to poison water wells and burn off agricultural fields rather than allow anything to fall into enemy hands. Few residents could bring themselves to carry out these orders, however.

Sherman continued his psychological warfare in South Carolina in early 1865. Although northerners came to believe his heroic efforts had broken the back of the Confederates and won the war, in the South, Sherman came to represent the cruelest aspects of the conflict.

Conclusion

Clearly, however, Sherman was true to the nature of warfare and at last prosecuted the war decisively, as Lincoln wanted. The war did extend into the West and into the sea but never to the same extent as that seen in other theaters.

Sources and Further Reading: Black, *Contesting Gettysburg: Preserving an American Shrine*; Brady, "The Wilderness War: Nature and Strategy in the American Civil War"; Kirby, *The American Civil War: An Environmental View*; Steinberg, *Down to Earth*.

OLMSTED AND VAUX DESIGN A CENTRAL PARK FOR NEW YORK CITY

Time Period: 1850–1860s
In This Corner: Landscape designers, social reformers, city officials
In the Other Corner: City developers
Other Interested Parties: New York citizens, American public
General Environmental Issue(s): Outdoor recreation, parks

With the completion of Central Park in 1862, America finally could boast a grand park to compete with those of Paris and London. It was, of course, a complete luxury to set land apart from development and then to spend resources on its preparation for recreation and aesthetic beauty. Although many Americans deemed it "waste" to use 700 acres of what would eventually become the center of Manhattan, the park's architects Frederick Law Olmsted and Calvert Vaux sought to create a pastoral landscape in the English Romantic tradition. Their hope was that such a park would serve the citizens of New York City while also serving as a symbol of intelligent planning for all Americans.

To describe the achievement of successfully incorporating utility with aesthetic appeal, designers, environmentalists, and others use the term "the genius of a place." For humans to live in balance with nature, they must access this genius and allow it to infuse decisions they make when altering a site. The genius of Central Park was human. Olmsted foresaw the city's development and designed a "middle ground" in which urbanites could recover their bearings.

Landscape architecture is not always linked to environmental awareness, yet it certainly possesses a clear environmental ethic. In fact, achieving the genius of the place reveals an awareness of humans' ecological interconnection to all else. Most architecture cannot worry about environmentalism as it goes about filling space. However, more than any other form of spatial design, landscape architecture pursues a natural aesthetic as it seeks to use organic forms to make outdoor spaces at once pleasing and useful. The best landscape architecture possesses an interest in applying the genius of the place to construct beauty. Through such an ethic or conviction, the creator and the created can come to possess at least a portion of a locale's genius, transformed into an environmental ethic.

While gardens and gardening have been a part of human society throughout history, it is particularly during the 1700s that they become an active buffer between society and the private space of the individual. This separation is based on the philosophy of Romanticism, which brought to Europe a sensibility that the human did not simply live at odds with nature. Arriving from Germany and elsewhere, this enlightened perspective met a new vision of progress in Great Britain where the pastoral was swiftly falling prey to the mechanical.

With James Watt's perfection of the steam engine in the late 1700s, the Industrial Revolution altered the fabric of human life in Great Britain. Mechanization changed ways of life, labor, and even physical surroundings. Industrial centers formed cities so that access could be had by laborers to workplace and factories to transportation. Aesthetic beauty lacked emphasis in such an existence. In reaction to this often brutal vision of progress, Romanticism served an escapist role in pastoral painting and literature and, eventually, on the physical landscape. Landscape designers or architects were commissioned to make a site resemble the pastoral beauty described in the imaginations of writers and artists.

Many American colonists brought with them ideal visions of beautiful gardens, particularly as they declared an individual's economic standing. The nature of the New World, however, was a brutal foe to settlement. Although many colonists constructed garden spaces, the proliferation of landscape architecture required the support of a resurgence of Romanticism.

In the young United States, this sensibility took the form of transcendentalism. From the pens and minds of Ralph Waldo Emerson, Henry David Thoreau, William Cullen Bryant, and others, transcendentalism linked spirituality (religion and God) to nature. In direct contrast to the aggressive enemy that the pioneer perceived in nature, these writers saw the opportunity for spiritual recovery. In their pastoral views, Hudson Valley painters of the early 1800s similarly created a natural environment replete with spiritual significance. Based on these landscapes and British models, designers planned the first "park-like" site in the United States, opened outside of Boston as Mt. Auburn Cemetery in 1831. This form became known as the "rural cemetery," a tradition of burial and park recreation that spread throughout the Northeast.

As the American field of landscape architecture spread from the cemetery model during the nineteenth century, Andrew Jackson Downing cut the greatest figure. A gardener and designer by trade, Downing designed home grounds for the wealthy of the Hudson Valley. He

This snowy view of Central Park in New York City is from 1906, nearly fifty years after Olmsted and Vaux made it America's first planned park. Library of Congress.

defined a new mode of living outside of urban areas (later called suburbs) by mimicking the country cottages of Great Britain. Most importantly, the well-connected Downing knew how to influence broader cultural tastes, and he began publishing plans and recommendations in magazines and books, culminating in his own influential periodical, the *Horticulturalist*.

After establishing himself as the national expert in landscape design during the 1840s, Downing accepted the commission to design the Mall area in Washington, DC, in 1850. It was planned that the new design would stand as a national model for rustic planning; instead, Downing died in a riverboat accident, and the mall was designed by those less imbued with the genius of place. The linearity and monumentality of today's site stands in direct contrast to Downing's plan for curved walkways accentuating privacy and sublimity.

Before his death in 1852, Downing publicly urged for the construction of a central park area within the growing metropolis of Manhattan. A few years later, his associate, Calvert Vaux, joined forces with the superintendent of the existing municipal park, Frederick Law Olmsted, and submitted the "Greensward Plan" for a 778-acre, multiuse park and pleasure ground within Manhattan.

Completed in stages during the early 1860s, Greensward revolutionized the American view of its landscape and represented a new relationship between Americans and the natural environment. The building of it required more than 20,000 workers, three million cubic yards of soil, and more than 270,000 trees and shrubs. By 1865, there were more than seven million visitors every year. Through Olmsted, the ethic introduced by Emerson and Thoreau found a physical form that specifically appealed to American ideals.

Sources and Further Reading: Rosensweig and Blackmar, *The Park and the People: A History of Central Park*; Schuyler, *Apostle of Taste: Andrew Jackson Downing, 1815–1852*.

OLMSTED HELPS TO DEFINE THE AMERICAN MOVEMENT FOR PARKS

Time Period: 1870–1910
In This Corner: Landscape designers
In the Other Corner: Utilitarian views of natural environment
Other Interested Parties: Upper-class Americans interested in outdoor activity, government
  officials
General Environmental Issue(s): Planning, parks, urban reform

Central Park in New York City became an emblem for a new era in the perception of the natural world. Many communities sought to plan natural spaces into their central areas. Also, Americans began to reconsider what should be done with areas designated as parks belonging to the entire nation: national parks. In each case, a single American thinker revolutionized the nation's spaces and, ultimately, its expectations for its natural environment.

Although the passion for nature in one's everyday life and the preservation interest in national parks each existed before 1900, they did not intersect and evolve into the modern parks movement until the 1910s. The growing interest in the natural environment did not immediately alter the places in which Americans chose to live. In fact, the American landscape urbanized at an increasingly rapid rate in the early 1900s.

With the population concentration in more urban regions, the impulse to preserve accessible areas of nature became even more imperative. These impulses drew a direct relation to those driving the design of Central Park, the nation's first planned park, and its designer, Frederick Law Olmsted. Olmsted urged Americans to appreciate the psychological and restorative power of nature. His plans for additional urban parks and the early suburbs brought nature nearer to the lives of most Americans. Olmsted also worked to inspire a national set of parks that would celebrate the nation's symbolic appreciation of its natural resources (Roper 1973, 122–40).

After making his name in New York City, Olmsted moved his practice to Brookline, Massachusetts, in 1883. Olmsted had begun work on a park system for the city of Boston, and eventually he focused much of his time on the area known as the Emerald Necklace. With his reputation reaching an international scale, Olmsted received one of the greatest assignments of his career: the grounds of the 1893 World's Fair in Chicago. In this remarkable site, Olmsted created a monumental landscape that served as the setting for the works of the world's greatest living architects. Although the entire landscape (including its buildings) would not be allowed to last beyond the term of the temporary fair, Olmsted's design created a spectacle that enhanced his reputation.

In each of his designs of this era, Olmsted sought to advance a shared sense of community among all of its members. In Olmsted's mind, landscape architecture provided a critical opportunity for the natural environment to shape healthy and productive ways of American life. Particularly in congested urban areas, Olmsted believed parks offered an antidote to stress and artificiality that would help prevent mental decay. In "Greenswards" and ornamental trees, Olmsted sought to spread calmness and democracy. He believed that, if properly designed, landscapes might enhance American ideals. His concept of "democratic recreation" was perfect for park spaces that sought to appeal beyond local needs.

It follows, therefore, that Olmsted took a pioneering role by defining the form of the emerging national park system. He wrote the initial report to establish Yosemite National

Park in 1865. Drawing on his Central Park experience, Olmsted viewed the valley's preservation as the creation of a work of art. Olmsted argued in terms of the psycho-sociological theory honed in the Central Park campaign:

> It is a scientific fact that the occasional contemplation of natural scenes of an impressive character, particularly if this contemplation occurs in connection with relief from ordinary cares, change of air and change of habits, is favorable to the health and vigor of men and especially to the health and vigor of their intellect.

> Without such recreation, in situations "where men and women are habitually pressed by their business and household cares," they are susceptible to "a class of disorders" that include such forms of "mental disability" as "softening of the brain, paralysis, palsy, monomania, or insanity." (Olmsted 1990, 17)

Through his work, Olmsted helped to transform American taste. Similar to Gifford Pinchot and others, Olmsted's work also altered the nation's concept of nature. The idea of parks as well as the growing number of natural areas set aside and placed under federal jurisdiction increased legislators' interest in formalizing the government's role in conservation. The effort to connect the growing interest in nature with the idea of national parks gained energy during the 1890s and early 1900s when Congress voted to create additional parks in Sequoia, Yosemite (to which California returned Yosemite Valley), Mount Rainier, Crater Lake, and Glacier (Fox 1981, 111–17).

During these same years, western railroads helped spur tourism to the new parks by building large hotels and rail access. Simultaneously, Congress added other types of sites to the national collection. Prehistoric Indian ruins and artifacts were first preserved by Congress at Arizona's Casa Grande Ruin in 1889. In 1906, Congress added Mesa Verde National Park and passed the Antiquities Act that authorized presidents to set aside "historic and prehistoric structures, and other objects of historic or scientific interest" in federal custody as national monuments.

Behind many of these initiatives was President Theodore Roosevelt. He used the act to proclaim eighteen national monuments, including El Morro, New Mexico, site of prehistoric petroglyphs and historic inscriptions, and natural features such as Arizona's Petrified Forest and the Grand Canyon. Congress later converted many of these natural monuments to national parks. Although these new federal sites reflected the changing interest in preserving nature and history, there remained no unified ethic to tie together these sites. This ethic emerged in 1916 with the passage of the National Park Service Organic Act, which established a separate federal agency to oversee the parks.

Sources and Further Reading: Roper, *FLO: A Biography of Frederick Olmsted*; Schuyler, *Apostle of Taste: Andrew Jackson Downing, 1815–1852*.

ARE DISAPPEARING SPECIES A PROBLEM OR AN ACCOMPLISHMENT?

Time Period: Late 1800s
In This Corner: Sportsmen, farmers, politicians, railroads
In the Other Corner: Sportsmen, native people, conservationists
Other Interested Parties: Upper-class women, western settlers
General Environmental Issue(s): Species depletion, hunting

The "problem" of human-caused species decline and, in a few cases, disappearance was not immediately obvious to many Americans in the late nineteenth century. Hunting and the use of guns had been part of the fabric of the nation during its first century of existence. In some cases, great pride was initially taken in the depletion of certain problem species.

Although the late 1800s made hunting unnecessary for most urban Americans to feed themselves, it remained culturally important for many wealthy men. In an era of change, hunting and posing for photos with a rifle allowed many urban-dwelling politicians and businessmen to maintain a connection to symbols of strength and virility as well as to the American past. This symbolism became more complicated, however, as some Americans began to consider wildlife conservation.

By the end of the nineteenth century, Americans had clear indications that their wasteful use of resources could result in their permanent loss. National forests were one demonstration of this shift in American thought. The trademark species of the continent, the American bison, became a symbol of the expansion westward and then of the gluttonous use of natural resources. The trade in buffalo began with the hide's use for robes in the early 1800s. Native Americans were involved in each part of the process, hunting and butchering, as well as tanning, which softened the robes. This process moved into factories after the Civil War.

Of course, this greatly intensified the number of buffalo killed. Typically, a hunting team included a shooter who sat in wait with a large-caliber rifle. The shooter first killed the dominant cow. If this shot was successful, the hunter's ability to kill others in the herd was limited only by his reloading and the amount of his ammunition. Two other men then worked as skinners. The carcasses were left behind. Manufacturers turned the hides into machine belts that ran factories throughout the nation (White 1991, 216–19).

This scene of America's far west shows a hunter shooting buffalo on the line of the Kansas-Pacific Railroad. The railroad brought trade opportunities to the West but could not coexist with the buffalo herd located there. Library of Congress.

The railroad altered the future of the bison herd forever. The tracks created a barrier to the bison and concentrated them in smaller areas. In 1884, the herd in eastern Montana numbered 75,000. An influx of approximately 5,000 hide hunters over the next year, however, reduced the herd to only a few hundred.

Extinction seemed a likely outcome. Many onlookers viewed a positive side to the loss of the bison: they felt that Native Americans could be more easily controlled without the bison. By 1900, the estimated thirty million bison of 1850 had been reduced to 2,500. An alternative sentiment that developed after 1900, however, marks one of the first focused efforts by the United States at wildlife conservation.

The first impulse to conserve the bison population came from a trained scientist: zoologist William T. Hornaday, who needed additional bison specimens for the National Museum in Washington, DC. Fully aware of their growing scarcity, Hornaday was still shocked that, during eight weeks in 1886, he could collect only twenty-five specimens in a region of Montana that just a few years prior supported tens of thousands of bison. He quickly became a leading proponent of the bison's imminent extinction, and his efforts came not a moment too soon. By 1893, the population of bison that had numbered nearly sixty million was estimated to have plummeted to just 300 (Isenberg 2001, 136–38).

Many upper-class Americans were ripe to hear Hornaday's conservation message. With consistent interest from urban areas in the northeastern United States, Hornaday worked with Theodore Roosevelt and others in 1905 to establish the American Bison Society. Their efforts combined with those of Roosevelt as U.S. president to help improve the bison's situation. During his presidency, Roosevelt persuaded Congress to establish a number of wildlife preserves. The Bison Society worked with a number of private ranch owners to raise bison that could be used to stock preserves and parks and to ultimately help the bison's numbers to rebound (Hays 1999, 141–44). The following is from Hornaday's report:

Of all the quadrupeds that have lived upon the earth, probably no other species has ever marshaled such innumerable hosts as those of the American bison. It would have been as easy to count or to estimate the number of lives in a forest as to calculate the number of buffaloes living at any given time during the history of the species previous to 1870. Even in South Central Africa, which has always been exceedingly prolific in great herds of game, it is probable that all its quadrupeds taken together on an equal area would never have more than equaled the total number of buffalo in this country forty years ago....

Between the Rocky Mountains and the States lying along the Mississippi River on the west, from Minnesota to Louisiana, the whole country was one vast buffalo range, inhabited by millions of buffaloes. One could fill a volume with the records of plainsmen and pioneers who penetrated or crossed that vast region between 1800 and 1870, and were in turn surprised, astounded, and frequently dismayed by the tens of thousands of buffaloes they observed, avoided, or escaped from. They lived and moved as no other quadrupeds ever have, in great multitudes, like grand armies in review, covering scores of square miles at once. They were so numerous they frequently stopped boats in the rivers, threatened to overwhelm travelers on the plains, and in later years derailed locomotives and cars, until railway engineers learned by experience the wisdom of stopping their trains whenever there were buffaloes crossing the track....

No wonder that the men of the West of those days, both white and red, thought it would be impossible to exterminate such a mighty multitude. The Indians of some tribes believed that the buffaloes issued from the earth continually, and that the supply was necessarily inexhaustible. And yet, in four short years the southern herd was almost totally annihilated....

Causes of Extermination

The causes which led to the practical extinction (in a wild state, at least) of the most economically valuable wild animal that ever inhabited the American continent, are by no means obscure. It is well that we should know precisely what they were, and by the sad fate of the buffalo be warned in time against allowing similar causes to produce the same results with our elk, antelope, deer, moose, caribou, mountain sheep, mountain goat, walrus, and other animals. It will be doubly deplorable if the remorseless slaughter we have witnessed during the last twenty years carries with it no lessons for the future. A continuation of the record we have lately made as wholesome game butchers will justify posterity in dating us back with the mound-builders and cave-dwellers, when man's only known function was to slay and eat.

The primary cause of the buffalo's extermination, and the one which embraced all others, was the descent of civilization, with all its elements of destructiveness, upon the whole of the country inhabited by that animal. From the Great Slave Lake to the Rio Grande the home of the buffalo was everywhere overrun by the man with a gun; and, as has ever been the case, the wild creatures were gradually swept away, the largest and most conspicuous forms being the first to go.

The secondary causes of the extermination of the buffalo may be catalogued as follows: (1) Man's reckless greed, his wanton destructiveness, and improvidence in not husbanding such resources as come to him from the hand of nature ready made. (2) The total and utterly inexcusable absence of protective measures and agencies on the part of the National Government and of the Western States and Territories. (3) The fatal preference on the part of hunters generally, both white and red, for the robe and flesh of the cow over that furnished by the bull. (4) The phenomenal stupidity of the animals themselves, and their indifference to man. (5) The perfection of modern breech-loading rifles and other sporting fire-arms in general....

Effects of the Extermination

The buffalo supplied the Indian with food, clothing, shelter, bedding, saddles, ropes, shields, and innumerable smaller articles of use and ornament. In the United States a paternal government takes the place of the buffalo in supplying all these wants of the red man, and it costs several millions of dollars annually to accomplish the task....

The Indians of what was once the buffalo country are not starving and freezing, for the reason that the United States Government supplies them regularly with beef and blankets in lieu of buffalo. Does any one imagine that the Government could not have regulated the killing of buffaloes, and thus maintained the supply, for far less money than it now costs to feed and clothe those 54,758 Indians? ...

Preservation of the Species from Absolute Extinction

There is reason to fear that unless the United States Government takes the matter in hand and makes a special effort to prevent it, the pure-blood bison will be lost irretrievably....

At least eight or ten buffaloes of pure breed should be secured very soon by the Zoo-logical Park Commission, by gift if possible, and cared for with special reference to keeping the breed absolutely pure, and keeping the herd from deteriorating and dying out through in-and-in breeding.

The total expense would be trifling in comparison with the importance of the end to be gained, and in that way we might, in a small measure, atone for our neglect of the means which would have protected the great herds from extinction. In this way, by proper management, it will be not only possible but easy to preserve fine living repre-sentatives of this important species for centuries to come .... (Hornaday's report)

Also in the Midwest, game markets took shape to harvest a number of flocking birds. The most intense hunting focused on the passenger pigeon. The demise of the pigeon at the end of the nineteenth century directly stems from the growth of consumer markets. In the great flocks of pigeon, game hunters saw a steady and significant food resource for eastern restaurants. In addition, however, nets could be used to capture flocks of live pigeons and return them to hunting clubs for the upper-class hunters. From their nesting areas in the northern Midwest, live pigeons were sent to hunting clubs in cities in the region as well as to Boston, New York, and Washington, DC.

Although the waste of hunters of the late 1800s seems terrible by modern sensibilities, hindsight reveals that it was the excessive killing that spurred collective action. Alarm over the disappearance of these species and others fueled additional calls for conservation action among the American public (Price 2000, 46–49). Eventually, this collective action altered American resource use and helped to create the modern environmental movement.

Sources and Further Reading: Hays, *Conservation and the Gospel of Efficiency*; Hornaday's report; Nash, *Wilderness and the American Mind*; Reiger, *American Sportsmen and the Origins of Conserva-tion*; Steinberg, *Down to Earth*; White, *It's Your Misfortune and None of My Own*.

SPORTSMEN HELP TO BUILD A CONSENSUS FOR CONSERVATION

Time Period: Late 1800s
In This Corner: Sportsmen, farmers, politicians, railroads
In the Other Corner: Sportsmen, native people, conservationists
Other Interested Parties: Upper-class women, western settlers
General Environmental Issue(s): Species depletion, hunting

To begin to acknowledge that it was possible to overhunt species, Americans needed to make basic changes in the fashion in which they viewed humans' role in the natural world. As more and more urban Americans grew less in tune with the natural world at the close of the 1800s, a surprising constituency took the lead in establishing an American conservation movement. The movement for conservation was initiated by those who were using natural resources: hunters and other sportsmen.

Sportsmen, particularly those living in urban areas, such as New York City, Chicago, and Boston, saw the impact of species depletion firsthand during their hunting trips. They also often interacted with members of society who had achieved wealth and stability that made hunting a leisure task instead of a matter of survival. When a few individuals (often with a background in the natural sciences) began talking about species depletion, these sportsmen listened. Ultimately, they also sought to preserve their ability to hunt by determining solutions to the problem. By the 1890s, the land and resource-use practices that emerged from this search for solutions became known as the American conservation movement.

Most historians agree that efforts to conserve wildlife began with fish. By 1870, hatcheries had been put in operation throughout the United States. Species ranging from shad, brook trout, and carp were reintroduced to depleted or polluted waters. In fact, seven years before George Perkins Marsh wrote *Man and Nature*, he was commissioned by sportsmen to study the feasibility of restoring fish populations lost from the Connecticut River (Reiger 1988, 53).

In 1871, the issue of fish populations resulted in the establishment of the U.S. Fish Commission, the first federal conservation agency. Although this agency produced some of the earliest studies of fish population, most of the action was initiated by sportsmen's organizations. Typically, fish population was not stressed by these groups for the fishes' sake; instead, efforts were organized to preserve "fish culture." Although the population of fish is a critical component of such culture, their worth comes from humans being able to catch them. Historian John Reiger wrote: "For sportsmen, [the fish culture idea] meant a restoration of angling opportunities; for farmers and ranchers, it meant a profitable sideline …, for commercial fishermen … it meant neverending profits …; and for the nation as a whole, it meant cheap food for the masses" (Reiger 1988, 53). Soon, other efforts to conserve supplies of wild game followed the early efforts to protect fish culture.

Another branch of the sportsmen's efforts were hunting preserves, which began appearing in the 1870s (Hays 1999, 114–16). "Deer parks" date back to the colonial era; however, in the 1870s, sportsmen aware of the depleted stocks of certain choice game established their own preserves and stocked them. Two of the earliest examples were the Bisby Club and the Blooming Grove Park. Located in Pike County, Pennsylvania, Blooming Grove spanned 12,000 acres. The founders of this preserve managed populations of game animals, fish, and forests to create the best environment for sportsmen. Reiger reported that it was Blooming Grove that inspired similar clubs in the Adirondacks beginning around 1890 (Hays 1999, 142).

Often, the efforts of early conservationists appear today to have been very self-serving. However, actions such as the designation of private hunting preserves helped to create a new cultural view of nature that was less exclusively utilitarian. The early regulations and laws that functioned to conserve wildlife and natural resources formed the groundwork of a new environmental ethic that took shape during the twentieth century. The interest of sportsmen helped to form the popular movement that formed a critical portion of the emerging environmental sensibility.

Sources and Further Reading: Hays, *Conservation and the Gospel of Efficiency*; Nash, *Wilderness and the American Mind*; Reiger, *American Sportsmen and the Origins of Conservation*; Steinberg, *Down to Earth*.

GEORGE BIRD GRINNELL SERVES AS A BRIDGE BETWEEN THE WEALTHY AND NATURE

Time Period: Late 1800s
In This Corner: Wealthy conservationists, reformers
In the Other Corner: Politicians, supporters of the status quo
Other Interested Parties: Sportsmen, upper-class women
General Environmental Issue(s): Conservation, popular environmentalism, birds

The issues and topics related to the new conservation movement of the late 1800s lay quite distant from the lives of many wealthy Americans. Writers and activists sought ways of fashioning a message that would interest wealthy patrons on behalf of nature. Using the name of one of the greatest American artists of nature, George Bird Grinnell distinguished himself from any other conservationist by initiating chapters of the Audubon Society throughout the nation. The Audubon chapters in major cities helped to spread revolutionary new ideas about nature throughout the United States. Some of these ideas were soon converted into some of the nation's first environmental policies.

Grinnell had known of Audubon's exploits all his life. In fact, as a child, Grinnell attended school in John James Audubon's family mansion in New York State. Similar to the great artist, Grinnell developed a love for birds at an early age. George, as well as his brothers and sisters, knew the Audubon family and were allowed to roam the grounds of their estate and play in the barn that held large collections of bird specimens and skins. Naturally, Grinnell grew into a budding naturalist and scientist.

Grinnell served as a naturalist on Custer's expedition to the Black Hills in 1874. While there, he developed an interest in what he could learn from the local Indian tribes. He was well known for his ability to get along with the tribe elders, and the Pawnee eventually adopted him into their tribe. He also studied other tribes, including the Gros Ventre and the Cheyenne, and his writings on the Indians during this time are considered some of the best in the anthropology field. He served as an advocate for Native Americans for the rest of his life and went on to work for fair and reasonable treaties with Indian tribes.

In addition to his studies on the Native Americans, Grinnell was also an advocate for environmental protection. He was editor of a weekly magazine for sportsmen and naturalists called *Forest and Stream* and used it to help him channel the dissatisfaction of outdoorsman with disappearing habitats and dwindling game populations into a fight to conserve natural resources. He advocated a game warden system to be financed by small fees from all hunters to ensure effective enforcement of game laws. His revolutionary regulation of hunting activity on the state level with financial support from the hunters themselves became a cornerstone for game management (Nash 1982, 152–53).

Grinnell was also active in the fight for habitat conservation. In 1882, he started an editorial effort to persuade Americans to efficiently manage timberlands. He was also involved with the preservation of Yellowstone National Park and launched a campaign to help protect

George B. Grinnell and his wife are shown on one of their frequent explorations. Through his influence, Grinnell inspired many Americans to consider conservation as well as their own efforts to explore the world around them. Library of Congress.

it against commercialization and to expose federal neglect. Theodore Roosevelt admired and supported Grinnell's efforts and joined Grinnell's battle for Yellowstone Park before he became president. Grinnell's conservation philosophy served as the basis of the American Conservation Program when Roosevelt became president in 1901.

One of his first battles, however, began in the late 1800s when he called for the formation of a society to fight the use of feathers, particularly in women's fashion. Hats decorated with feathers, bird parts, and even entire stuffed birds were a mainstay of the upper-class woman in France, Britain, and England. Eight to ten warblers were used on a single hat. Additionally, the long plumes of egrets or even the taxidermied heads of owls defined the well-dressed. Style reached broader audiences through Grinnell when he announced the formation of the Audubon Society in *Forest and Stream* on February 11, 1886:

> Very slowly the public are awakening to see that the fashion of wearing the feathers and skins of birds is abominable. There is, we think, no doubt that when the facts about this fashion are known, it will be frowned down and will cease to exist. Legislation of itself can do little against the barbarous practice, but if public sentiment can be aroused against it, it will die a speedy death ...

The reform in America ... must be inaugurated by women.... [But] Something more than this is needed. Men, women, and children all over our land should take the matter in hand, and urge its importance upon those with whom they are brought in contact. A general effort of this kind will not fail to awaken public interest, and information given to a right-thinking public will set the ball of reform in motion....

We propose the formation of an association for the protection of wild birds and their eggs, which shall be called the Audubon Society. (Rieger 1988, 68–69)

Grinnell's Audubon Society tapped into a budding interest of many Americans. The Society was an instant success, attracting 50,000 members by 1888. Overwhelmed, Grinnell needed to choose between Audubon and *Forest and Stream*. In 1889, he disbanded the Audubon Society; however, others would form new chapters of the society within the decade. The new Audubons began in 1896 and had the social status and financial power of urban elite women behind them, starting in Boston and then extending to New York and other cities throughout the nation (Price 2000, 75). Although critics may refer to these as upper-class social clubs, Audubons became an important mechanism for altering the American idea of nature

Sources and Further Reading: Hays, *Conservation and the Gospel of Efficiency*; Nash, *Wilderness and the American Mind*; Price, *Flight Maps*; Reiger, *American Sportsmen and the Origins of Conservation*; Steinberg, *Down to Earth*.

TRANSFORMING WHALING INTO AMERICA'S LEADING ENERGY INDUSTRY

Time Period: Early 1800s
In This Corner: Quaker business leaders in New Bedford, Massachusetts
In the Other Corner: Fishermen and whalers of Nantucket and elsewhere
Other Interested Parties: Illumination consumers worldwide
General Environmental Issue(s): Species management, energy supplies

The energy to create light promised to revolutionize American life. Limited supply marked the greatest difficulty to each of the early illuminants. For American whalers, the limited capabilities of Nantucket were replaced by grander undertakings in the early 1800s. Increasing whale oil supplies required the use of more ships traveling the sea for longer periods. Orchestrating this shift in whaling's scale and scope would enable it to become one of the first great energy industries. The emergence of New Bedford, Massachusetts, coincided with the nation's growing ability to take full advantage of its economic resources. The city would take shape as a multifaceted industrial center based almost completely around whale oil. Although the resource had not changed at all, the nature of the enterprise had been revolutionized.

By the early 1800s, whaling ships remained at sea for three to five years. On their journeys, they reached through the Pacific and into the Arctic. In their Pacific travels, these early tankers discovered many tiny islands that were considered worthless at the time but that subsequently became essential to American expansion in the region. Whalers also extended into the Arctic Ocean in 1846 when an American ship drifted in a dense fog northward through the Bering Strait. Proximity to the new grounds and the advent of steam whalers (1879–1880) made San Francisco a major whaling port. It would also lead to some of the first American relations with the Hawaiian Islands.

The vast majority of these whaling vessels hailed from New Bedford, Massachusetts, the commercial experiment of a few of Nantucket's most successful whalers. The community that took shape along the Acushnet in the early 1800s did not replicate other sea towns. The added responsibilities of New Bedford demanded additional facilities. Specifically, factories and manufacturing centers rapidly appeared between Water and Third Streets, immediately adjacent to the waterfront. New Bedford's complexity grew with candle manufacturers and factories for refining whale and sperm oil.

At the apex of American whaling, nearly half of the recorded voyages left from New Bedford. Mid-century, New Bedford's fleet numbered 320, with the next "competing" port possessing only sixty-five. Unlike a traditional seaport, New Bedford served as much more than entrepôt. Although the docks and berths allowed large ships to use New Bedford as a port, the town's one-dimensional infrastructure clearly positioned it as a fusion of a site of oil transshipment, such as Valdez, Alaska, and of oil refinery, such as locations in the American South and West. Despite maintaining the Seamen's Bethel, counting house, and other hallmarks of maritime society, this infrastructure was less designed around the sailor and more conditioned for industrial workers, such as refiners and factory workers.

By the 1850s, sprawling sites for processing oil, known as oil works, had been added just outside of the riverfront section of town. The industrial nature of the enterprise derived from the need to refine both whale and sperm oil. First, the blubber oil and the head oil were combined in large tanks. Pipes lined the tanks, and steam was introduced to heat the oil to a temperature of about 212 degrees Fahrenheit for six to ten hours. This boiled off any remaining water and melted any solid blubber that remained. The oil that went into the tank then was typically light yellow, with a mild odor and taste; when it came out, it was without color, odor, or flavor. After the oil had cooled, it was pumped into fifty-gallon barrels. The oil would then be cooled to freezing before being reheated to various levels to create different "grades," including winter oil, taut pressed oil, and spring oil. The remaining substance was pure spermaceti and could be used in candle and soap making.

With the industry's scale and scope expanded, America's energy future also changed significantly. Domestic and industrial use of energy products very swiftly exceeded the capabilities of whalers. When the global whalers began using cannons and other means to increase the supply of whales, American whalers looked to other energy resources altogether. New Bedford's heyday, however, helped Americans step toward a new era in energy production and consumption.

Sources and Further Reading: Black, *Petrolia: The Landscape of America's First Oil Boom*; Creighton, *Rites and Passages*; Hohman, *The American Whaleman*; Labaree, *Americans and the Sea*.

ENERGY TRANSITION LEAVES WHALING BEHIND

Time Period: 1871
In This Corner: Whalers, whaling interests
In the Other Corner: Petroleum developers

Other Interested Parties: Energy consumers
General Environmental Issue(s): Energy, consumption

The significant economic stakes rarely allow transitions from one energy source to another to proceed gradually. In the case of the United States in the late 1800s, cultural demand for additional energy supplies drove a search that brought dramatic changes to everyday American life. To fill this demand, production of the primary resource for home energy use moved from the ocean and onto the land. Whaling, which had been one of the nation's leading industries in the early 1800s, rather swiftly became a thing of America's past during the 1870s.

Ultimately, the American Civil War played a key role in the demise of American whaling. The war wreaked havoc on the fleet, reducing the number of ships out of New Bedford by one-quarter. More importantly, however, petroleum discovered in 1859 offered a much cheaper illuminant (kerosene). Considering physical risks and pecuniary hazards, the whaling industry was never extremely, or consistently, profitable. However, during the early era of industrialization, the limitations of whaling did not preclude it from becoming one of the world's most important energy industries.

The Civil War had already significantly impacted the whaling industry. By early 1861, Confederate leaders identified whaling as one of the Union's most important industries. Confederate efforts to thwart Union commerce were not based on future promise but on current reliance, and so commenced the whalers' problems of the 1860s. In summary, during the space of five years, the fleet was cut in half. Between January 1, 1861 and January 1, 1866, whaling tonnage fell from 158,746 to 68,536, and the number of vessels fell from 514 to 263 (Hohman 1928, 296–300). In addition, the federal government purchased forty older whaling ships, laden with stones, and deliberately sunk them off Charleston and Savannah harbors in an effort to make the navigable channels unsafe for blockade runners. Although the government purchased the "Great Stone Fleet" vessels outright, their destruction left an immediate gap in the ranks of the whaling fleet that was never filled. Ironically, the next front of the whalers' Civil War arrived after the actual fighting had ceased.

On June 28, 1865, more than a month after the surrender by General Robert E. Lee at Appomattox, Virginia, the CSS *Shenandoah* wreaked havoc on whaling ships in the Arctic. Two ships were allowed to leave to transport the crews. Eight whaling ships, however, burned amid the Arctic ice as the prizes of *Shenandoah* and the defunct Confederacy (Hohman 1928, 290). Repeatedly, Commander Thomas Waddell's captives attempted to convince him of Lee's surrender. Waddell, however, refused to trust such claims. Finally, on June 23, Waddell's log verifies that one captured vessel used newspaper accounts to convince the Confederate raider that Lee had surrendered at Appomattox. Despite the danger of being declared a pirate, Waddell chose to press on. In his log, Waddell emphasized that, without orders to discontinue his efforts, he felt that he must continue as an officer of the Confederate Navy. With the number of prisoners rapidly growing, *Shenandoah* led an odd ocean-going train of whaleboats (twelve was the largest number).

Despite his success, the pressure on Waddell's conscience soon grew too great. He commanded his crew to abandon their raid and flee to safe harbor in Liverpool, and the reign of terror on the Arctic whalers was over. Waddell had taken more than 1,000 prisoners, destroyed twenty-five ships, converted four others for prisoner transport, and seized cargo valued at about $1.5 million. Added to other ships' losses in the war, forty-six whaling ships were taken from service. From a New Bedford fleet of 186 whaling ships in 1861, these losses added to the forty in the Great Stone Fleet to seriously diminish the industry.

After the Civil War, whale oil's market share weakened just as petroleum's surged ahead. In an effort to compete with the more abundant and easier gotten petroleum, whalers remained at sea longer and pushed more deeply into the volatile Arctic waters than ever before. This effort in 1871 spelled the end of the American whale fishery. As the whalers watched the ices close in during the spring, they resisted the urge to leave. The Arctic whales pursued by *Taber* and others provided a variety of materials to commercial markets. Baleen whales offered stores of oil for illumination but also bone for a variety of uses, including the ribs in umbrellas and skirt hoops and other materials used for buggy whips. The oil, however, remained the preeminent product of the Arctic whale hunt. The expansion northward, in essence, was the New Bedford financiers bow to changing times; of course, it would turn out that competitive whaling needed more than an expansion of grounds.

On September 1, 1871, the thick ice pressed in on twenty-six whaling vessels. The *Comet* succumbed first. One by one, the ships gave in to the monstrous ice as it squeezed their hulls and lifted them out of the water before simply shattering many of their stern. Those that could flee by and large elected instead to continue to take whales. After two more days of whaling, the fleet's fate began to unfold. Of the forty-two New Bedford whalers in the region, thirty-three would not return. Like the tentacles of a giant sea beast, the ices came faster than expected. Rapidly, the bulk of the fleet stood stock-still in the sea, as if suddenly transmuted into a maritime painting adorning a collector's wall.

The whalers pressed forward, despite Native Alaskan warnings that the weather was going to be unusually poor. As the ice locked the ships in place, heavy snow also added to the whalers' difficulty, packing them in more solidly than before. In a moment of desperation, the whaler overlooked the limits under which he worked.

The American fleet would not recover. When other nations integrated new technologies to expand whaling, Americans, instead, moved ashore and found their energy elsewhere. As the American appetite for energy grew with new supplies such as petroleum, whaling appeared more and more as an industry of a bygone era.

Sources and Further Reading: Black, *Petrolia: The Landscape of America's First Oil Boom*; Creighton, *Rites and Passages*; Hohman, *The American Whaleman*; Labaree, *Americans and the Sea*.

DISCOVERY OF PETROLEUM IN PENNSYLVANIA

Time Period: 1859
In This Corner: Financiers of early petroleum exploration, regional interests
In the Other Corner: Whalers, existing energy industrialists
Other Interested Parties: American consumers, illumination and otherwise
General Environmental Issue(s): Energy, consumption

The culture of industrialization altered the ways that Americans conceived of any and all of their natural resources. Sometimes, this meant that known resources such as coal became integral to new processes, such as making steel. However, the industrial sensibility also inspired a new approach to known resources. The best example might be petroleum, which was a well-known nuisance in places where it seeped to the earth's surface, such as western Pennsylvania.

Although the first occurrence of oil floating in streams was in New York in 1627, Pennsylvania was the place most identified with oil seepage. These observations often noted the presence of oil as a way of setting off a less-than-desirable location for agriculture. The intricacy of petroleum to American life in the 1990s would have shocked nineteenth-century users of "Pennsylvania rock oil." Most farmers who knew about the oil in the early 1800s knew seeping crude as a nuisance to agriculture and supplies of drinking water. These observers were not the first people to consider the usefulness of petroleum, which had been a part of human society for thousands of years. Its value grew only when European-Americans offered the resource their skills of commodification.

Crude oil was found and used in some manner in ancient societies of Asia. However, the area credited with first noticing the value of petroleum lays just beyond North America's Appalachian Plateau in western Pennsylvania, nearly one hundred miles above Pittsburgh. The oil occurring along Oil Creek was named initially for the Seneca people, who were the native inhabitants of this portion of North America at the time of European settlement. However, there were earlier users of this same supply.

Northwestern Pennsylvania had served as a temporary home to the mound-builder society living centuries before the Seneca. Paleo-Indians of the Woodland period, before 1400, ventured from their original homelands in the Ohio Valley and along the Great Lakes on frequent journeys to Oil Creek, where they collected oil on a fairly large scale for use in their religious rituals. Although no written accounts remain, it was well known that initial European explorers in the area found long, narrow troughs that had been dug along Oil Creek. Early use of the crude oil reveals interesting contrasts between native and European cultures. The Seneca skimmed the oil from the water's surface, using a blanket as a sponge or dipping a container into the water, and then used the collected crude as ointment or skin coloring. European explorers designated this Pennsylvania stream as Oil Creek beginning in 1755. Tourists and soldiers passing through the area were known to soak aching joints in the surrounding oil springs and even to imbibe the crude as a castor oil variation.

As the oil's reputation grew, settlers to the region gathered oil from springs on their property by constructing dams of loose stones to confine the floating oil for collection. In the mid-1840s, one entrepreneur noticed the similarity between the oil prescribed to his ill wife and the annoying substance that was invading the salt wells on his family's property outside Pittsburgh, Pennsylvania. He began bottling the waste substance in 1849 and marketed it as a mysterious cure-all available throughout the northeastern United States. Although he still acquired the oil only by skimming, Samuel Kier's supply quickly exceeded demand because there was a constant flow of the oil from the salt wells. With the excess, he began the first experiments with using the substance as an illuminant, or substance that gives off light. The culture of expansion and development was beginning to focus on petroleum.

Other interested parties soon began to experiment with oil's use as an illuminant. Dr. Francis Brewer, a resident of New Hampshire, traveled to Titusville in 1851 to work with a lumbering firm of which he was part owner. During the visit, Brewer entered into the first oil lease ever signed with a local resident. Instead of drilling, however, the lessee merely dug trenches to convey oil and water to a central basin. On his return to New England, Brewer left a small bottle of crude with Dixi Crosby, a chemist at Dartmouth College, who then showed it to a young businessman, George Bissell.

Petroleum's similarity to coal oil immediately struck Bissell. He signed a lease with Brewer to develop the petroleum on the lumber company's land, but first Bissell needed to attract financial backing of $250,000. This would not be easy because neither he nor anyone else knew what utility petroleum would have. Some of the risk could be assuaged by scientific explanation of the odd curiosity, petroleum. Benjamin Silliman, Jr. of Yale University provided such backing in his report released in April 1855. Silliman estimated that at least 50 percent of the crude could be distilled into a satisfactory illuminant for use in camphene lamps and 90 percent in the form of distilled products holding commercial promise. On September 18, 1855, Bissell incorporated the Pennsylvania Rock Oil Company of Connecticut, the first organization founded solely to speculate with the potential value of the oil occurring naturally beneath and around the Oil Creek valley.

From this point forward, petroleum's emergence became the product of entrepreneurs, except for one important character: Edwin L. Drake of the New Haven Railroad. In 1857, the company sent Drake to Pennsylvania to attempt to drill the first well intended for oil. The novelty of the project soon worn off for Drake and his assistant Billy Smith. The townspeople irreverently heckled the endeavor of a "lunatic." During the late summer of 1859, Drake ran out of funds and wired to New Haven, Connecticut, for more money. He was told that he would be given money only for a trip home, that the Seneca Oil Company, as the group was now called, was done supporting him in this folly. Drake took out a personal line of credit to continue, and, a few days later, on August 29, 1859, Drake and his assistant discovered oozing oil.

E. L. Drake and Peter Wilson in foreground, and James Smith, Wm. Smith, Jr., and Elbridge Lock in background, before the first well ever drilled for petroleum oil in the United States. Library of Congress.

What happened next made petroleum the world's next great energy resource and defined a century of human life. The manner in which oil pooled under the surface of the earth's crust combined with the drilling technology used to access it to create the practice that became known as "the rule of capture." Once oil was struck by one well, others were immediately struck by either the same company or other lessees.

This need to rush to known supplies created a cycle of boom and bust that would define the petroleum industry. The boom and bust cycle was even underwritten by the courts in the case of *Brown v. Vandergrift* (1875), which established the laissez-faire development policy that became known as "the rule of capture." The oil could be owned by whomever first pulled it from the ground, or captured it. The rush to newly opened areas became a race to be the first to sink the wells that would bring the most oil up from its geological pockets.

Sources and Further Reading: Black, *Petrolia: The Landscape of America's First Oil Boom*; Darrah, *Pithole: The Vanished City*; Yergin, *The Prize: The Epic Quest for Oil, Money & Power*.

PITHOLE, PENNSYLVANIA: BOOMTOWNS HARVEST RESOURCES AND THEN MOVE ON

Time Period: 1860 to the present
In This Corner: Oil companies, developers
In the Other Corner: Local and regional residents
Other Interested Parties: Energy consumers
General Environmental Issue(s): Energy management, sustainability

Whether the petroleum wells are found in the jungle of Nigeria, the ices of Siberia, or off the coast of Mississippi, today's oil development continues to use at least a few of the devices perfected on the oil frontier of Pennsylvania. One of the most recognizable is the boomtown. Originally created as a service community for resource extraction, boomtowns still function effectively to support the energy industry; today, however, some critics have come to question the ethics of establishing communities intentionally designed not to last.

The oil fields of Pennsylvania relied on the rapid development of many boomtowns. Such towns became part of the mystique of the public's infatuation with the petroleum boom. None, however, stirred as much interest as Pithole, which was established just after the Civil War, around 1866. Pithole received a national reputation as the greatest of all boomtowns and the symbol of the progress embodied by the developing oil industry. From an industrial work camp, this place suddenly warranted comparisons with the great cities of the Northeast, and it was even claimed that it was to be Pennsylvania's second city after Philadelphia. This was an astounding transition from the farm that formerly filled this space. Within six months of the discovery of oil here, a town had taken form, the largest boomtown in the region, with a population of 10,000. This population would spring from zero to its reported high of 15,000 in less than eight months.

The 1859 striking of the first petroleum in Pennsylvania forever altered American life and industry. During this era of change, Pithole marked a new model for the industrial boomtown. Located a few miles from Titusville, the town of Pithole was created by the United States Petroleum Company in 1864. By 1865, petroleum production of 1,200 barrels a day brought "boomers," particularly Civil War veterans, from throughout the northeast. Overnight, a town was constructed, companies founded, and wells sunk. In September 1865, the population reached 15,000, and the daily production exceeded 6,000 barrels. If oil development was an organic part of nature, then this was the first flower that it produced: a human community entirely based on petroleum.

No residents intended to call Pithole home for long. Most workers lived in boarding-houses. With open flames part of life for lighting and heat, fires occurred very often. Pithole burned repeatedly. With no water supply, little effort was made to stop a fire's spread. When town leaders met to discuss starting a fire company, residents resisted paying any money to save a community that they did not really think would last very long.

Their suspicion proved correct. Within six months, Pithole's supply of oil began to wane. Soon, workers began to leave to find work in oil fields elsewhere. Some residents began setting intentional fires on their property to collect insurance money. Every resident of Pithole, however, gave up on its future and stopped rebuilding the town. Evidently, Pithole's only reason for being was its petroleum. Without it, Pithole became a deserted city. Although its post office was Pennsylvania's third most active in early 1865, by year's end, it no longer existed.

The image of the boomtown showed humans in such control of their enterprise that they were able to exist outside of their primary cultural construction: the town. Loosened from the limiting strictures of community and heritage, boom was about the future: expansion, growth, and greed. This dynamic became petroleum's enduring claim to iconic appeal. Obviously, instant wealth intrigued the onlooker in 1866. Petroleum wealth is a fleeting possibility. Pithole demonstrated the other attraction of petroleum: boom, of course, includes the possibility of bust.

The intrigue with boom came to permeate petroleum's image in popular culture. The little known film *Boomtown* (1940), starring Spencer Tracy, Claudette Colbert, Hedy Lamarr, and Clark Gable, uses Texas wildcatting of the 1930s to pull at every portion of petroleum's myth, including its corruptive capability. Nothing is sacred in East Texas during boom: one of the film's most memorable scenes takes place when Tracy's character stands with the church minister in the sanctuary of his defunct church observing the derrick that the congregation had voted to construct in the midst of their site of worship. Gable's character rides the boom mentality to fortune all over the world and becomes a global tycoon. Tracy's character pursues happiness instead.

The greed inherent in the expansive oil industry is also the main plot structure of James Dean's final film *Giant* (1956). In one of the most memorable scenes, director George Stevens has Dean's character, Jett Rink, strike a flowing well that erupts from the earth and drenches him in crude. He leaps into his truck to drive directly to the home of Bick Benedict, played by Rock Hudson, and his wife, played by Elizabeth Taylor. The Benedicts have proudly resisted Texas's drive for oil in favor of the more stable, albeit less profitable, cattle industry. Their cowhand, Rink, comes to stand before them to show them what they are missing. When Benedict is unimpressed, Rink attempts to fight Benedict and all he stands for. Through the two characters, visions of western progress clash, just as they still do today.

In the closing scene of *Giant*, we see many of the same implications of growth with which twentieth-century America has contended. Rink has ridden his single well and expanded it to corporate heights. He constructs his own city and then shows up drunk to its dedication, where he once again fights Benedict. The morality play is complete: greed can't win; oil equals greed. Thanks to Rink and J. R. Ewing of television's *Dallas*, such images are so familiar that the plot may seem trite.

Of course, for many years, the West has possessed significant resonance in many forms of popular culture. Most often, popular depictions of western development have fueled an exploitative ethic. The boom mentality grew from human's experience with petroleum in the 1860s to predominate western resource management throughout the twentieth century. This is most obvious if one looks at depictions of a single resource, such as petroleum. The twanging mountain tune rings familiar as the happy, bouncing voice sings:

> Well, listen to the story of a man named Jed, a poor mountaineer trying to keep his family fed.

> When one day he was shooting at some food, and up came a fountain of bubbling crude—OIL that is, black gold, Texas tea.

> Well, the first thing you know, ole Jed's a millionaire and kin folks said, "Jed, move away from there. California's the place you ought to be."

> So they loaded up the truck and moved to Beverly.

The opening of *The Beverly Hillbillies* combines with Jed Klampet's instant and uncomfortable fortune to perpetuate the myth of petroleum wealth well beyond 1950. The experience of Jed Klampet, the hero of *The Beverly Hillbillies*, indicates the resonance of such images into the twentieth century. In film and literary examples, the iconography of petroleum remained remarkably consistent with that seen first in the 1860s.

These images mean something about the actual ethics exhibited in American resource management as well as by the viewing public. Images of boom contribute to an ethic of exploitation. If one views western resource development in the twentieth century, it becomes very obvious that fictitious depictions have actual impacts on the physical environment. The amount of control that the human can exert over the construction of such ethos is still being determined. Petroleum use has been defined by this ethic rooted in the oil fields of Pennsylvania and now extending into the twenty-first century.

Sources and Further Reading: Black, *Petrolia: The Landscape of America's First Oil Boom*; Darrah, *Pithole: The Vanished City*; Yergin, *The Prize: The Epic Quest for Oil, Money & Power*.

ROCKEFELLER, STANDARD, AND THE CONSTRUCTION OF "BIG OIL"

Time Period: 1870s to the present
In This Corner: Standard Oil Trust
In the Other Corner: Other oil companies
Other Interested Parties: Federal government, energy consumers
General Environmental Issue(s): Energy management

After the American Civil War, the petroleum industry consistently moved toward the streamlined state that would allow it to grow into the world's major source of energy and lubrication during the twentieth century. Oil was a commodity with so much potential that it attracted the eye and interest of one of the most effective businessmen in history, John D.

Rockefeller. Through his management of the world's supply of oil and his attitude toward business, Rockefeller expanded the impact of this resource originally found in the streams of western Pennsylvania.

Having made his living as a fruit trader in Cleveland, Rockefeller heard early the great uproar about oil found in the ground just over the state border in Titusville. The new industry certainly appealed to Rockefeller's interest in growing wealthy; however, its boom-to-bust cycles also appealed to his personal interest to bring efficiency to disorder and chaos. Without getting tangled in the unpredictable nature of speculating for oil, Rockefeller would focus his attention on the oil once it had reached the earth's surface.

Working within the South Improvement Company for much of the late 1860s, Rockefeller laid the groundwork for his effort to control the entire industry at each step in its process. Rockefeller formed the Standard Oil Company of Ohio in 1870. Oil exploration grew from the Oil Creek area of Pennsylvania in the early 1870s and would expand from Pennsylvania to other states and nations during the next decade. By 1879, Standard controlled 90 percent of the U.S. refining capacity, most of the rail lines between urban centers in the northeast, and many of the leasing companies at the various sites of oil speculation. Through Rockefeller's efforts and the organization he made possible, petroleum became the primary energy source for the nation and the world.

Standard's Ethics Brought into Question

Rockefeller's Standard Oil first demonstrated the possible domination available to those who controlled the flow of crude oil—what Americans call "big oil" today. Rockefeller's system of refineries grew so great at the close of the nineteenth century that he could demand lower rates and eventually even kickbacks from rail companies. One by one, he put his competitors out of business, and his own corporation grew into what observers in the late 1800s called a trust (what today is called a monopoly). Standard's reach extended throughout the world, and it became a symbol of the "Gilded Age" when businesses were allowed to grow too large and benefit only a few wealthy people. Reformers vowed things would change.

President Theodore Roosevelt, who took office in 1901, led the Progressive interest to involve the federal government in monitoring the business sector. In the late 1890s, "muckraking" journalists had written articles and books that exposed unfair and hazardous business practices. Ida Tarbell, an editor at *McClure's*, who had grown up the daughter of a barrel maker in Titusville, took aim at Rockefeller. Her "History of the Standard Oil Company" produced a national furor over unfair trading practices. Roosevelt used her information to enforce antitrust laws that would result in Standard's dissolution in 1911. Rockefeller's company had become so large that, when broken into subsidiaries, the pieces would grow to be Mobil, Exxon, Chevron, Amoco, Conoco, and Atlantic, among others.

Conclusion: Big Oil as a National Problem

Even after Standard's dissolution in 1911, the image of its dominance continued. Standard led the way into international oil exploration, suggesting that national borders need not limit the oil-controlling entity. Throughout the twentieth century, large multinational corporations or singular wealthy businessmen attempted to develop supplies and bring them to market. In

An antitrust cartoon from 1901 asserts that Rockefeller's Standard Oil is mugging the common people. Library of Congress.

the 1960s, however, many petroleum nations would draw from Rockefeller's model to devise a new structure. Still, massive international companies managed the import and export of oil regardless of the nation of origin. The importer, often companies in western, industrialized nations, was most in control of supply and demand and, therefore, the prices.

This situation began to shift in the late 1950s. The Eisenhower Administration decided to implement quotas on the import of crude oil. Quotas, when only a specified amount of oil could be imported from outside the country, were designed to protect the sale of domestic oil. Begun in 1959, such quotas infuriated oil-producing countries throughout the 1960s. By September 14, 1960, a new organization had been formed with which to battle companies making money by extracting oil around the world. The Organization of Petroleum Exporting Countries (OPEC) had a single clear intention: to defend the price of oil. OPEC would from this point forward insist that companies consult them before altering the price of crude. They also committed themselves to solidarity, and they aspired to a day when oil companies and western nations would come to them to negotiate. "Big oil" had come to mean something else, as well.

Sources and Further Reading: Black, *Petrolia: The Landscape of America's First Oil Boom*; Chernow, *Titan: The Life of John D. Rockefeller*; Yergin, *The Prize: The Epic Quest for Oil, Money & Power*.

SOCIAL REFORMERS SET SIGHTS ON URBAN PROBLEMS

Time Period: 1870s to 1910s
In This Corner: Social reformers, journalists, political activists
In the Other Corner: Entrenched corporate interests, government leaders
Other Interested Parties: Government reformers, American public
General Environmental Issue(s): Environmental reform, pollution, urban life

The advance of modern efforts to use technology to solve problems clearly overwhelmed American life in the late 1800s. Historian Alan Trachtenberg and others have written that American culture tolerated social and environmental injustices if they were accompanied by a commensurate economic benefit. In short, environmental or health impacts were of little concern in the nineteenth century.

Reforming this ethic required basic cultural changes in American society. The calls for these reforms came from many different types of Americans. One of the first issues to galvanize reformist interest was urban squalor. Whereas some observers argued that the priority of cities needed to be facilitating the industries that fed the economy, reformers factored in new information about influences on the health of urban residents to demand change.

Demanding Change in New York City

One of the most important new understandings of the late nineteenth century was the origin of diseases. Concentrated in cities, diseases such as cholera became epidemics during the early 1800s because of a lack of understanding. For instance, when New York City was struck by an epidemic of cholera in 1832, ordinary citizens pointed to the prevalence of immoral behavior as the cause of this latest malady. In such instances, cholera was seen as God's judgment. Historian Charles Rosenberg wrote: "Cholera was a scourge not of mankind but of the sinner ... [and most] Americans did not doubt that cholera was a divine imposition."

When the disease returned to New York in 1849, medically trained observers focused on specific parts of the city such as the tenements inhabited by poor immigrants. Some nativists used the moral argument to call for controls on immigration. By the mid-1860s, public health workers or sanitarians observed a connection between dirt and pollution and the outbreak of disease. Soon, city inspectors used this logic on New York tenements to argue that cleaner rear yard areas indicated healthy occupants.

The concept of germs was not yet understood, but discerning viewers were growing more able to pick out indicators of good or ill health. That said, most ordinary Americans showed little concern for the implications of how they did basic things such as cooking or storing food, going to the bathroom, or acquiring their drinking water. Thus, many of the first efforts to reform urban areas occurred from the top down: cleaning streets, emptying privies, disinfecting tenement buildings, and inspecting food and beverage manufacturing all helped to mitigate the dangers posed by unsanitary conditions.

One of the nation's first organizations devoted to urban reform was the Citizen's Association of New York Council on Hygiene and Public Health, which was founded in 1865. Under the supervision of John Griscom, the Citizen's Association set out to document the living conditions of the working class and poor. Comprising wealthy merchants and city

New York City street sweepers (here shown around 1896) were one of the first applications of the new understanding of the impact that filth had on human health. Library of Congress.

leaders, it was in the best interest of the Citizen's Association to improve urban squalor. Many urban elite feared that disease that began in tenements would wind up in their neighborhoods as well.

Widely distributed in a variety of forms, the 1865 Citizen's Association report laid bare the links between poverty, unsanitary living conditions, and ill health and was used to compel city and state governments to create a permanent health department. The Metropolitan Board of Health used the reports to exert new controls over the urban environment. With new authority to quarantine and disinfect unsanitary houses and rear yards, the Metropolitan Board of Health cleared more than 160,000 tons of manure from vacant lots, cleaned and disinfected more than 4,000 yards, emptied 771 cisterns, and cleaned more than 6,418 privies (Melosi 1999).

By the late 1800s, many reformers were willing to listen to the findings of scientific professionals. Collections of data such as John Shaw Billings's Vital Statistics of New York City and Brooklyn became important tools for determining strategies for reforming urban life.

The Organic City

Amid the squalor of the organic city, cities such as Chicago, Illinois, were often considered to be state of the art for using waterways to take wastes away from the city and into larger bodies of water. Few people comprehended that such practices simply spread the problem over a broader area.

Waste management became a reality first in the cities, where large-scale changes were more easily carried out. Historian Martin Melosi traces the start of a waste management effort to the work of George Waring (Melosi 1999, 46). After 5,000 residents died from yellow fever, Memphis, Tennessee, hired Waring in 1880 to develop a system for disposing of its sewage. His design became the state of the art, and he defined the new role of the "sanitation engineer." Melosi wrote that Waring was committed "to the positive implications of environmental santitation." Waring clarified the important connection between disease and mismanaged urban waste. Melosi wrote, "Despite his advocacy of an outdated theory of disease, Waring instinctively recognized many of the potential dangers of unsanitary surroundings" (Melosi 1999, 50).

He became an advisor to the federal government and worked with the National Board of Health. In 1894, political shifts in New York pushed out the Tammany Hall machine and the new mayor brought in Waring, the world's leading authority on waste management. Under Waring, the Department of Street Cleaning unleashed thousands of military-like workers to clean up New York City and make it a national leader in sanitation. A corps of 1,450 sweepers cleaned each of the city's 433 miles of paved road one to five times per day (Melosi 1999, 56).

By the 1910s, approximately 50 percent of the nation's cities operated municipally owned collection systems for solid waste. The Waring system was adopted in most cities; however, many cities remained a patchwork of collection practices. In terms of water, by 1910, seven of ten cities of more than 30,000 residents had constructed municipally owned wastewater disposal infrastructures (Tarr 1996).

The influence of Waring helped to diminish the "out of sight, out of mind" mentality that allowed many communities to dump waste into nearby bodies of water. Most often, dumps were sited in less-desirable areas of the city (Melosi 1999, 150–53). Crematories and incinerators, a practice imported from Britain, became popular in the 1900s. Landfills became more popular by 1920. As cities grew after 1920, Melosi argued that the organization of waste management systems significantly increased. Although industrialization created more wastes with which to deal, it also helped to created an engineering mindset in which solutions were sought through planning and technology.

Living with Technology

As industrialization increased in the late 1800s, radically new ways of doing things changed many very basic parts of American life. Historian Thomas Hughes referred to the emergence of these new technologies as "the American genesis." Inventors, he wrote, "persuaded us that we were involved in a second creation of the world" (Hughes 1989, 3). For instance, the number of patents issued annually more than doubled between 1866 and 1896 (Hughes 1989, 14). Humans began to interact closely with technology at both work and home. Food, travel, lighting, and heating are just a few of the essential parts of life that changed in the 1890s. Those who seized these new opportunities also had the capability to make some of the greatest fortunes the world had ever seen in nonroyalty.

Although the ethics of many of these wealthy Americans earned them the name "robber barons," the era of big business additionally helped to shape a middle class of managers and business people. This younger generation also began to ask serious questions of the American model of progress. Overall, however, this model of progress continued to prioritize economic expansion in the late nineteenth century. Industrialists of the era viewed mechanization as a boon because they could use cheaper labor. This social organization, of course, created a distinct working class. The ethics with which the few successful tycoons managed their work forces are considered extremely dubious by today's standards; although they amassed some of the greatest fortunes in human history, these businessmen would become known as "robber barons."

During the Gilded Age, many capitalists dismissed considerations of workplace safety and human welfare to create the greatest possible profits. For instance, child labor became a norm in factories, mines, and other extremely dangerous environments, largely because children required the least pay. All industrial growth surged after the Civil War. This growth and a general faith in economic development allowed a few corporations to gain control of entire commodities and their production. Called "trusts," these conglomerates were near monopolies during an era when government and society had not yet defined such an entity as evil. With such power concentrated in a single entity, any efforts for reform faced a mighty challenge.

American cities had, by and large, borne the brunt of rapid industrial development. In the late 1800s, cities were more often viewed as entities for producing economic development than as places to live. Reformers called for change beginning in the early 1900s.

Today, environmental policies regulate many different aspects of the world around us. This is particularly true of industries and factories that produce air, water, and other types of pollution. Today, we even know that there are different types of pollution: point, which effects the area immediately adjacent to the pollution's creation, and nonpoint, which effects a wider region, possibly even distant portions of Earth such as Antarctica. Scientists have given us a much clearer idea about how such pollution damages the surrounding environment as well as the human body.

These ideas, of course, are radically different from those of the nineteenth century, when few questions were asked of economic development. Personal injuries or health problems were considered by many to be the price of having a job. This expectation began to change during the Progressive Era of the early twentieth century. Muckraking journalists alerted the public and politicians to examples in which industry exploited resources, including the natural environment and the human worker. Unions called for more attention to be paid to workers' rights. In each case, the focus of concern became the massive factories that could be found concentrated in many American cities.

Reforming the City

Against a culture in which these ethics were the norm, a younger generation came of age in the 1890s and began to demand reform. Initially, one of the most common outlets for voicing discontent was journalism. The impassioned pleas of concerned journalists then found receptive ears among the elite women of the era. Activists such as Jacob Riis, who wrote *How the Other Half Lives* in 1890 to describe life in New York City slums, and Jane Addams, who started Hull House to aid immigrant acclimation to American culture, led a movement for progressive reform. Ironically, the wealth of some robber barons would contribute to the evolution of a public consciousness on issues such as ghettoes, environmental degradation, and unfair labor practices. In New York City, Riis, observed that "three-fourths of [the city's residents] live in the tenements, and the nineteenth-century drift of the population to the cities is sending ever-increasing multitudes to crowd them.... We know now that there is no way

Similar to the work of Jacob Riis, this image shows "how the other half lives." This scene from around 1907 shows a crowded Hebrew district on the Lower East Side of New York City. Library of Congress.

out; that the 'system' that was the evil offspring of public neglect and private greed has come to stay, a storm-centre forever of our civilization" (Riis, *How the Other Half Lives*).

Riis used his writing and drawings to paint a picture of urban life that was largely unknown, or at least unacknowledged, by wealthy Americans. In *How the Other Half Lives*, he set out to overcome this oversight:

… The name of the pile is not down in the City Directory, but in the public records it holds an unenviable place. It was here the mortality rose during the last great cholera epidemic to the unprecedented rate of 195 in 1,000 inhabitants. In its worst days a full thousand could not be packed into the court, though the number did probably not fall far short of it. Even now, under the management of men of conscience, and an agent, a King's Daughter, whose practical energy, kindliness and good sense have done much to redeem its foul reputation, the swarms it shelters would make more than one fair-sized country village …

… It is curious to find that this notorious block, whose name was so long synonymous with all that was desperately bad, was originally built (in 1851) by a benevolent Quaker for the express purpose of rescuing the poor people from the dreadful rookeries they were then living in. How long it continued a model tenement is not on record. It could not have been very long, for already in 1862, ten years after it was finished, a sanitary official counted 146 cases of sickness in the court, including "all kinds of infectious disease," from small-pox down, and reported that of 138 children born in it in less than three years 61 had died, mostly before they were one year old. Seven years later the inspector of the district reported to the Board of Health that "nearly ten per cent of the population is sent to the public hospitals each year." When the alley was finally taken in hand by the authorities, and, as a first step toward its reclamation, the entire population was driven out by the police, experience dictated, as one of the first improvements to be made, the putting in of a kind of sewer-grating, so constructed, as the official report patiently puts it, "as to prevent the ingress of persons disposed to make a hiding-place" of the sewer and the cellars into which they opened. The fact was that the big vaulted sewers had long been a runway for thieves—the Swamp Angels—who through them easily escaped when chased by the police, as well as a storehouse for their plunder. The sewers are there to-day; in fact the two alleys are nothing but the roofs of these enormous tunnels in which a man may walk upright the full distance of the block and into the Cherry Street sewer—if he likes the fun and is not afraid of rats. Could their grimy walls speak, the big canals might tell many a startling tale. But they are silent enough, and 80 are most of those whose secrets they might betray. The flood-gates connecting with the Cherry Street main are closed now, except when the water is drained off. Then there were no gates, and it is on record that the sewers were chosen as a short cut habitually by residents of the court whose business lay on the line of them, near a manhole, perhaps, in Cherry Street, or at the river mouth of the big pipe when it was clear at low tide. "Me Jimmy," said one wrinkled old dame, who looked in while we were nosing about under Double Alley, "he used to go to his work along down Cherry Street that way every morning and come back at night." The associations must have been congenial. Probably "Jimmy" himself fitted into the landscape. (Melosi 1999)

Combining his writing and drawing talents with care and concern, Riis brought the plight of the poor to many wealthy Americans. He made it more difficult for Americans to look the other way, and, as a result, reformers began to take effective action. Soon, city, state, and federal government had little choice but to listen to their calls for reform and to take action.

Sewage and Water Technology

The technology for managing water and wastewater in urban areas changed steadily from the decentralized privy vault-cesspool system of the early 1800s. By 1850, centralized water-carriage sewer systems had been placed in some urban areas. By the end of the nineteenth century, centralized management systems had become the preferred urban wastewater management method, and most communities installed them by the end of the twentieth century (Melosi 1999).

The primary reason for improving water treatment was new understanding about the role that mistreated water played in breeding disease. The new technologies did not necessarily consistently prevent contamination of nearby surface water or groundwater. By the mid-nineteenth century, engineers, public health officials, and the general public were searching for alternative wastewater management options.

Imported from Europe, the new concept of centralized water-carriage waste removal entailed planning a coordinated system of conduits and channels that used water to convey the wastes away from the sources to a central disposal location. The primary ingredient was piped-in water, which allowed each home's water closet to send out its waste into the system. The water closet marked a significant shift in sewage by increasing both the quantity of the fecal matter treated and the quality of the treatment. In an odd twist, this increase in discharge actually made the risk of disease transfer greater.

The transition slowed at this point when the increased wastewater levels overwhelmed the privy vault-cesspool system. Few municipalities were able to plan for the new needs. Left to handle the problem on their own, many residents either continued to use the existing privy vault or cesspool or created illegal connections to storm sewers or gutters along streets. Each of these illegal options, of course, created additional problems.

The population in the United States surged more than fourfold from 1850 to 1920. This population increase was accompanied by an increase in the number of cities with populations greater than 50,000 (from 392 to 2,722). During the same time period, the percentage of total U.S. population in urban areas increased from 12.5 percent to 51 percent. By the end of the nineteenth century, sewage technology began to have a noticeable effect on inner-city areas. By that time, most major U.S. cities had also constructed some form of a sewer system. In 1909, cities with populations over 30,000 had approximately 24,972 miles of sewers, of which 18,361 miles were combined sewers, 5,258 miles were separate sanitary sewers, and 1,352 miles were storm sewers. In larger cities (populations over 100,000), there were 17,068 miles of sewers, of which 14,240 miles were combined sewers, 2,194 miles were separate sanitary sewers, and 634 miles were storm sewers (Melosi 1999).

By the early 1900s, the most common technologies for treating wastewater were dilution, land application and irrigation of farmlands, filtration, and chemical precipitation. Each of these practices was performed with untreated wastewater. By 1905, more than 95 percent of the urban population discharged their wastewater untreated to waterways. Little changed over the first quarter of the twentieth century, and, in 1924, more than 88 percent of the population in cities of over 100,000 continued to dispose of their wastewater directly in waterways (Melosi, 1999).

Federal regulations demanded change beginning with the Water Pollution Control Act of 1948, which demanded and provided funds for planning, technical services, research, financial assistance, and enforcement. The renewal of this act through 1965 set water quality standards for every community. Although plans of the late 1960s continued to prioritize the protection of public health, they also began to stress the need for preserving the aesthetics of water resources and protecting aquatic life. In 1972, the Water Pollution Act set the unprecedented goal of eliminating all water pollution by 1985 and authorized expenditures of $24.6 billion in research and construction grants. A new environmental era had descended on America's water and wastewater management program.

Taming Trash

Ideas for organizing and properly disposing of waste continued to evolve in the twentieth century. A frequent sight in the 1950s, burning dumps were used by many urban areas to condense waste. The premise, of course, was that burning reduced the volume of refuse received at the dump and therefore extended the life of the site. However, the burning dumps' impact on local air quality was a primary reason that early efforts after World War II to address the problems of dumps were directed toward putting out the fires. On a smaller scale, many homeowners also had burning pits in their backyards. Overall, open burning of refuse stopped in the 1950s.

Discontinuation of the fires created additional problems at dumps throughout the United States. Size increased more rapidly, and often the trash was more likely to breed disease and threaten human health. These needs stimulated a national movement in solid waste management during World War II toward the ideal that became known as the "sanitary landfill."

The model for such landfills derived from the military. Because of the tremendous growth of new military bases during the Cold War, studies determined that the sanitary landfill was adaptable to changing conditions and would accommodate varying quantities of refuse with little significant change in equipment need or operating procedures. The essential element was a heavy piece of equipment called a "bull clam." Resembling a bulldozer, the bull clam carried a moveable flap or blade that could form a bucket or basket to hold significant quantities of refuse or push material around the landfill to cover the surface. Almost constantly, the bull clam moved material and compacted it more and more. Finally, it moved earth-cover material over the surface of the fill. As landfills became common place at military bases, they also influenced civilian refuse operations. By the end of 1945, almost one hundred cities in the United States were using sanitary landfills, and, by 1960, some 1,400 cities were using sanitary landfills (Hickman and Eldredge).

Historians H. Lanier Hickman, Jr. and Richard W. Eldredge wrote, "The significance of the cessation of open burning at dumps and backyards, as well as commercial and industrial open burning, cannot be overstated. The issue of stopping open burning at landfills was the first real national effort to change the management of refuse" (Hickman and Eldredge 2000). Of course, the downside was that now municipalities had much more refuse with which to deal. By professionalizing their approach to the landfills, however, engineers perfected more sanitary methods.

Trash Timeline

1842   A report in England links disease to filthy environmental conditions.

1874   In Nottingham, England, the "destructor" burns garbage and produces electricity. Eleven years later, the first American incinerator opens in New York.

1898   The first energy recovery from garbage incineration in the United States starts in New York City.

1900s Pigs are used to help get rid of garbage in several cities. One expert said 75 pigs could consume one ton of garbage a day.

1904   First major aluminum recycling plants open in the United States.

1920s Landfilling becomes the most popular way to get rid of garbage.

1959   The first guide to sanitary landfilling is published.

1965   Congress passes the first set of solid waste management laws.

1987   A garbage barge circles Long Island with no place to unload its cargo. Americans perceive a new garbage crisis.

1989   *The Solid Waste Dilemma: An Agenda for Action*, an Environmental Protection Agency report, advocates recycling as a waste management tool.

1997   First "America Recycles Day."

(from U.S. Environmental Protection Agency, *Milestones in Garbage*)

Conclusion: Minimizing the Human Footprint

Humans effect the environment around them. It then makes sense that, where humans are most concentrated, in urban areas, the impact is most acute. During the 1900s, Americans came to realize that they impacted the environment around them. In addition, these impacts created health implications for each of us. Therefore, over the course of the twentieth century, Americans took positive action to minimize in many ways their impact on the world in which they lived. Trash and waste are very likely the most undeniable part of the human life cycle.

Sources and Further Reading: Sewer History, Tracking Down the Roots of Our Sanitary Sewers, http://www.sewerhistory.org; Burian et al., "Urban Wastewater Management in the United States: Past, Present, and Future"; Hickman and Eldredge: http://www.forester.net/ msw_9909_brief_history.html; Lower East Side Tenement Museum, *Health and Disease*, http:// www.tenement.org/encyclopedia/diseases_cholera.htm; Kraut, *Silent Travelers: Germs, Genes, and the Immigrant Menace*; Melosi, *Sanitary City*; Pittsburgh Survey; Rosenberg, *The Cholera Years: The United States in 1832, 1849, and 1866*; Rosner, *Hives of Sickness: Public Health and Epidemics in New York City*; Rosner, *The Living City: Engineering Social and Urban Change in New York City, 1865 to 1920*; Tarr, *The Search for the Ultimate Sink*; Tomes, *The Gospel of Germs: Men, Women, and the Microbe in American Life*; U.S. Environmental Protection Agency, *Milestones in Garbage*, http://www.epa.gov/msw/timeline_alt.htm.

RUSSELL SAGE STUDIES URBAN PROBLEMS IN PITTSBURGH

Time Period: 1910s to 1930s

In This Corner: Social reformers, Russell Sage sociologists

In the Other Corner: Business leaders, government leaders

Other Interested Parties: Government leaders, American public

General Environmental Issue(s): Social reform, urban life, pollution

It was no easy task to convince Americans that industrial cities, with the economic opportunity that they offered, were generating social difficulties for those living in them. By the early 1900s, sociologists began to actively study cities to quantify the claims being made by activists and reformers. It is not surprising that one of the first sources of study was Pittsburgh, Pennsylvania, one of the most active industrial cities on Earth.

Rivers were the corridor that first initiated settlement at Pittsburgh. The confluence of three mighty rivers, Pittsburgh hosts the Monongahela, Ohio, and Allegheny Rivers. Travel westward on the rivers also initiated a great deal of industry in the city. Building and outfitting boats became Pittsburgh's first big business in the early 1800s. This industry was concentrated primarily on the shores of the Monongahela River near Pittsburgh's Point Park. During this era, inexpensive flatboats were powered solely by river current and steered with

Iron ore and coke cars in Pittsburgh, Pennsylvania, c. 1905. The presence of the railroad (foreground) and the rivers (background) allowed industry to flourish there. Heavy industry made Pittsburgh famous as one of the nation's most polluted cities in the nineteenth century. Library of Congress.

a thirty-to forty-foot oar at the back. This traffic traveled only in one direction. Once they arrived at points in the west, settlers would break up the boats and use the wood as building material.

In 1852, however, new railroads assured efficient connections to the east. Pittsburgh became the leading city not only in western Pennsylvania but also in many nearby states as well. Even so, shipping by water remained much cheaper than sending raw material via rail. Pittsburgh offered the confluence of these transportation technologies. Pittsburgh became a city in which trains and rivers worked together to lay the foundation for the city's industrial future. Often, trains shifted their loads to barges to make the trip downriver. Then barges might empty their loads onto trains to send finished goods into the countryside.

The heavy industries of the 1800s also centered on Pittsburgh's transportation connection. When the coke-burning blast furnace was developed, iron makers moved their operations close to the rivers. This way, the coke could be delivered by barge or rail to a riverside furnace near rolling mills and other ironworking operations whose engines and processes demanded water.

By 1850, Pittsburgh possessed a remarkably diversified economy. With railroads to deliver materials to Pittsburgh's factories and carry off finished products to markets in other cities, other industries began to flourish as well. For instance, by 1857, five large textile mills employed more than 3,000 workers. In addition, during the 1860s, Pittsburgh also became the world's greatest petroleum-refining center. When western Pennsylvania brought the world its first supply of oil, the boats full of crude oil headed down the Allegheny River toward Pittsburgh. The refinery period was brief, however, because John D. Rockefeller's Standard Oil Company attracted the petroleum shippers to Cleveland (Opie 1998, 276–79).

By the 1870s, however, Pittsburgh's involvement in industry had just begun. The growth of the steel and iron industry changed the city's population, economy, and environment forever. Thanks to its transportation infrastructure and its access to raw materials, Pittsburgh became one of the world's greatest symbols of the new industrial era (Tarr 2003, 23).

When sociologists working for the Russell Sage Foundation looked for a city in which to study the human effects of industrialization and pollution, they immediately found Pittsburgh to be the best case study. Here is how Paul U. Kellogg described the process:

> The main work was set under way in September, 1907, when a company of men and women of established reputation as students of social and industrial problems, spent the month in Pittsburgh. On the basis of their diagnosis, a series of specialized investigations was projected along a few of the lines which promised significant results. The staff has included not only trained investigators but also representatives of the different races who make up so large a share of the working population dealt with. Limitations of time and money set definite bounds to the work, which will become clear as the findings are presented. The experimental nature of the undertaking, and the unfavorable trade conditions which during the past year have reacted upon economic life in all its phases, have set other limits. Our inquiries have dealt with the wage-earners of Pittsburgh (a) in their relation to the community as a whole, and (b) in their relation to industry. Under the former we have studied the genesis and racial make-up of the population; its physical setting and its social institutions; under the latter we have studied the general labor situation; hours, wages, and labor control in the steel industry; child labor, industrial education, women in industry, the cost of living, and industrial accidents. (Pittsburgh Survey)

The findings of the Pittsburgh Survey remain one of the best representations of the ills of unregulated industrialization. This study was an important step in the efforts of reformers to help make everyday life in American cities safer from industrial hazards. Such an attitude for reform, however, directly contradicted many Americans' view of what was responsible corporate behavior (Tarr 2003, 64–66).

Sources and Further Reading: Tarr, *Devastation and Renewal*; Tarr, *The Search for the Ultimate Sink*; Tomes, *The Gospel of Germs: Men, Women, and the Microbe in American Life*.

FORCING THE AMERICAN TRADITION OF NATIONAL PARKS

Time Period: 1860s to 1890s
In This Corner: Property rights advocates and western developers
In the Other Corner: Western interests, railroads, conservationists
Other Interested Parties: Native Americans, politicians, preservationists, romantics
General Environmental Issue(s): Preservation, national parks

Do Americans need nature in their lives? Romantic writers and painters said yes. However, theirs was a wondrous, overwhelming, and sublime nature. Other intellectuals took this impulse and used it to create a model of a useable natural form that became known as parks. Still, many Americans viewed the idea as excessive waste.

Although other nations had established parks and planned natural areas for leisure use, the United States coined the model of setting specific areas aside from development for no reason related to religion or historic importance. This cultural tradition became known as preservation. The name given to these federally owned treasures became national parks. In fact, this term was used before anyone even knew what it meant and against the wishes of some Americans who feared the dangerous precedent of taking land out of private hands and locking it away. Not only did many Americans disagree with setting land aside, but many also argued that the federal government had much more important things to do than to administer and care for such parks.

The idea for national parks was a blend of intellectual traditions, including ideas of romanticism particularly as they took physical form in the work of artists and photographers. The exact moment of origin for the idea of national parks, however, remains the stuff of legend. In national park folklore, it is said that the idea originated in September 1870 among the members of the Washburn-Doane Expedition (a largely amateur party organized to investigate tales of scenic wonders in the area).

The legendary origin myth proceeds like this. During an evening campfire discussion near the Madison Junction, where the Firehole and Gibbon Rivers join to form the Madison River in present-day Yellowstone National Park, the explorers recalled the natural spectacles of the day. Americans should see these wonders, all the campers agreed. But how long could they last before developers realized what great profits could be made by exploiting the natural attractions? The concerns of the group convinced them that action needed to be taken to protect unique sites such as Yellowstone. Everyone around the fire agreed that Yellowstone's awe-inspiring geysers, waterfalls, and canyons should be preserved as a public park. In short, it was Yellowstone's oddity that fueled the designation of the world's first national park.

Convincing politicians to enforce this designation was something else entirely. Although painters and photographers could use their skills to portray the physical beauty of areas such as Yellowstone to the American public, nineteenth-century Americans needed to know something basic about the areas to be set aside from development. In an era of rapid development and growth, nineteenth-century Americans were unable to envision the luxury of leaving natural resources unused. For this reason, before they could consider setting Yellowstone aside, they needed to first establish that it was "worthless." In presenting the idea of preservation of the Yellowstone area to Congress, proponents of the park needed to describe its beauty and also establish that the area contained no resources of value. (Runte, 1987)

The description would have been convincing regardless of its author. However, Hayden's reputation was such that his report proved very influential. To convince Congress of the need to create the park, Hayden convinced followers that they needed to establish the park's uselessness for all but scenic enjoyment. George Edmunds of Vermont opened the brief debate by declaring that Yellowstone was "so far elevated above the sea" that it could not "be used for private occupation at all." He therefore assured his colleagues they did "no harm to the material interests of the people in endeavoring to preserve" the region.

Critics countered that the only rebuttal of significance came from Senator Cornelius Cole of California who stated: "I have grave doubts about the propriety of passing this bill." Although he was convinced of there being "very little timber on this tract of land," he could not believe that it was off limits to grazing and agriculture. He and other critics argued that preservationists overstated the threat to the area. For instance, he argued, what harm would come to the geysers and natural curiosities if they fell into private control?

Throughout the Rocky Mountains, argued Cole, were many areas that would make a splendid public park; however, Yellowstone was a place "where persons can and would go and settle and improve and cultivate the grounds, if there be ground fit for cultivation." Further guarantees by Senator Edmunds that Yellowstone was "north of latitude forty" and "over seven thousand feet above the level of the sea" failed in the least to quiet Cole's objections. "Ground of a greater height than that has been cultivated and occupied," he retorted before asking, "But if it cannot be occupied and cultivated, why should we make a public park of it? If it cannot be occupied by man, why protect it from occupation? I see no reason in that" (Sellars 1979). The argument proved moot. In just a matter of a few years, evidence of Yellowstone's vulnerability to development appeared.

This proposal moved through high political circles and was approved within a year and a half. In 1872, Yellowstone National Park was established, making it the world's first national park. The actual congressional act included the following text:

> … is hereby reserved and withdrawn from settlement, occupancy, or sale under the laws of the United States, and dedicated and set apart as a public park or pleasuring-ground for the benefit and enjoyment of the people; and all persons who shall locate or settle upon or occupy the same, or any part thereof, except as hereinafter provided, shall be considered trespassers and removed there from.

The act also sought to establish basic guidelines for the use and management of the new park:

> SECTION 2. That said public park shall be under the exclusive control of the Secretary of the Interior, whose duty it shall be, as soon as practicable, to make and publish such rules and regulations as he may deem necessary or proper for the care and management of the same. Such regulations shall provide for the preservation, from injury or spoliation, of all timber, mineral deposits, natural curiosities, or wonders within said park, and their retention in their natural conditions. The secretary may in his discretion, grant leases for building purposes for terms not exceeding ten years, of small parcels or ground; at such places in said park as shall require the erection of buildings for the accommodation of visitors; all of the proceeds of said leases, and all other revenues that may be derived from any source connected with said park, to be expended under his direction in the management of the same, and the construction of roads and bridle-paths therein. He shall provide against the wanton destruction of the fish and game found within said park, and against their capture or destruction for the purposes of merchandise or profit. He shall also cause all persons trespassing upon the same after the passage of this act to be removed therefrom, and generally shall be authorized to take all such measures as shall be necessary or proper to fully carry out the objects and purposes of this act. (Forty-Second Congress. Session II Ch. 21–24. 1872. March 1, 1872. CHAP. XXIV)

Although it had created the park, Congress did not approve funding for Yellowstone until 1877 and even that was insufficient to manage and protect the reserve. In 1884, additional problems came about from the proposal to construct an access railroad across the northeast corner of the park. Proponents argued that the railroad was the only way to remove gold-bearing ores from Cooke City, just east of the park, to the Northern Pacific Railway at

Gardiner Gateway, Yellowstone's northern entrance. Although Congress turned down the plan, wrote Sellers, "the project was denied more because of what the mines lacked rather than what the tracks would have threatened" (Sellers, 1979). The Cooke City mines actually never lived up to expectations. Historian Sellers reported that, "In truth, Dr. Ferdinand V. Hayden had been vindicated; his assessment in 1871 that few of Yellowstone's volcanic formations contained precious metals was correct" (Sellers, 1979). Still, the willingness of many parties to consider the railway demonstrated that Yellowstone's status still hinged on its worthlessness.

At Yellowstone and soon at Yosemite, Americans learned that a new ethic could be applied to land. In the case of Yellowstone, the actions of those seated around the fire have allowed Americans nearly 150 years later to see at least a version of what confronted them. The Hayden explorers attributed the name Yellowstone to the Native Americans living in the area at the time of the Hayden Expedition. Because of the high yellow cliffs surrounding the waterway that passed through this region, they referred to it as the Yellowstone River. Eventually, this geological oddity also was used to name the nation's first national park. Today, the park spans 2.2 million acres. Although most of the acreage is contained in northwest Wyoming, the park also reaches into Montana and Idaho. Initially, the odd geological landforms within the park, including geysers and fissures in the earth, attracted the imaginations of many Americans. After initiating Americans' conception of national parks, Yellowstone's role as a symbol for all parks has never diminished. Today, Yellowstone continues to serve as an active battleground as Americans strive to define the meaning of preservation and wilderness.

The other great symbol of the American movement for national parks can be found in northern California; however, its origin story differs significantly from that of Yellowstone. Originally referred to as America's Switzerland, the Yosemite Valley possessed great mountains, waterfalls, and forests that, similar to Yellowstone, struck preservationists as unique oddities. The great girth and height of the Giant Sequoias, for instance, made the Mariposa Grove area one of the nation's most fabled landscapes by the 1870s. President Abraham Lincoln signed the bill to set aside the Yosemite Valley and the Mariposa Grove during the Civil War in 1864. This bill, however, did not designate the area a national park; instead, it granted the land to the state of California as a public trust.

One of the earliest observers of Yosemite was America's leading proponent of parks, Frederick Law Olmsted. Before the establishment of the National Park Service (NPS) in 1916, planners such as Olmsted were the leading authorities on the American idea of parks and park development. In an 1865 report, he cautioned that the great features of the park might be exploited by development if government restrictions were not placed on the park:

> It was during one of the darkest hours, before Sherman had begun the march upon Atlanta or Grant his terrible movement through the Wilderness, when the paintings of Bierstadt and the photographs of Watkins, both productions of the War time, had given to the people on the Atlantic some idea of the sublimity of the Yosemite, and of the stateliness of the neighboring Sequoia grove, that consideration was first given to the danger that such scenes might become private property and through the false taste, the caprice or requirements of some industrial speculation of their holders; their value to posterity be injured. To secure them against this danger Congress passed an

act providing that the premises should be segregated from the general domain of the public lands, and devoted forever to popular resort and recreation, under the administration of a Board of Commissioners, to serve without pecuniary compensation, to be appointed by the Executive of the State of California.

By no statement of the elements of the scenery can any idea of that scenery he given, any more than a true impression can be conveyed of a human face by a measured account of its features. It is conceivable that any one or all of the cliffs of the Yosemite might be changed in form and color, without lessening the enjoyment which is now obtained from the scenery. Nor is this enjoyment any more essentially derived from its meadows, its trees, streams, least of all can it be attributed to the cascades. These, indeed, are scarcely to be named among the elements of the scenery. They are mere incidents, of far less consequence any day of the summer than the imperceptible humidity of the atmosphere and the soil. The chasm remains when they are dry, and the scenery may be, and often is, more effective, by reason of some temporary condition of the air, of clouds, of moonlight, or of sunlight through mist or smoke, in the season when the cascades attract the least attention, than when their volume of water is largest and their roar like constant thunder.

There are falls of water elsewhere finer, there are more stupendous rocks, more beetling cliffs, there are deeper and more awful chasms, there may be as beautiful streams, as lovely meadows, there are larger trees. It is in no scene or scenes the charm consists, but in the miles of scenery where cliffs of awful height and rocks of vast magnitude and of varied and exquisite coloring, are banked and fringed and draped and shadowed by the tender foliage of noble and lovely trees and hushes, reflected from the most placid pools, and associated with the most tranquil meadows, the most playful streams, and every variety of soft and peaceful pastoral beauty.

This union of the deepest sublimity with the deepest beauty of nature, not in one feature or another, not in one part or one scene or another, not any landscape that can be framed by itself, but all around and wherever the visitor goes, constitutes the Yosemite the greatest glory of nature. (Olmsted 1865)

Although Olmsted, similar to romantics, believed that Americans needed to interact with nature to keep from becoming overcivilized, he also sought to exploit and develop this interest of the wealthy into a bone fide American tradition of landscape design.

Dominating the design of urban parks for a generation, Olmsted was uniquely suited to establish how a federally sponsored system of larger park areas would need to be administered and constructed. In addition to planning the roads and grounds of Yosemite, Olmsted wrote widely on humans' need to maintain a connection with nature in an era of increasing industry and urbanity. Later in his report on the Yosemite Valley, Olmsted wrote, "It is a scientific fact that the occasional contemplation of natural scenes of an impressive character, particularly if this contemplation occurs in connection with the relief from ordinary cares, change of air and change of habits, is favorable to the health and vigor of men and especially to the health and vigor of their intellect beyond any other conditions which can be offered them, that it not only gives pleasure for the time being but increases the subsequent capacity for happiness and the means of securing happiness" (Olmsted 1865).

Olmsted's forecast proved correct by the 1870s when a tourist landscape of roads, hotels, and cabins, teamed with incursions by cattle and hogs, threatened to overtake Yosemite's majesty. The park was moved to federal authority in 1890, which eventually improved the misuses. Together, Yellowstone and Yosemite defined an American original in the human relationship with nature: the national park.

In the wilderness setting and with a backdrop of the vast, dramatic landscape of the western frontier, the origin of the national park idea seemed fitting and noble. However, the reality of America's national parks is not quite this simple. The effort to establish the first parks forced Americans to consider the idea of preservation, but no one defined what exactly that meant. Future debates were needed to establish the logic of the American national parks.

Sources and Further Reading: Olmsted, *Yosemite and the Mariposa Grove*; Runte, *National Parks: The American Experience*; Sellars, *Preserving Nature in the National Parks: A History*.

TRANSFORMING THE PLAINS WITH NEW AGRICULTURAL TECHNOLOGY

Time Period: 1800–1930
In This Corner: Agricultural interests in the West, federal government
In the Other Corner: Natural limitations of the region
Other Interested Parties: Scientists; environmentalists
General Environmental Issue(s): Agriculture, the West, biotechnology

Although few can debate the benefits of efforts by pioneers to transform the Plains to farmland, historians and ecologists have learned a great deal about the impact of those changes on the region and its occupants. Some historians use these impacts to include this transformation as a major portion of a "Legacy of Conquest." Others refuse to separate pioneering efforts from the heroic imagery with which they were depicted by stories of the "Old West."

Historic Farming Technology

On the typical pioneer farm, there was no end to the work that needed to be done. In addition to farming, the household needed to shoe and feed horses and oxen and repair the plows and other equipment. As the number of farmers in the midwest grew, businessmen found that there was significant opportunity available to service the equipment of the rapidly increasing number of agriculturalists. One of these entrepreneurs was John Deere.

While servicing equipment, Deere learned of the most serious problem that pioneer farmers encountered in trying to farm the midwestern soil: the cast-iron plows brought from the east were designed for the light, sandy New England soil; the rich midwestern soil was much thicker. The dirt clung to the plow bottoms, clogging their forward motion. Plowmen were forced to slow their progress by stopping repeatedly to scrape the sticky soil from the plow. The plowing that was necessary to pull up the deeply rooted grasses became one of the settlers' most dreaded and laborious tasks. The difficult plowing made many homesteaders consider moving on or returning to the eastern United States before they had improved their land.

Deere became convinced that a plow with a highly polished and properly shaped moldboard and share would scour itself as it turned the furrow slice. He fashioned such a plow in 1837, using the steel from a broken saw blade. He successfully tested the plow and then began selling it to farmers of the west and midwest. Deere's steel plow proved to be the answer for the unique soil found west of the Mississippi. However, Deere's efforts to sell his product are what truly changed agriculture in the western United States. Additionally, in a market in which most plows were made to order, Deere produced a surplus. Once he had produced a full railcar load, Deere took the plows to rural areas and offered his John Deere's "self-polishers" for sale. For the first time, marketing treated rural communities like any other.

Of course, once the vast fields of grain had been plowed and grown, they needed to be picked or cut. The McCormick Reaper, which was perfected in Virginia in the 1830s, began large-scale use in the West after Cyrus McCormick moved his factory to Chicago in 1848 (Opie 1998, 237–39).

Describing his innovation, one historian wrote, "McCormick used seven fundamental principles in the reaper he invented. A divider separated the wheat so it could be cut. The reel pushed the wheat toward the knife and then pushed the cut wheat onto the platform. A straight reciprocating knife cut the wheat. Fingers held the wheat while it was cut. The platform held the cut wheat. The main wheel and gears drove all the moving parts of the reaper. The front-side draft traction provided the reaper a firm grip on the ground" (Wong 1992).

By the time that his Chicago factory opened, McCormick was estimated to have manufactured about 1,300 reapers. Using his new factory and marketing efforts, between 1848 and 1850 he manufactured and sold 4,000 reapers. The influence of the plow and the reaper could be seen throughout the western United States.

Ranching and "Bonanza Farms"

Liberated from the limits of self-sufficiency and connected to larger markets by the railroad, western farmers began to use the vast open spaces to create massive farming establishments. Usually referred to as bonanza farms, these farms were most popular in states such as Minnesota in which the Northern Pacific Railroad sold off huge acreage.

George W. Cass, president of the Northern Pacific, and George Cheney, a board member, needed to prove to potential buyers that the land was fertile and productive. The bigger the farm, the more convincing their case. So, the first "bonanza farms" were composed of massive amounts of acreage and typically grew only wheat. The Northern Pacific investors hired wheat expert Oliver Dalrymple to develop and manage their demonstration farm. The Cass-Cheney farm (which was also referred to as the Dalrymple farm) was located about 20 miles west of Fargo, North Dakota. It grew from 5,000 acres in 1877 to 32,000 acres by 1885 and yielded as much as 600,000 bushels of wheat per year. The bonanza farm needed 600 men to plant seed and 800 to harvest. The scale of the mechanical enterprise was similarly massive, including 200 plows, 200 self-binding reapers, 30 steam threshers, and 400 teams of horses or mules. One manager typically oversaw a single 2,500-acre tract (Opie 1998, 234–39).

Other huge farms, ranging from 3,000 to more than 30,000 acres, soon appeared in the Red River Valley. The soil, topography, and climate in the upper midwest and west were ideal for large-scale farming. The necessary workforce was drawn from new immigrants from Northern Europe as well as from off-season loggers from Minnesota's timber industry. Typically, workers were hired on a limited term contract. Although most owners of these vast farms did not live in the area, the absentee landowners hired local managers to run the farms.

The Industrial Farmscape

Through the creation of bonanza farms, Minnesota and North Dakota, and specifically the Red River Valley, became one of the country's largest wheat-producing areas. Between 1875 and 1890, the bonanza farms that were modeled after Dalrymple's became highly profitable, but the scale of such enterprise could not be sustained. Eventually, this intense use of the soil resulted in exhaustion of the necessary minerals. Many of the farms were no longer profitable by the 1890s. However, regional industry helped keep this style of agriculture going into the 1900s.

These great farms helped to support a new midwestern grain industry that possessed awe-inspiring scale. Minneapolis was one of the many cities to prosper in milling. By the 1880s, twenty-seven mills produced more than two million barrels of flour annually, making Minneapolis the nation's largest flour manufacturing center. This primarily occurred through corporate consolidation. In 1876, seventeen firms operated twenty mills; in 1890, four large corporations produced almost all of the flour made in Minnesota.

Leading these corporate entities, the Pillsbury Company completed its gigantic A Mill in 1880. Containing two identical units, it had a capacity of 4,000 barrels of flour per day when it opened. By 1905, the mill had tripled its output. Its owners claimed that it was the largest flour mill in the world (Opie 1998, 234–39).

By the early twentieth century, the day of the bonanza farm had passed. Run as large-scale businesses, the farms operated on a very narrow profit margin. The price of land began to rise after 1900, as did corporate taxes. Costs of labor and equipment rose as wheat prices fell. The huge farms were gradually broken up, but the geography of the valley dictated that the new farms still would encompass thousands of acres.

Creating a Future for Grain: The Elevator

In the 1850s, many midwestern cities began using a device found on many farms that grew grain: the silo. The urban model, however, was much larger and became known as an elevator. Intimately linked into the system of growing grain and transporting it by railroad, the elevator became a revolutionary device when used to store grain for market-driven reasons. Historian William Cronon wrote, "Chicagoans began to discover that a grain elevator had much in common with a bank—albeit a bank that paid no interest to its depositors" (Cronon 1991, 120). Primarily, the elevator stored the grain and helped growers maintain a supply throughout the year. It functioned as kind of a holding tank.

As if carrying a precious mineral, farmers brought their wheat or corn to the elevator operator. The operator gave the grower a receipt that could be redeemed for grain when the original grower wanted to make a withdrawal. Early on, grain was measured, traded, and sold in sacks. To simplify its trading, grain operators adopted a "liquid" form of measuring it that liberated the grain from the sack. This was the change that allowed the elevator to take over the landscape of each midwestern city.

In cities such as Chicago, the supply of grain seemed endless. Supplying much of the nation's needs, the Chicago elevator operators were grain brokers. Much as a bank, the brokers bought and sold grain. The elevator, however, allowed this financial market to become a "futures" market. Cronon wrote, "Grain elevators and grading systems had helped

Wheat wagon at railroad depot and grain elevator in Offerle, Kansas, c. 1916. Technologies of many levels combined to help make the western United States agriculturally successful. Library of Congress.

transmute wheat and corn into monetary abstractions, but the futures contract extended the abstraction by liberating the grain trade itself from the very process which once defined it: the exchange of physical grain" (Cronon 1991, 126).

Bounding the Grid: Barbed Wire

Commodifying the open space of the West allowed animals to become eating machines. However, a necessary technology in this configuration was a fencing material that could stretch over large expanses and use little wood. Before 1863, several individuals created forms of fencing that could be considered early versions of barbed wire. However, none of these products reached the mass market. In 1863, Michael Kelly developed a type of fence with points affixed to twisted strands of wire, but, again, he was not able to create a large-scale market for his product. In fact, it took ten more years for another inventor to file a patent to begin the barbed wire industry, and it was not Kelly.

The initial spark for the innovation came at the county fair in DeKalb, Illinois, in 1873. The invention, however, was not barbed wire. Henry M. Rose exhibited a new fence material that could be added to an existing wooden fence. Rose's additional wooden rails contained a series of sharp spikes that would prick animals to keep them from breaking through the fence. Rose's model fence impressed Joseph Glidden, Jacob Haish, and Isaac Ellwood, who visited the fair. Each man had a very similar idea to improve Rose's fence: get completely rid of the wood and attach the barbs directly to pieces of wire. Each took his idea and left the fair without knowing that they each shared this inspiration.

Glidden reportedly used his invention to enclose his wife's vegetable garden. To create his product, he bent shorter wires around long strands of straight wire. To wind the wire together, Glidden modified a coffee mill. When the crank was turned, two pins that he had installed on each end twisted the wire to form a loop. The wire was then clipped off approximately one inch on each end at an angle to form a sharp point. Glidden then placed barbs on one of two parallel strands of wire. Then the two strands of wire were attached to a hook on the side of an old grinding wheel. As the barbs were positioned, the grinding wheel turned to twist the two strands and to lock the barbs in place.

During this time, one of the other inspired fair visitors, Isaac Ellwood, had been unsuccessful in his own efforts. Glidden was awarded a patent on November 24, 1874 for his fencing material, which he called "The Winner." Because Ellwood was a hardware merchant, he and Glidden joined forces and formed the Barb Fence Company. The other inspired fair visitor, Jacob Haish, had also patented a wire invention, but he had not yet marketed it. He filed additional patents and tried to thwart the efforts of the other men to market their own fencing product. Legal battles continued for the next few years. Ultimately, however, Glidden was given credit for being the inventor of barbed wire.

This fencing material provided the flexibility that was needed for western ranchers. Delivered on a large spool, the wire could be hung on a variety of posts while being unwound from a wagon. In this fashion, ranchers enclosed vast acreage, allowing the open spaces of the West to be made profitable through grazing.

1930s Dust Bowl as Outcome of Aggressive Agriculture

The disappearance of the buffalo was one outcome of massive changes in the area known as the Great Plains. Between 1878 and 1887, nature welcomed a massive influx of new settlers to the West with some of the wettest weather on record. Just in the western third of Kansas, the population rose from 38,000 in 1885 to 139,000 in 1887. Many settlers assumed that, indeed, "rain did follow the plow." Throughout the Plains, settlers used new innovations in agriculture, including plows, combines, and tractors, to farm more land than ever before. Wheat production, for instance, became prevalent throughout the Dakotas and into California. Wheat required more water than the average rainfall in many of these regions; however, the wet years misled settlers into thinking that this was not the case.

However, in 1887, rainfall diminished and settlers endured one of the most difficult periods of drought in American history. Many settlers simply left the land. In some areas, as much as three-quarters of the population left the land. Farmers who remained attempted dry-land farming, which emphasized the use of drought-resistant grain crops, deep plowing, and quick cultivation after a period of rain.

These techniques helped Great Plains farmers become successful again by the end of the 1909. Success, and the crisis of World War I in the 1910s, fueled further expansion of wheat farming. This expansion helped to usher in the dust bowl of the 1930s (Worster 1979, 5–8).

Severe dust storms took place during the 1930s to create an agricultural, ecological, and social crisis in the Great Plains. This event is now referred to as the Dust Bowl. The causes of the crisis begin with the agricultural practices of American settlers who were unfamiliar with the ecological limitations of the region.

This photo of Oklahoma taken in April 1936 shows an abandoned farm surrounded by drifts of soil blown by "Dust Bowl" winds. The winds combined with agricultural practices to create an ecological disaster. Library of Congress.

In addition, there were four periods of extreme drought. The first of these periods was 1930–1931, which marked the beginning of the Dust Bowl. Then there was another period of drought in 1934. This period also included extreme heat and numerous dust storms. The third period was 1936, which had even more extreme heat. Finally, the last period was 1939–1940, which marked the end of the Dust Bowl.

Most important, however, severe wind erosion occurred in the dry conditions. Wind erosion is essentially the wind blowing loose topsoil off of the ground. This was intensified during the 1930s for a number of reasons. The choice of settlers to grow wheat led to fine particles of soil blowing free in the wind. Little effort was made by farmers to rotate crops, so wheat was continually planted in the same fields, resulting in soil depletion. Also, with the land in cultivation, there were seasons in which the soil was covered by no vegetation. This left the uncovered fields vulnerable to wind erosion. With the increase of cropland, the threat for wind erosion in this respect also increased.

The climatic event that stripped much of the midwestern United States of topsoil is called the Dust Bowl. This ecological disaster awakened Americans to the need for a change in the ethic with which it used and managed its supply of natural resources. The dust storms were partly weather phenomena; however, as historian Donald Worster wrote, they were also a product of a culture working perfectly. He wrote, "It cannot be blamed on illiteracy or overpopulation or social disorder. It came about because the culture was operating in precisely the way it was supposed to" (Worster 1979, 4). This expansionist ethic had largely governed settlement and agriculture on the Great Plains in the early 1900s. The expansionist

Auto camp north of Calipatria, California. Approximately eighty families from the Dust Bowl camped here during March 1937. They paid fifty cents per week. Library of Congress.

zeal behind the 1893 Colombian Exposition led a nation to tear away the grasses that naturally preserved the rich soils of the Midwest to clear the way for agriculture.

By World War I, the nation's agriculture had settled into basic patterns based primarily on region and climate but also containing important elements of cultural traditions. In areas such as the upper Midwest, this may mean dairy farming because ancestors had begun the tradition. In the Great Plains from Texas to the Canadian border, however, agricultural traditions grew at least partly from national cultural patterns. Here, in the country's breadbasket, settlers had been assured that vast fields of wheat were the best way of making a profit from the vast expanse of flat lands. Their crop relied on railroads to carry grain to distribution centers such as Chicago, Illinois. Most often, the crop required more moisture than what fell from the sky (Worster 1979, 34).

Although American agriculture was extremely productive relative to that of other nations, most farmers were not prosperous. In 1900, the income per worker in agriculture was only $260 annually compared with $622 for nonagricultural workers. Tenancy increased from 25 to 35 percent between 1880 and 1900. World War I brought a great boon to wheat farmers as the federal government urged that "Wheat will win the War!" The demand for farm commodities increased and land values rose. In addition, the federal government offered farmers subsidies to help pay for tractors and other new agricultural technology that could increase production. Nineteen hundred to 1920 went down as a golden age for Great Plains agriculture (Opie 1998, 357–59).

What followed, however, was worse than anyone could have imagined. Postwar price drops hit Great Plains farmers particularly hard. Although federal assistance was considered, it did not arrive until the 1933 Agricultural Adjustment Act that inaugurated a wide range of federal assistance programs, including the stabilization of the prices of basic agricultural commodities, encouragement of soil conservation, and subsidization of exports. It became clear in the early 1930s that the golden 1920s had taken a toll on Plains farming.

The introduction of the gasoline tractor that was discussed above brought a close to the era of farming with horses and other animals. In addition, the development of bigger and better machines, such as the grain combine harvester and the mechanical cotton picker, continued to reduce the amount of labor needed in agriculture. In each case, technology overcame farmers' unintended limitations: they could now grow more and spread their crop over larger areas. These changes set the stage for a fairly typical climatic event of the region: drought. During the 1930s, the Great Plains experienced a lack of rainfall that created perfect conditions for an agricultural calamity. Federal efforts to help began during the 1930s and continued throughout the twentieth century. Government payments to farmers in 1934 totaled $134 million; by 1961, they had increased to $1.5 billion, and by 1987, to $22 billion (Schlebecker 1975, 295–300).

During the 1930s, author John Steinbeck wrote about the dreams and the difficulties of the thousands of American farmers who, no longer able to earn livings in their former homes, began a mass exodus to California where they hoped to find work as migrant field hands. In March 1938, he toured migrant worker camps in California with a photographer. His assignment was to write a photo essay for *Life* magazine, but he decided the story was too important and he wrote a novel instead, *The Grapes of Wrath*, which became an American classic.

Conclusion: Committing to Aquifers

Although a few critics have called for the end of agriculture in the Plains, the region remains with more than 90 percent of the land in farms and ranches and 75 percent cultivated. In this agricultural paradigm, there exist five major production systems: range livestock, crop fallow, river valley irrigation (snowmelt-dependent), confined livestock feeding, and groundwater irrigation (aquifer-dependent).

Today, farmers realize that the precipitation of the western United States is different from other portions of the country, approximately 50 centimeters (twenty inches) per year. To continue its rate of agriculture after the 1930s, farmers turned to irrigation.

Irrigation waters for many of the Great Plains states (North Dakota south to Texas) are taken from the 180,000-square-kilometer High Plains aquifer, which extends under South Dakota to west Texas. Today, approximately 170,000 wells draw water from the aquifer, making the region the largest area of irrigation-sustained cropland in the world.

Groundwater from the aquifer was first widely used in the 1930s in Texas. Experts believe that, over the twentieth century, approximately 11 percent of the total groundwater supply has been extracted. The most significant declines in the aquifer have occurred in southwest Kansas and the Texas Panhandle. Western Kansas has consumed 38 percent of the groundwater in the underlying aquifer.

These depleted regions are considered to be examples of groundwater overdraft, where groundwater extraction occurs more rapidly than recharge. If managed correctly, aquifers will replenish themselves naturally. The sand and gravel aquifer is unconfined (open) and is recharged by water infiltrating from above. Limited recharge occurs from precipitation and stream outflow.

Climate scientists believe that issues of global warming are likely to effect regions such as the Great Plains most acutely. Climate change may also result in greater crop damages attributable to increased drought stress resulting from higher growing season temperatures. The loss of soil from these croplands may be enhanced by the lack of plant cover. Ranchers in the region may not be able to support the current number of animals on the existing range lands because of reduced dry-land pasture production and lack of water resources for their animals.

One U.S. geology study group anticipates the following trends for the Great Plains: increased size and reduced number of farms; increased importance of technology to achieve those increases; increased livestock and mixed cropland-livestock operations; continued water competition between agricultural and urban areas; and population shifts out of rural areas.

Sources and Further Reading: Opie, *Nature's Nation*; U.S. Global Change Research Program, *U.S. National Assessment of the Potential Consequences of Climate Variability and Change*, http://www.usgcrp.gov/usgcrp/nacc/background/meetings/forum/greatplains_summary.html; White, *It's Your Misfortune and None of My Own*; Worster, *Dust Bowl: The Southern Plains in the 1930s*.

POWELL AND EFFORTS TO EXPLORE THE CONTOURS OF THE WEST

Time Period: 1878
In This Corner: Powell and scientific understanding of the West, Mormons, native peoples
In the Other Corner: Settlement interests, federal government, railroads
Other Interested Parties: Settlers
General Environmental Issue(s): Aridity, the West

Settling western land into agricultural units required that the area first be known or better understood. Although this process may seem unlikely to inspire debate, there was significant disagreement over at least two primary issues: proper treatment of native inhabitants and whether or not the federal government should finance the surveying process. The findings of these scientific excursions only created more discussion, particularly about issues such as the West's supply of water.

To most efficiently disperse western lands, the government undertook a number of "Great Surveys" to chart the paths by which new transportation routes might be established to knit together a growing empire. These excursions detailed the natural environment of the West but also laid the foundation for the exaggerations that would become known as the West's mythic importance to all Americans.

Aridity in the West

Possibly no explorer better grasped the complications of the natural environment of the West than John Wesley Powell, who led one of these surveys into the Southwest. Powell's legendary exploration of the Colorado River provided him with unprecedented and celebrated insights into the geology of the West before almost any other American had set his or her eyes on the place. One of Powell's first observations, however, inspired a generation of disagreement and debate. After studying the water situation in the West, Powell argued that the lands of the West were significantly unlike those of the East and, therefore, should be treated differently (Worster 1979, 337).

The primary distinction between regions began with moisture. In the areas to the east of the 100th meridian (about mid-Kansas), in excess of fifteen inches of rainfall per year helped to create soil and access to reliable sources of surface water that permitted conventional agriculture over surprisingly large areas. In this area, the logic of the grid and the organization exerted by the Homestead Act made sense for supporting agricultural families. However, without this minimum amount of rain, western lands beyond the 100th meridian would have much more difficulty supporting farmers.

In the far West, rainfall was scarce and unreliable. In areas where annual rainfall dipped below fifteen to twenty inches, rains also might come in very different amounts from year to year. A wetter year might allow farmers to prosper, but drought could be around the corner. For farmers carrying mortgages for equipment, seed, and animals, one bad year could put an end to their enterprises. Led by railroad companies, land developers advertised to settlers that "rain followed the plow." The claim was that more settlers and agriculture meant more rain would eventually come. A geologist by training, Powell learned in his travels of the West that "average" rainfall declines as one moves West and, more importantly, variability increases to where rain-fed agriculture is an extremely risky, and ultimately futile, proposition in the far West.

In actuality, the high plains that stretched west to the Rockies might expect dry years in three or four of every ten years. Farther west, growing crops without irrigation was largely impossible. Grazing livestock was a much better option for these vast open areas of little rainfall. However, it might take more than forty acres of sparse vegetation to support one cow and a calf in these places. In a very real sense, then, it became a matter of increasing cultivated acreage as rainfall decreased, trading land for rain.

With these ecological parameters in mind, it made no sense to limit homesteads to 160 acres. This also meant that the West might accommodate fewer people and, thus, attract less of the capital needed for development. This, of course, was not exactly what boosters of the West wanted to hear.

John Wesley Powell Urges Control of Western Settlement

After traveling through much of the western United States, Powell argued that mining would likely be a better option to initiate development in the region. If mineral wealth was not available, Powell argued that climate and regional topography should determine what activity took place there. If rainfall could not support agriculture, argued Powell, irrigation could be used to supplement the lack of moisture. Powell studied the native communities of the southwest and also the Mormons of Utah and urged that such a "hydraulic" society should be planned in advance.

Powell still believed that, in higher elevations, where rainfall was greater but soils poorer, lands could be made profitable if they were earmarked primarily for timber harvest and grazing. However, in much of the southwest, Powell pointed out, settlement should not be carried out without irrigation.

Powell's ideas, of course, contrasted significantly with the description of railroads, land development companies, and even many politicians. He gathered and expressed these opinions just as proponents of western development attempted to create a land rush by painting the rosiest of depictions of western life. Any effort to distinguish the far western lands from those of the midwest, politicians feared, would make settlers less likely to go there.

When Powell filed his report with Congress in 1878, it argued that the homestead model of settlement, which worked well in the wetter forests and prairies of the central part of the country, should be reexamined and restructured for the West. For the far West, Powell recommended two revolutionary proposals, each based in the nature of the place. First, believing that water was the key to development, he argued for land management units organized around watersheds. This would require scrapping the "township and range" survey system that imposed a rigid systematic grid pattern on the land. Second, the government must perform surveys to establish the potential value of the land and make survey results known to the public. For the reservoirs and canals that would inevitably be necessary then, Powell argued that, rather than relying on individual initiative, communities should undertake development of western "watershed commonwealths" (Worster 1979, 377–80).

Initial Reactions to Powell

Powell's ideas, of course, directly contradicted the spirit of the Jeffersonian ideal of democracy based on individual independent farmers that was at the root of the American movement westward. By prioritizing communities, he also directly challenged much of the settlement efforts of some of the most influential capitalists of the era. If his model were followed, farmers would be allowed to purchase eighty acres of land that could be irrigated, and then they would be given collective stewardship over remaining range and forest lands. A new, decentralized model of the West—and of planning in general—would have emerged.

Congress largely ignored Powell's report. Some of his ideas reverberated around Washington over the next sixteen years, but few lawmakers bought into his recommendations. Primarily, by proposing to develop and distribute information on the economic potential of western lands, Powell undercut speculators who relied on settlers' ignorance, many of whom helped to perpetuate such ignorance. His plans cut capitalists out of the process of land dispersal and also took potential lumber supplies from timber interests. By insisting that lands be withheld from entry until they were surveyed and described, he stymied the developers who sought any sort of advantage in gaining access to the most favored lands. Furthermore, his ideas complicated the use of the Homestead Act because he demanded a survey of the land before their dispersal. In short, Powell directly went against the culture of westward expansion that had evolved during the early and mid-1800s. By the 1870s, many influential thinkers had bought into the mythical West, and they wanted to know little of Powell's "real" West.

Conclusion: Lessons for the Arid West

Although proponents of western settlement urged for a rapid influx of population without limits, Powell's report presented the obvious reality of the natural limitations of the region. It was a wise way of perceiving the ecology of the West that did not jibe with the culture of the era.

Sources and Further Reading: Opie, *Nature's Nation*; White, *It's Your Misfortune and None of My Own*; Worster, *A River Running West*.

WHO PAYS FOR RECLAMATION AS A FEDERAL POLICY FOR THE WEST?

Time Period: Early 1900s
In This Corner: Western interests, railroad companies
In the Other Corner: Residents of other regions, state governments
Other Interested Parties: Federal government, engineers and planners, conservationists
General Environmental Issue(s): Aridity, river management, the West

Eventually, lawmakers realized the prescience of Powell's perspective on the aridity of the western United States. As the technology emerged to hydraulically manage rivers and water supplies, it became clear that the effort to manipulate the water supply of these states would require vast sums of capital. Although specific states would benefit most, one could clearly argue that, much like the railroad, the entire nation would gain valuable resources from such development in western states. Therefore, as a major part of the Progressive Era of the early 1900s, lawmakers debated the extent to which federal funds should be used for dam building and water management in the West.

By the second decade of the 1900s, federal money would be channeled into revolutionary projects to aid in making the western lands more easily inhabitable by settlers. One can argue, however, that this federal support was made even more essential because settlers had moved to the West and now were stuck. When Powell made his observations, of course, many of the arid lands that he critiqued had not yet been inhabited by American settlers.

California marks one of the earliest examples of the application of such irrigation planning and development. Before the turn of the nineteenth century, officials realized that the Los Angeles River could only accommodate 300,000 people. The mayor of Los Angeles, Fred Eaton, originally proposed the Owens Valley as a water supply in 1893, but William Mulholland, the head of the Water Department, thought it was unrealistic. Eaton decided to purchase land options in the valley on his own, and, when it was finally chosen as a new supply, Eaton decided to make money for himself from the city. Mulholland and the city attorneys put Eaton off for three years and finally made a deal to purchase the land for one-quarter of Eaton's original offer (Reisner 1993, 64–65).

Marc Reisner wrote, "The Owens River created Los Angeles, letting a great city grow where common sense dictated that one should never be" (Reisner 1993, 106). Following the model of politically controlling water supplies to dictate which areas could be developed, much of southern California followed during the next fifty years. The formerly arid landscape became some of the greatest agricultural land in the world by managing its water supply through a system of aqueducts and management pools. Eventually, this same technology enabled the growth of metropolis cities as well.

The City of Los Angeles received its biggest boost in 1913 with the construction of the Los Angeles Aqueduct. By delivering the Owens River water to Los Angeles, Mulholland had made it possible for the city to prosper. However, this growth demanded additional water supplies. Combined with drought, this unprecedented growth in homes and businesses that spread across the Los Angeles basin created water shortages for the Owens Valley in 1923.

In need of a dam or a storage reservoir to control the flow of the Owens River above the aqueduct intake at Independence, developers searched for a proper location for such a

Los Angeles Aqueduct piping water through the desert to supply the burgeoning metropolis of L.A. Bringing water to the desert of Southern California in the late 1800s allowed the nation's largest city to take shape during the following century. Library of Congress.

facility. The best site, Long Valley, was still owned by Fred Eaton, and Mulholland believed he was asking too much for the land. As an alternative to increase the water supply, the city began pumping groundwater into the irrigation canals that ran throughout the area.

This action brought outcry from farmers who feared the drop in the water tables that supplied their irrigation supplies for agriculture. As the city purchased areas north of Independence to acquire groundwater rights, a series of confrontations took place. Financial leaders of the county banded together and formed a unified opposition movement in the form of their own irrigation district. Farmers also illegally diverted water, leaving the canal empty.

On May 21, 1924, forty men dynamited the Lone Pine aqueduct spillway gate. The city continued to indiscriminately purchase land and water resources. The two sides had reached a violent stalemate. Increasingly, many valley residents thought that Los Angeles should buy out the entire area.

Although this may have been best for farmers in the area, the city believed the wholesale purchase of the district was unnecessary to meet its water needs. Instead, in October 1924, the city proposed a plan to leave 30,000 acres in the Bishop area free of city purchases and to allow its expansion into the other regions. The directors of the Owens Valley Irrigation District rejected the proposal. They demanded the outright farm purchase and full compensation for all the townspeople. Supporters of the directors staged an occupation of the Alabama Gates on November 16 and closed the aqueduct by opening the emergency spillway. More attacks on the aqueduct occurred in April 1926.

Los Angeles demonstrated that, in the West, political power and water went hand in hand. In arid regions, urban growth demanded control and development of adequate water supplies.

Newland's Act Brings Federal Funds to Reclamation

For the entire western region, the true admission of Powell's wise reading of the nature of the West, however, came in the Newland's Reclamation Act. The Corps of Engineers typically used river conservation plans to assist or maintain existing human communities. When progressive politicians began to see the capabilities of dams, it did not take long for their gaze to turn to the arid regions of the American West. With the Newland's or Reclamation Act of 1902, river engineering was connected to a federal effort to overcome the rainfall deficiencies of the American West and allow it to be "reclaimed" for human development. Although these intrusive projects significantly manipulated western environments, they are typically included under the term "conservation" because they necessitated the management and maintenance of natural resources.

The Newland's Reclamation Act of 1902 was predominantly relevant to irrigation developments in the American West, but it also set the precedent for the Bureau of Reclamation and U.S. Corps of Engineers to use federal funds for hydraulic development of the West, including dams to generate electricity (Reisner 1993, 110–15). In addition, President Roosevelt also appointed an Inland Waterways Commission in 1907, which subsequently issued a report advocating a national policy of planned development of water resources, including both navigable streams and non-navigable streams on public lands. Multiple uses of the nation's waterways were proposed and were to be financed through the sale of electric power. A conference on natural resources held at the White House in 1908 stressed the necessity of exploiting more fully the nation's hydroelectric power capacity of more than thirty million horsepower to conserve nonrenewable fossil fuels.

Originally, the Bureau of Reclamation came to the field of hydropower simply to create revenue while achieving its larger goal of managing the water resources in the arid West. Government planners realized that Reclamation dams could provide inexpensive electricity, which might also stimulate regional growth. Reclamation's first hydroelectric power plant was built to aid construction of the Theodore Roosevelt Dam. Even before fully constructing the dam, the bureau installed small hydroelectric generators to manufacture energy for building the dam and running equipment. Surplus power was sold to the community, which helped citizens to quickly fall in line to support expansion of the dam's hydroelectric capacity.

In 1909, the Theodore Roosevelt Power Plant became one of the first large power facilities constructed by the federal government. Initially, the plant provided the Phoenix area with 4,500 kilowatts before being expanded to more than 36,000 kilowatts. Power, first developed for building Theodore Roosevelt Dam and for pumping irrigation water, also helped pay for construction, enhanced the lives of farmers and city dwellers, and attracted new industry to the Phoenix area (Reisner 1993, 84–89).

By the early 1900s, hydroelectric power accounted for more than 40 percent of the United States' supply of electricity. In the 1940s, hydropower provided about 75 percent of all the electricity consumed in the West and Pacific Northwest and about one-third of the total United States' electrical energy. With the increase in development of other forms of electric power generation, hydropower's percentage has slowly declined and today provides about one-tenth of the United States' electricity.

In 1902, this legislation gave the federal government responsibility for planning and financing the replumbing of the American West:

That all moneys received from the sale and disposal of public lands in Arizona, California, Colorado, Idaho, Kansas, Montana, Nebraska, Nevada, New Mexico, North Dakota, Oklahoma, Oregon, South Dakota, Utah, Washington, and Wyoming, beginning with the fiscal year ending June thirtieth, nineteen hundred and one, including the surplus of fees and commissions in excess of allowances to registers and receivers, and excepting the five per centum of the proceeds of the sales of public lands in the above States set aside by law for educational and other purposes, shall be, and the same are hereby, reserved set aside, and appropriated as a special fund in the Treasury to be known as the "reclamation fund," to be used in the examination and survey for and the construction and maintenance of irrigation works for the storage, diversion, and development of waters for the reclamation of arid and semiarid lands in the said States and Territories.…

In the Colorado Compact of 1922, legislation got even more specific about entrusting the water development of the West to federal agencies. Meeting in New Mexico, representatives of the seven Colorado River Basin states literally divided up the flow of the distant West's most important river. California's rapid growth made it a major negotiator, although it added little to the river in terms of flow. In addition to California, the negotiating parties included Arizona, Colorado, Nevada, New Mexico, Utah, and Wyoming.

The text makes it absolutely clear that the flow of the river was now clearly understood as a significant tool for American development of the entire region:

The major purposes of this compact are to provide for the equitable division and apportionment of the use of the waters of the Colorado River System; to establish the relative importance of different beneficial uses of water, to promote interstate comity; to remove causes of present and future controversies; and to secure the expeditious agricultural and industrial development of the Colorado River Basin, the storage of its waters, and the protection of life and property from floods. To these ends the Colorado River Basin is divided into two Basins, and an apportionment of the use of part of the water of the Colorado River System is made to each of them with the provision that further equitable apportionments may be made.…

(a) There is hereby apportioned from the Colorado River System in perpetuity to the Upper Basin and to the Lower Basin, respectively, the exclusive beneficial consumptive use of 7,500,000 acre-feet of water per annum, which shall include all water necessary for the supply of any rights which may now exist. (b) In addition to the apportionment in paragraph (a), the Lower Basin is hereby given the right to increase its beneficial consumptive use of such waters by one million acre-feet per annum.

Hoover Dam Symbolizes the Future of the Hydraulic West

The new era of hydro-development in the West received its greatest symbol in 1935, when the Hoover Dam was completed. The large dam and the completeness with which it turned the unruly river to human good sent shockwaves around the world. In fact, Hoover Dam served as a symbol of the efficiency of modern technology in general, not just hydro-electric development. The dam, which has long since repaid the $165 million cost for construction, is a National Historic Landmark and has been rated by the American Society of

Aerial view of Hoover Dam, a symbol of the great federal efforts and expense devoted to "reclaiming" the arid western United States. Library of Congress.

Civil Engineers as one of America's Seven Modern Civil Engineering Wonders. The structure contains more than four million cubic yards of concrete, which, if placed in a monument one hundred feet square, would reach 2.5 miles high, higher than the Empire State Building.

As proposed in the 1910s, the mammoth Boulder Dam (as it was first referred to) served as the linchpin of a western land-use policy designed to "reclaim" dry, barren regions by applying human ingenuity. This ingenuity would be applied to the region's few existing waterways, including the Colorado River. Most of the flow, including the electricity made at Hoover Dam, would be managed by the Six Companies contractors to power development more than 300 miles away in Southern California. Today, the majority of Hoover Dam's power is passed over wires to Los Angeles.

The symbolic significance of this immense structure became obvious immediately, which led developers to name it after President Herbert Hoover (an engineer who had been a great supporter of the project). On its completion in 1935, Hoover Dam became a symbol of America's technological prowess, firmly placing the United States with the great civilizations in world history. More importantly, however, conservationists had adopted a policy format that included scientific management based in ecological understanding. This perspective viewed technology, such as dams, as a tool of conservation.

Conclusion: Dams for the West

Although certain urban areas, such as Los Angeles and Las Vegas, Nevada, used private funding and financing to initiate their water development, the idea of reclamation placed the federal government squarely behind the idea of funding western water development. There was little doubt that Americans considered populating and developing the West to be in the nation's best interest.

Sources and Further Reading: Los Angeles Department of Water and Power, *The Story of the Los Angeles Aqueduct*, http://www.ladwp.com/ladwp/cms/ladwp001006.jsp; Opie, *Nature's Nation*; Reisner, *Cadillac Desert*; White, *It's Your Misfortune and None of My Own*; Worster, *Rivers of Empire*.

LITTLE BIGHORN AND NATIVE POLICY IN THE WEST

Time Period: 1876–1930
In This Corner: American government and military
In the Other Corner: Native Americans
Other Interested Parties: Western settlers
General Environmental Issue(s): Native American rights, the West

In rethinking and considering the history of the American West, a primary element of the story is the conquest of native peoples residing in the region. Of course, the Anglo and American conquest of the region was not viewed as inevitable by its native residents. At different junctures in history, different tribes confronted white settlement. In addition, as the U.S. government settled into a generation of changing policy toward native people, these shifts grew from the dynamic conceptions of the American public. Were the Indians the same type of human as Americans? Were they lesser beings? And, therefore, did they merit less consideration in policy negotiations?

Crazy Horse Leads Fight against American Expansion

Many native people resisted European settlement. Their efforts took many forms and created lasting heroes for their people. One of the most famous was Crazy Horse, also known as Tashunca-uitco. He is recalled most for his ferocity in battle. This inspired fighting, however, came from his passion to defend his people and their heritage. Crazy Horse was recognized among his own people as a visionary leader committed to preserving the traditions and values of the Lakota way of life.

As a young man, Crazy Horse inspired legend with his abilities and fearlessness. Before he was thirteen, he stole horses from the Crow Indians. Before he turned twenty, he had led his first war party. When Oglala chief Red Cloud led a fight against American settlers in Wyoming in 1865–1868, Crazy Horse led his war party into the fray. In 1867, it was Crazy Horse who played a significant role in destroying William J. Fetterman's brigade at Fort Phil Kearny.

Crazy Horse refused to conform to American ways of life, which he viewed as a challenge to native tradition. For instance, throughout his life, Crazy Horse refused to allow himself to be photographed, and this determination inspired his reluctance to the mantel of leadership for the Lakota's effort at resistance. Closely allied to the Cheyenne through his first marriage to a Cheyenne woman, he gathered a force of 1,200 Oglala and Cheyenne at his village and turned back General George Crook on June 17, 1876, as Crook tried to advance up Rosebud Creek toward Sitting Bull's encampment on the Little Bighorn. After this victory, Crazy Horse joined forces with Sitting Bull and then carried out his most infamous fight. On June 25, he led his band in the counterattack that destroyed Custer's Seventh Cavalry. The band of Indian warriors flanked the Americans from the north and west as Hunkpapa warriors led by Chief Gall approached from the south and east. The bloodiest day for American sol-diers would define American-Indian relations for the rest of the century.

After the Lakota victory at the Little Bighorn, Sitting Bull and Gall retreated to Canada. Crazy Horse, however, refused to leave the fight. The public and military commitment to the fight against the Indians intensified after Little Bighorn, however. U.S. General Nelson Miles arrived to tirelessly pursue the Lakota and their allies throughout the winter of 1876–77. Combined with the decline of the buffalo population, the military harassment and the decline of the buffalo population eventually forced Crazy Horse to surrender on May 6, 1877.

Although most Native Americans already considered Crazy Horse a legend, his image grew when he was killed while in federal custody on September 6, 1877. In the minds of many Native Americans today, the life and death of Crazy Horse parallels the treatment of natives in the American West.

Federal Efforts to Control the "Indian Problem"

The existing human resources of the American West were also under siege during the late 1800s. In 1850, Commissioner of Indian Affairs Luke Lea proposed a system of reservations that would place all tribes on reserves where they could be controlled and possibly educated and converted into Christians. Reservations were established as permanent, limited areas where native peoples would live in fixed homes and eventually have their own land. These areas were set aside west of the Mississippi. Soon, of course, resistance took a variety of forms.

To ensure the safety of settlers, the U.S. cavalry established forts throughout the West. From these forts, parties were sent out to intimidate native groups who were not peacefully giving way to European settlers. Most famous, General George Armstrong Custer led a party into the Dakota territories under the guise of the discovery of gold in Sioux territory near the Black Hills of the Dakotas (White 1991, 104–6).

Discussed above for its impact on Crazy Horse's life, Little Bighorn became a watershed moment in native revolt against the expansion of settlers into their lands. Expecting a small group of Sioux, Custer and his 600 cavalry were met by approximately 2,000 warriors. All the U.S. soldiers were killed. Although the Sioux and other tribes may win single battles such as that at Little Bighorn, they would not win the war. In fact, victories such as these helped to swing American public opinion more vociferously against native concerns and treatment (White 1991, 106). Native resistance continued, though.

Another form of resistance came from Chief Joseph of the Nez Perce, who was known to his people as "Thunder Traveling to the Loftier Mountain Heights." He led his people in an attempt to resist the takeover of their lands in the Oregon Territory by white settlers. In 1877, the Nez Perce were ordered to move to a reservation in Idaho. Chief Joseph agreed at first, but after members of his tribe killed a group of settlers, he tried to flee to Canada with his followers, traveling more than 1,500 miles through Oregon, Washington, Idaho, and Montana. Along the way, they fought several battles with the pursuing U.S. Army. Chief Joseph spoke these words when they finally surrendered on October 5, 1877:

Tell General Howard I know his heart. What he told me before, I have it in my heart. I am tired of fighting. Our Chiefs are killed; Looking Glass is dead, Ta Hool Hool Shute is dead. The old men are all dead. It is the young men who say yes or no.

He who led on the young men is dead. It is cold, and we have no blankets; the little children are freezing to death. My people, some of them, have run away to the hills, and have no blankets, no food. No one knows where they are—perhaps freezing to death. I want to have time to look for my children, and see how many of them I can find. Maybe I shall find them among the dead. Hear me, my Chiefs! I am tired; my heart is sick and sad. From where the sun now stands I will fight no more forever. (Chief Joseph, Thunder Traveling to the Loftier Mountain Heights, 1877)

Policy Shifts Resulting from Little Bighorn

In reaction to the fighting at Little Bighorn and other forms of resistance, the federal government abandoned reservation policy and adopted a new approach. Federal Indian policy during the period from 1870 to 1900 marked a departure from earlier policies that were dominated by removal, treaties, reservations, and even war. The new policy focused specifically on breaking up reservations by granting land allotments to individual Native Americans. Very sincere individuals reasoned that they were helping native people by assisting them in adopting the culture of Americans. Many truly believed that they were helping the native peoples by making them responsible for their own farm and assimilating them into American life. Today, many critics argue that such paternalism destroyed native culture and life.

Creating policies that would at once break down native culture and free up western lands for American settlement, the U.S. government made these ideas official on February 8, 1887 when Congress passed the Dawes Act. Named for its author, Senator Henry Dawes of Massachusetts, the law is also referred to as the General Allotment Act. Essentially, it gave the U.S. president the right to break up reservation land, which was held in common by the members of a tribe. The act allowed the president to break the common ownership into small allotments that were then parceled out to individuals. Primarily, however, this act made it easier for Indian lands to be sold off to American settlers. The basic structure of the allotments included the following: each head of family would receive one-quarter of a section (120 acres); each single person over eighteen or orphan child under eighteen would receive one-eighth of a section (sixty acres); and other single persons under eighteen would receive one-sixteenth of a section (thirty acres).

Initially, the act exempted the Cherokees, Creeks, Choctaws, Chickasaws, Seminoles, and Osage, Miamies and Peorias, and Sacs and Foxes, who were located in the Indian Territory as well as lands belonging to the Seneca Nation of New York and the Sioux Nation in Nebraska. In later negotiations, however, most of these groups agreed to allotment if they would dissolve their tribal governments. To receive the allotted land, tribal members needed to enroll with the Bureau of Indian Affairs, which placed the individual's name on the "Dawes rolls." At this point, the individual (who now had no tribal affiliation) was eligible to receive land.

At the time, the federal government argued that the Dawes Act protected Indian property rights, particularly during the land rushes of the 1890s. Most often, however, the new landowners received land of little worth: desert or near-desert lands unsuitable for farming. Even if the land proved capable of being farmed, agricultural techniques used by settlers were often quite foreign to Indian understanding. Often, the landowners could not afford the tools, animals, seed, and other supplies necessary to get started.

Conclusion: A Legacy of Conquest

From one line of thought, American development relied on ridding the West of native residents. Although the policy was erratic, over the course of a century, the federal government cleared settlers of this impediment. Unlike other nations in world history, the United States did not use mass exterminations or organized genocide.

The other line of thinking begins with the founding principles of the United States. In a nation founded on democracy and fairness among diversity, natives were mistreated throughout the period of settlement, but, say many critics, the most problematic issue may have been the complete inconsistency and deception with which the U.S. government carried out its agenda.

Sources and Further Reading: Krech, *The Ecological Indian: Myth and History*; Opie, *Nature's Nation*; White, *It's Your Misfortune and None of My Own*.

BUFFALO EXTERMINATION SPELLS DISASTER FOR NATIVE PEOPLE

Time Period: Late 1800s
In This Corner: Native groups of the Midwest and West
In the Other Corner: Military, federal government, hunters
Other Interested Parties: Settlers
General Environmental Issue(s): Species management, bison, hunting, the West

The great, wandering herd of American bison marked one of the most distinctive characteristics of the continent when Europeans arrived around 1500. For centuries, native peoples defined their life by the pursuit of this creature, creating a mobile society that could be certain to always retain access to the bison herd. When European settlers sought a more sedentary style of trade and habitation, the free-ranging bison was no longer practical or acceptable. Although the near extermination of the bison allowed European-Americans to achieve successful settlement of the western United States, it spelled disaster for cultures relying on the herd. The size of the herds was affected by predation (by humans and wolves), disease, fires, climate, grazing competition from horses, and other factors. Throughout the nineteenth century, the story of the great bison decline ran parallel with the demise of many native tribes.

Tribes of the American Plains, including the Blackfeet, Gros Ventre, Assiniboin, Crow, Cheyenne, Shoshoni, Arapaho, various Sioux or Dakota people, Comanche, and others, had come to rely on the bison the way that coastal tribes depended on the Great Lakes or ocean. Many cultures consumed almost every portion of the bison, including blood, milk, marrow, meat, fat, organs, testicles, nose gristle, nipples, and the fetus. An average bison weighed 700–800 pounds. From this massive animal, native users could butcher 225–400 pounds of meat. Given their massive size, bison were often pursued in communal hunts that resulted in dozens of kills at once. Most native groups preferred cows over bulls. Humps, tongues, and fetuses were considered delicacies by more than one group. Some observers were appalled to see the great waste of some native hunters, who removed tongues or other delicacies and abandoned the rest of the carcass.

Table 2.
Native American Uses of the Bison

Hides (with bison hair off)	Moccasins, leggings and other clothing, tipi covers and linings, shields, maul covers, cups and kettles, parfleches (carrying cases)
Robes (with bison hair on)	Winter clothing, gloves, bedding, costumes (ceremonial and decoy)
Hair	Ropes, stuffing, yarn
Sinew	Thread, bowstrings, snowshoe webbing
Horns	Arrow points, bow parts, ladles and spoons, cups, containers (for tobacco and medicine)
Hoofs	Rattles, glue
Tibia and other bones	Brushes, awls, fleshers, other tools
Ribs	Arrow straighteners
Brains	To soften skins
Fat	Paint base
Dung	Fuel to polish stone
Teeth	Ornaments
Paunch and large intestine	Containers
Gallstones	Yellow pigment
Penis	Glue

Source: Krech 1999.

Eating was only one of the native uses of the bison. For most of these groups, the animal's body possessed many additional uses. Historian Shepherd Krech refers to the bison as "the era's Wal-Mart."

The Impact of Markets

Despite their reliance on the bison, natives did not overly stress the continuity of the animal's population. So, regardless of the impact of drought, horses, or fires, the true culprit of the bison's demise can only be the markets connected to the expanding American economy. Buffalo tongues, skins, meat, and robes were in demand all over the world. Once the American railway system accessed the Plains for trade purposes, bison parts could be sold widely.

Clearly, the railroad brought new strain to the bison herd, specifically between 1867 to 1884. In addition to importing more farmers and increasing the connection to outside markets, the railroads also increased the amount of waste that would be tolerated, which one scholar estimates to be by three to five times. Thus, more animals were taken without regard to any need to use the entire animal. In 1873, more than 750,000 hides were shipped on the Atchison, Topeka, and Santa Fe Railroad alone, and it is estimated that more than 7.5 million buffalo were killed from 1872 to 1874 (Haines 1995, 179). Millions were killed each year during the early 1880s until the commercial hunt ended in 1883.

The herd was also impacted by settlement's increase in the possibility of fire. Started by railroad engines or settlers, many fires swept the grasslands, killing or disrupting bison herds. To make matters worse, drought became most pronounced in this portion of North America from 1840 to 1880. The bison population was stressed even further.

By this time, most western tribes had been confined to reservations. With limited ability to intervene or pursue the remaining bison, many tribes suffered from hunger. By 1884, wrote Krech, "With very few exceptions, the buffalo was gone and bone collectors scooped up all the remains they could find for shipment east where they were processed into phosphate fertilizer." Colonel Homer W. Wheeler, of the U.S. Cavalry, observed the following:

> Millions of Buffalo were slaughtered for the hides and meat, principally for the hide. Some of the expert hunters made considerable money at that occupation....
>
> Some of the habits of the Buffalo herds are clearly fixed in my memory. The bulls were always found on the outer edge, supposedly acting as protectors to the cows and calves. For ten to twenty miles one would often see solid herds of the animals. Until the hunters commenced to kill them off, their only enemies were the wolves and coyotes. A medium-sized herd, at that time, dotted the prairie for hundreds of miles, and to guess at the number in a herd was like trying to compute the grains of wheat in a granary …
>
> … Buffalo hunting was dangerous sport. Although at times it looked like murder, if you took a buffalo in his native element he had plenty of courage and would fight tenaciously for His life if given an opportunity. Like all other animals, the buffalo scented danger at a distance and tried to escape by running away, but if he did not escape he would make a stand and fight to the last, for which every one must respect him.
>
> The stupidity of the buffalo was remarkable. When one of their number was killed the rest of the herd, smelling the blood, would become excited, but instead of stampeding would gather around the dead buffalo, pawing, bellowing and hooking it viciously. Taking advantage of this well-known habit of the creature, the hunter would kill one animal and then wipe out almost the entire herd. (Wheeler 1925, 80–82)

Conclusion: Was There an Official U.S. Policy to Exterminate the Bison?

The bison extermination brought great suffering and change to natives of the Plains. Conveniently, this provided the groundwork for the efficient expansion of the American population westward. Some scholars have pondered whether or not it was too convenient. They have speculated that there may have been an official policy to exterminate the buffalo to undermine natives of the region.

Records show that U.S. Generals Sheridan and Sherman recognized that eliminating the buffalo severely reduced the Indians' capacity to continue an armed struggle against the United States. The editors of the *Army and Navy Journal* supported the proposition, comparing such an effort with Civil War campaigns against Confederate supplies and food sources (Wooster 1988, 171).

Forts provided the infrastructure for settlement but also provided outposts for hunters and traders. In fact, often soldiers were hunters (primarily for sport). General George Custer observed the following:

> To find employment for the few weeks which must ensue before breaking up camp was sometimes a difficult task. To break the monotony and give horses and men exercise, buffalo hunts were organized, in which officers and men joined heartily. I know of no better drill for perfecting men in the use of firearms on horseback, and

thoroughly accustoming them to the saddle, than buffalo-hunting over a moderately rough country. No amount of riding under the best of drill-masters will give that confidence and security in the saddle which will result from a few spirited charges into a buffalo herd. (Custer 1966, 111)

For these reasons, most scholars agree that the U.S. Army undoubtedly participated in the demise of the bison; however, there does not appear to be an official policy dedicated to this pursuit.

The unofficial policy, then, is very clear in Secretary of the Interior Delano's 1864 testimony before Congress in which he stated, "The buffalo are disappearing rapidly, but not faster than I desire. I regard the destruction of such game as Indians subsist upon as facilitating the policy of the Government, of destroying their hunting habits, coercing them on reservations, and compelling them to begin to adopt the habits of civilization" (Wooster 1988, 171). General Sheridan added, "If I could learn that every buffalo in the northern herd were killed I would be glad" (Wooster 1988, 172).

By 1873, several U.S. officers protested the wanton destruction of the bison to Henry Bergh, president of the America Society for the Prevention of Cruelty to Animals. The Army, while anxious to strike against the Indians' ability to continue their resistance, did not make the virtual extermination of the American bison part of its official policy; in some cases, individual officers took it upon themselves to try and end the slaughter (Wooster 1988, 171).

Sources and Further Reading: Custer, *My Life on the Plains*, 1966; Flores, "Bison Ecology and Bison Diplomacy: The Southern Plains from 1800–1850"; Haines, *The Buffalo: The Story of American Bison and Their Hunters from Prehistoric Times to the Present*; Inventory of Conflict and Environment, *The Buffalo Harvest*, http://www.american.edu/TED/ice/buffalo.htm; Isenberg, *The Destruction of the Bison: Social and Ecological Changes in the Great Plains, 1750–1920*; Krech, *Buffalo Tales: The Near Extermination of the American Bison*, http://www.nhc.rtp.nc.us/tserve/nattrans/ntecoindian/essays/buffalo.htm; Krech, *The Ecological Indian: Myth and History*; Library of Congress, *The Extermination of the American Bison*, http://memory.loc.gov/learn/features/timeline/riseind/west/bison.html; Utley, *Cavalier in Buckskin: George Armstrong Custer and the Western Military Frontier*; Wheeler, *Buffalo Days: Forty Years in the Old West: The Personal Narrative of a Cattleman Indian Fighter, and Army Officer Colonel Homer W. Wheeler*; Wooster, *The Military and United States Indian Policy 1865–1903*.

WHITE CITY REDEFINES THE EXPECTATIONS OF ENVIRONMENTAL PLANNING

Time Period: 1893
In This Corner: Planners and engineers, conservationists
In the Other Corner: Defenders of status quo
Other Interested Parties: Policy officials and American citizens
General Environmental Issue(s): Planning, landscape

The "White City," as the spectacle was popularly called, spread out in front of the world in 1893 and demonstrated where the United States wanted to go in the future. Standing at the dawn of the twentieth century—what *Time* magazine founder Henry Luce would later dub the "American century"—the White City helped to define the way Americans did basic things, including leisure and planning. Through its use of technology and planning, in particular, the White City established a new instrumentalized role for the natural environment in the life of most Americans. Referred to as modernism or as the modern, great minds all over the world spent the next century using technological innovation to solve problems. But was this always wise? Can technology in its own right always solve problems without creating new ones?

Defining the American Future at the White City

The grand landscape of the Columbian Exposition was one of the first planned, multiuse sites in American history; despite this tremendous feat, however, it remained a landscape ripe with irony. Intended as a celebration of Columbus's voyages 400 years before, the Exposition attracted more than twenty-seven million visitors in its single year of existence. That was the irony of the great spectacle: this great example of modern design was not intended to last. The material that held together the exterior of many buildings resembled papier-mâché and began disintegrating after one year. Designers and architects made buildings that could be dismantled and moved elsewhere or whose infrastructure could be used for other purposes. Ironically, this aspect of modern planning became even more pronounced in the "disposable society" that followed World War II.

The symbol of a great American past and future, the White City also became a symbol of the nation's precarious present. The industrial, and increasingly electrical, revolutions were transforming America at a remarkable pace. Most Americans no longer earned their living on farms but worked in factories and lived in urban centers. Historian Alan Trachtenberg calls this an era of the "incorporation of America," the shift of social control from the people and government to big business (Trachtenberg 1982, 3–10). Overall, the United States was in the process of shifting from a producer society to one of consumers. This required a culture that accepted and even expected to consume. As strikes, economic depression, and social issues such as unemployment and homelessness marred the early 1890s, the nation collectively looked to a more positive future, a glimpse of which could be found in the White City.

Many American historians have dubbed the World's Columbian Exposition the beginning of the consumer-based society that would prosper in the twentieth century. Although some of the designers and conceivers of the fair might have entertained such aspirations, the real success of the World's Columbian Exposition derived from its timing. Most important, the 1890s witnessed the emergence of revolutionary new technologies that radically altered every-day human life. The fair offered an opportunity for the unveiling of such innovations. In addition, modern thought was reconsidering a variety of ideas relating to social progress and the laissez-faire treatment of corporations. Altogether, the fair created symbols of American aspirations and ideals, even if some of these were icons of consumption, including Juicy Fruit Gum, Pabst Blue Ribbon beer, ragtime music, and Quaker Oats.

Financially, the World's Columbian Exposition was an overwhelming success. The world, now equipped with modes of transportation that facilitated tourism, had never seen anything like it. Although August attracted 3.5 million visitors, this more than doubled in October with more than 6.8 million paid visitors. The concession stands brought in more than $4 million! The final calculations revealed in excess of approximately $1 million surplus for its 30,000 stockholders. More important, however, in the stretch of Chicago known most for

Exposition Grounds at World's Columbian Exposition, Chicago. Known as the White City, this grand event of the 1890s ushered in the modern era. Library of Congress.

its worthless swampland, the United States had grasped a glimpse of its future (Columbian Exposition, *The World's Columbian Exposition: Idea, Experience, Aftermath*).

Shaping a New Vision of Nature

The White City, of course, was no city at all. It had no residences or utilities; however, it was a fully designed space. The greatest landscape planner of the era, Frederick Law Olmsted, who had designed New York's Central Park in 1862, gave the Columbian Exposition a unifying ground plan. In its unity, the landscape taught visitors about beauty and living among art while also possessing state-of-the-art amenities and technologies.

The scene's aesthetic beauty was a coordination of Olmsted's management of water, ornamental plants, and landscape art with the structures that had been organized by the director of works, Daniel H. Burnham, one of the first great American architects. As an overall design, Burnham wanted a neoclassical style set in overwhelming whiteness. Seven hundred acres of swampy land had been dredged and filled to create artificial canals, plazas, promenades, lagoons, and even a forest preserve. This setting then held 400 buildings, each with an intricate design and specific purpose (i.e., Manufactures, Agriculture, Electricity, and Women's Progress). The idea of a landscape that both performed and functioned well while also appearing beautiful seemed a fantasy to visitors who resided in farms and cities thick with the grit and grime of life in 1893. Trachtenberg wrote, "For a summer's moment, White City had seemed the fruition of a nation, a culture, a whole society: the celestial city of man set upon a hill for all the world to behold" (Trachtenberg 1982, 230).

In laying out this new city form, Burnham relied on overall organization and symmetry. This was most obvious in the Court of Honor, where the elaborate buildings balanced one another around Olmsted's system of pools, trees, and bridges. To most visitors, Machinery Hall was the grandest attraction of all. It covered seventeen acres and housed an elevated, moveable crane. Two dynamos powered a 2,000-horsepower engine and lit nearly 200,000 lights. Although the structure was made of steel and iron, it was covered with the white plaster-like material called "staff" that covered most of the structures in the court. It was this plaster veneer that added a fantasy-like aura to the landscape, transforming the diverse structures into the spectacle known as the White City. Trachtenberg observes that these buildings were "composed … as pictures of art, thus establishing the place of culture in relation to the activities of society embodied by the exhibitions within" (Trachtenberg 1982, 215–16).

Although there were many significant speakers at the Exposition, the words of one have proven to have remarkable repercussions: the lecture given by Frederick Jackson Turner, which is now referred to as the "Frontier Thesis." Using demographic data from counties in the western territories, Turner proclaimed that the American West was populated and, therefore, "settled." He worried that Americans now had to live without the emotional and demographic "safety valve" of the empty western lands. Turner spoke of a spiritual frontier as well as a physical one, and he urged Americans to now seek out new challenges (Smith 1978, 15–20).

Historian Roderick Nash argued that Turner's formulation forced Americans to reconfigure the expansive spirit that had driven the nation's development thus far. Some energy would be channeled abroad, providing the United States with a new prominence in global affairs. In addition, the American effort to tame the wilderness would be channeled to a new effort to preserve remaining wild areas. The preservation impulse grew from panic concerning overdevelopment as well as a growing aesthetic appreciation for the beauty of raw nature. This impulse resulted in the movement for national parks as well as in efforts to conserve natural resources. Turner's general theme about limits, however, could be found at the core of each (Nash 1982, 24).

Using Technology to Construct a Middle Ground

Behind the pastoral, park-like façade of the White City lurked the primary agent of change: new technological advances. The early 1900s involved such significant change that these years moved Karl Marx to describe them as follows:

> All fixed, fast-frozen relations, with their train of ancient and venerable prejudices and opinions, are swept away, all new-formed ones become antiquated before they can ossify. All that is solid melts into air, all that is holy is profaned, and men at last are forced to face … the real conditions of their lives and their relations with their fellow men. (Berman 1988, 21)

At the Exposition, technology was symbolically portrayed as the nation's successful future, but there was a cost to technological development. Trachtenberg observed the other side of this equation when he wrote, "If the machine seemed the prime cause of the abundance of new products changing the character of daily life, it also seemed responsible for newly visible poverty, slums, and an unexpected wretchedness of industrial conditions" (Trachtenberg

1982, 38). The realities of the vicious change wrought on society by the machine were most acute at the beginning of the century when factories concentrated labor into the cities that they polluted. Reform and regulation informed better urban planning and eventually helped create new understandings of environmental health.

The pace of growth stemming from the technology emerging at the start of the twentieth century cannot be overstated. Historian John McNeil estimates that, during this century, the world's gross domestic product increased by more than ten times. The world population also increased incredibly: demographic historians estimate that, if one took all the years lived by the 80 billion hominids that have ever been born in the past four million years, 20 percent of those years took place during the twentieth century (McNeil 2001, 8–9). The engine behind this century of rapid change was the United States, which became the international industrial leader by the end of World War I.

Technology came to permeate every aspect of American life. For instance, the literary scholar Leo Marx argues that American culture is defined by two alternating pulls: the machine toward technological advancement and the garden toward aesthetic and natural beauty (Marx 1964, 2–10). In this formulation, however, Marx makes room for a middle ground between technology and nature. City planner and writer Lewis Mumford believed that the application of modernist design helped to construct such a middle ground in urban spaces (Mumford 1963, 321–25). This was one of the primary tasks of planners and designers (such as Mumford) who emerged after 1900. Just as Burnham and Olmsted had created a new reality at the Exposition, urban planners rationalized human existence and planned its most suitable environment. Possibly the most startling intellectual revolution of the twentieth century was the ability to entirely diminish the human's role in the natural world. In other words, just as humans mastered very unnatural forms such as the urban skyscraper, they alternatively realized what was being lost and initiated efforts to set aside other areas from development.

In the same general time period that Americans constructed the Empire State Building, Hoover Dam, and some of the early highways, a scholar in the Wisconsin marsh developed an ethic that would inspire environmentalists for generations. Referred to as the "Land Ethic," the words written by Aldo Leopold contained a new way of viewing the human's place within the natural environment. He wrote, "In short, a land ethic changes the role of *Homo sapiens* from conqueror of the land-community to plain member and citizen of it. It implies respect for his fellow-members, and also respect for the community as such" (Leopold 1987, 240).

Although this perspective draws inspiration from romanticism of the nineteenth century, the Land Ethic is grounded in new scientific understandings of the mid-twentieth century. Therefore, just as engineers taught new, innovative ways to solve everyday problems, scientists, writers, and naturalists took the opportunity to better explain and consider the complexity of the human condition, particularly as it relates to the natural world.

Armed with a modern sensibility such as that exhibited at the Columbian Exposition, Americans redefined many portions of their everyday lives. Cities took shape in difficult climates and locations suspended by only the contrivances of modern technology. During the twentieth century, nature brought momentary reminders of our tenuous relationship with nature in some of these locations. In 2005, for example, Hurricane Katrina wrought a reminder of nature's power that promises to last more than a moment. Although the heavy manipulation of nature brought heightened impacts on the natural world, it also helped to usher in the beginning of an environmental sensibility. During the twentieth century, both perspectives, the use of technology and the appreciation of the natural environment, each gathered the interest of the American public. In this era of modernity, we can locate a few symbols of the compromise that became necessary.

Conclusion: "Fallingwater" as a Symbol for a New Technological Era

Nestled in the White City, one could find a stunning Japanese cottage built by the unknown protégé of one of the greatest American architects, Louis Sullivan. In the Asian styling and

Frank Lloyd Wright's "Fallingwater," a symbol of the "middle ground" between nature and technology. Library of Congress.

the use of natural forms, one could make out the fusion of modernist sensibilities with an awareness of natural forms and materials. Four decades later, the organic architecture of the young architect reached its maturity in the Laurel Highlands of Pennsylvania. Frank Lloyd Wright's design in Pennsylvania was the renowned structure known as "Fallingwater." Using modern technology to cantilever the structure like a diving board from a hillside, Wright merged the building's form with the rushing water of the mountain stream (Smith 1993, 142–45).

Today, Fallingwater remains the preeminent example of designing with nature as builders and landscape designers continue to try to infuse their forms with a similar ethic. In this effort for reconciliation, Fallingwater presents a suitable representation for Americans' relationship with nature during the entire twentieth century.

Undoubtedly, during the twentieth century, Americans became the most technologically advanced humans that the world had ever seen. Although the details of these technical feats are fascinating as well, the most remarkable point might be that massive technological development contributed to perpetuating and complicating Americans' relationship with nature. In short, during the twentieth century, Americans' fascination with nature did not diminish; instead, Americans used new technologies and scientific understanding to define for themselves a new place in the natural world.

Fallingwater might demonstrate to us the next phase in our technological relationship with nature. In an era when science provides a microscopic and macroscopic understanding of the change caused by our activities and innovations, we have begun to ask new questions of our technology. Similar to the White City, this approach embraces human ingenuity and technology while also demanding an awareness of the environment.

Sources and Further Reading: Columbian Exposition, *The World's Columbian Exposition: Idea, Experience, Aftermath,* http://xroads.virginia.edu/~MA96/WCE/title.html; Smith, *Making the Modern;* Trachtenberg, *The Incorporation of America.*

WORKING FOR WORKERS' RIGHTS

Time Period: 1850–1970
In This Corner: Workers, labor organizations
In the Other Corner: Industrial leaders
Other Interested Parties: Federal government, political parties
General Environmental Issue(s): Labor, human rights, economic development

The industrial era marked incredible accomplishments in human progress; however, many of these innovations forced the human worker into a new low in depersonalization. The strict hierarchy of the economy during the Gilded Age combined with the use of nonskilled laborers to make workers seem superfluous to some employers. In the late 1800s, of course, there was always new immigrant labor arriving in the United States who were willing to work for lower wages. Many employers resisted workers' call for better treatment as well as their right to organize into labor unions. The changes in American ideas about factory and worker management required a wholesale ethical shift that could only be done gradually beginning in the early 1800s.

Organizing Workers

By the 1820s, various unions involved in the effort to reduce the working day from twelve to ten hours began to show interest in the idea of federation, of joining together in pursuit of common objectives for working people. Starting in the 1830s and accelerating rapidly during the Civil War, the factory system accounted for an ever-growing share of American production. Although this new way of organizing work and the workplace brought great wealth to a few, it also provided significant new incentive to exploit the rights of laborers.

Trade unions began to take shape that were connected to specific occupations and activities. With workers recognizing the power of their employers, the number of local union organizations increased steadily during the mid-nineteenth century. In a number of cities, unions in various trades joined together in citywide federations. One of the earliest examples was the National Trades' Union, which was formed in 1834 in five cities.

By 1866, several national associations of unions functioning in one trade—printers, machinists, and stone cutters, to name a few—sent delegates to a Baltimore meeting that brought forth the National Labor Union. In the case of each of these early attempts at unionization, financial fluctuations such as depressions could destroy them overnight.

The first union with staying power was the Knights of Labor of the 1870s, which was an all-embracing organization committed to a cooperative society; in some circles, it seemed more a social club than a labor organization. Membership was not limited to wage earners; it was open to farmers and small business people—everybody, that is, except lawyers, bankers, stockbrokers, professional gamblers, and anyone involved in the sale of alcoholic beverages. The Knights achieved a membership of nearly 750,000 during its first few years of existence.

In 1881, a meeting in Pittsburgh brought together delegates from many fields including carpenters, cigar makers, printers, merchant seamen, and steel workers, as well as from a few city labor bodies and a sprinkling of delegates from local units of the Knights of Labor. Together, they formed the Federation of Organized Trades and Labor Unions, which was guided by a constitution inspired by that of the British Trades Union Congress. The chairman was thirty-one-year-old Samuel Gompers of the Cigar Makers Union, who would become the American leader of the movement for unionization.

Growing through many different phases over the next decade, the American Federation of Labor (AFL) evolved around 1886. Gompers and the other founders of the AFL expressed their belief in the need for more effective union organization: "The various trades have been affected by the introduction of machinery, the subdivision of labor, the use of women's and children's labor and the lack of an apprentice system—so that the skilled trades were rapidly sinking to the level of pauper labor," the AFL declared. "To protect the skilled labor of America from being reduced to beggary and to sustain the standard of American workmanship and skill, the trades unions of America have been established."

The strike would emerge as the primary tool of the new AFL with its 300,000 members in twenty-five separate unions. Earlier in 1886, railroad workers in the Southwest had been involved in a losing strike against the properties of Jay Gould, one of the more flamboyant of the great industrial leaders now referred to as robber barons. On May 1, 1886, some 200,000 workers struck in support of the effort to achieve the eight-hour day.

Although the national eight-hour-day strike movement was generally peaceful, violence erupted in Chicago that would define a watershed event for unionization. The McCormick Harvester Company in Chicago, learning in advance of the planned strike, locked out all its employees who held union cards. Fights erupted and the police opened fire on the union members, killing four of them.

A public rally at Haymarket Square to protest the killings drew a large and peaceful throng. As the meeting drew to a close, a bomb exploded near the lines of police guards, and seven of the uniformed force were killed, with some fifty persons wounded. The police began to fire into the crowd; several more people were killed, and about 200 were wounded. Eight anarchists were arrested and charged with a capital crime. Four were executed; four others were eventually freed by the Illinois governor after he concluded that the trial had been unfairly conducted.

Upheaval at Homestead

For a dramatic example of the new way of viewing human labor, many historians point first to the iron and steel town outside of Pittsburgh known as Homestead. What in 1892 had begun as an interest in unionization by steelworkers with hopes of higher pay became a type of last stand by Andrew Carnegie, owner of the company and one of the great symbols of the robber baron era. Although Carnegie had amassed the world's largest fortune, he did not want to lose control over the rate by which he paid his work force.

To prevent workers from forming a union, Carnegie had his assistant, Henry Clay Frick, lock the 3,800 workers out of his factory. Two days later, frustrated workers seized the mill and sealed off the town from strikebreakers. Frick summoned a private police force, the Pinkerton Detective Agency, to protect the nonunion workers he planned to hire. When the 300 detectives arrived, they were met by stones and bullets flying through the air. Steelworker William Foy and the captain of the Pinkertons fell wounded. What had begun as a simple disagreement over wages between the nation's largest steelmaker and its largest craft union, the Amalgamated Association of Iron and Steel Workers, had taken a tragic turn (Demarest 1992, vi–xi).

A fierce battle raged for twelve hours. Outgunned by the Pinkertons' Winchester rifles, Homestead's citizens pressed into service everything from ancient muzzle loaders to a twenty-pound cannon. Wheelbarrow loads of ammunition came from a local hardware store. Meanwhile, news of the battle reached nearby Pittsburgh. By 6:00 A.M., more than 5,000 curious spectators lined the riverbanks. Indeed, the story had reached the entire nation, creating strong disagreements everywhere between Americans who favored workers' rights and those who sided with the right of owners to do as they pleased.

Finally, the Pinkertons attempted to land again before 8:00 A.M. the following day. From across the Monongahela, workers blasted the cannon at the Pinkertons' barges with little luck. Workers lit rafts and railroad cars on fire and sent them toward the barge. Each fell harmlessly short.

However, below deck on their barge, the terrified Pinkertons cowered in fear. Four times the Pinkertons raised a white flag to surrender. Each time, workers shot it down. Finally, at 5:00 P.M., the workers accepted the Pinkertons' surrender. In the melee, three workers and seven Pinkertons had been killed, and the conflict was not yet over.

At Frick's request, the Pennsylvania governor sent 8,500 troops to Homestead. Within twenty minutes, the guardsmen had secured the mill. Homestead was placed under martial law, and, by mid-August, the mill was in full swing, employing 1,700 nonunion replacement workers.

In mid-November, the union conceded and asked for their jobs back. Three hundred locked-out workers were rehired once they agreed not to unionize. Many others were placed on blacklists and not employed. With the union crushed, Carnegie slashed wages, imposed twelve-hour workdays, and eliminated 500 jobs.

Pullman Starts a Trend

Workers had little or no recourse to improve their situation. However, collective action, including unionization and strikes, brought national attention to the needs of workers by the end of the 1870s, yet the overall organization of society was slow to change. In some complicated examples, industrialists inspired by social Darwinism and other approaches attempted change only to create more problems. Outside Chicago, one finds a primary example of this in the model town of Pullman.

George Pullman, who manufactured railroad cars, created a model company town, organized around a church and store. However, in 1894, his insistence on cutting wages and implementing oppressive labor practices fueled a large-scale strike, with its epicenter in Pullman. In sympathy, a national railway workers' boycott was directed against the handling of trains carrying Pullman cars.

Acting at the behest of his attorney general, a former railroad attorney named Richard Olney, President Grover Cleveland appointed a special counsel to deal with the strike on the grounds that U.S. mails were being impeded. Although this was true, many railroad workers were intentionally attaching Pullman cars to mail trains to spur strikers. The attorney for the Milwaukee Railroad hired 4,000 strikebreakers, armed them, and deputized them. Great masses of sympathetic workers, particularly in the Chicago area, responded by attacking the trains. There were casualties and burned trains. Ultimately, approximately half the U.S. Army was deployed (12,000 federal troops). Although they were sent to keep the peace, it was also hoped that they would break the boycott. Clearly, the Pullman experience showed a federal government willing to use military force to uphold the moral order imposed by business owners. Most often, business owners used injunctions, issued by compliant judges on the request of government officials or corporations, as their prime legal weapon against union organizing and action.

Just as the strikes and riots at Haymarket and Homestead demonstrated, the power in American society during the Gilded Age did not lie with political leaders. Barons of enterprise, with their extreme commitment to profit, ruled the day. With the federal government on the side of business leaders, workers' groups needed to go elsewhere to stir support for change. As the public became more aware of the plight of the workers, political leaders began to respond. This shift was most obvious when the executive branch of the federal government began to exert a different view of such situations.

When 100,000 members of the United Mine Workers (UMW) went on strike in May 1902, they kept the mine closed all summer. When the mine owners refused an UMW proposal for arbitration, President Theodore Roosevelt intervened. In October, he appointed a commission to arbitrate the situation and to establish a compromise. The miners returned five days later. After deliberating, the presidential commission awarded them a 10 percent wage increase and shorter work days.

Triangle Shirtwaist Tragedy

On March 25, 1911, a fire, which broke out on the top floors of the ten-story Asch Building in lower Manhattan, New York, created the type of lesson that made every American reconsider the ethics with which factories were governed. The Triangle Shirtwaist Company was a fairly typical "sweat-shop" factory in the heart of Manhattan, at 23–29 Washington Place, at the northern corner of Washington Square East.

Workers suffered with low wages, excessively long hours, and unsanitary and dangerous working conditions. If they registered a complaint, they were typically fired and replaced with cheaper workers. In this case, when the fire broke out, the exits had been blocked to keep workers from going outside on breaks. Unable to flee, many of the workers died in the flames. The fire killed 146 of the 500 employees of the Triangle Shirtwaist Company in one of the worst industrial disasters in the nation's history.

These factory workers, mostly young female immigrants from Europe working long hours for low wages, died because of inadequate safety precautions and lack of fire escapes. As a result of the Triangle fire, the International Ladies' Garment Workers Union stepped up its organizing efforts and fought to improve working conditions for garment workers. Also a public outcry prompted the New York State Legislature to appoint a commission to investigate the causes of the fire. The commission's investigation, and union organizing, eventually led to the introduction of fire-prevention legislation, factory inspections, liability insurance, and better working conditions for all workers.

A state factory investigation committee headed by Frances Perkins [she was to become President Franklin Delano Roosevelt's (FDR) secretary of labor in 1933, the first woman cabinet member in history] paved the way for many long-needed reforms in industrial safety and fire-prevention measures.

Legislating for Workers' Rights

The lesson of these early events brought a steady stream of legislative change during the twentieth century. Congress, at the urging of the AFL, created a separate U.S. Department of Labor with a legislative mandate to protect and extend the rights of wage earners. This bureau also included a Children's Bureau that could specifically work to protect the victims of job exploitation. In addition, the LaFollette Seaman's Act required urgently needed improvements in the working conditions on ships of the U.S. Merchant Marines.

A watershed day for workers came in 1914 when the Clayton Act made explicit the legal concept that "the labor of a human being is not a commodity or article of commerce" and hence not subject to the kind of Sherman Act provisions that had been used in earlier cases decided in favor of business owners. At long last, this act provided a legal basis in the federal jurisdiction for strikes and boycotts and peaceful picketing and dramatically limited the use of injunctions in labor disputes.

The effort to achieve the eight-hour work day received its legislative approval in 1916 with the passage of the Adamson Act. Followed up in 1936 with the Public Contracts Act and in 1938 with the Labor Standards Act, most occupations took important steps toward capping work hours at forty per week.

Cesar Chavez Speaks for Workers

The rights of workers continued to form an active debate in American society throughout the twentieth century. Toward the end of the century, many of these issues included concerns over environmental health. This was one of the most important portions of the famous work carried out by the United Farm Workers Union (UFW) and its founder Cesar Chavez on behalf of California's agricultural workers.

After working as a migrant employee on farms in the western United States, Chavez, in 1952, joined the Community Service Organization and became a community organizer. By 1965, the UFW had become the main conduit of Chavez's activities. In this same year, migrant grape pickers had gone on strike, demanding a raise from the dollar an hour they were paid. More and more workers joined the Huelga (Spanish for "strike"), even in the face of threats from farm owners and labor contractors. Chavez worked tirelessly in support of the strike even leading a 250-mile march from Delano to Sacramento. In 1968, he began a twenty-five-day hunger strike, organized more rallies and demonstrations, and called for a national boycott of grapes. By 1970, the grape growers agreed to a contract with the UFW, which gave the workers healthcare benefits and a raise in pay.

There would be many other examples before his death in 1993. Chavez's *Prayer of the Farm Workers' Struggle* shows the continuity of the workers' rights movement throughout American history:

Show me the suffering of the most miserable;
So I will know my people's plight.
Free me to pray for others;
For you are present in every person.
Help me take responsibility for my own life;
So that I can be free at last.
Grant me courage to serve others;
For in service there is true life.
Give me honesty and patience;
So that the Spirit will be alive among us.
Let the Spirit flourish and grow;
So that we will never tire of the struggle.
Let us remember those who have died for justice
For they have given us life.
Help us love even those who hate us;
So we can change the world
Amen

Conclusion: The Legacy of Unions

By the end of the twentieth century, union membership had dropped to its lowest point in more than a century. To many observers, unions' time had come and gone. Only time will tell if that is true. Clearly, though, efforts to organize workers altered the expectations with which every American approaches the workplace.

Sources and Further Reading: Trachtenberg, *The Incorporation of America*; United Farm Workers, *Address by Cesar Chavez*, http://www.ufw.org/_page.php?menu=research&inc=history/10.html.

POPULISM AND THE GRASSROOTS OF THE WEST

Time Period: 1880–1910
In This Corner: Workers and farmers of the United States
In the Other Corner: Industrial elites
Other Interested Parties: Federal government, American voters
General Environmental Issue(s): Agriculture, politics

Although the exploits of Buffalo Bill and others romanticized portions of life on the frontier, in reality, life in the West was anything but romantic. Historian Bill Cronon has referred to economic development, particularly agriculture, in the West as methods for "annihilating space," which was the region's greatest natural resource. This approach to development and land use, often without the towns used in other regions, often seemed impersonal and inhuman. Some historians and writers have even referred to it as machinelike particularly because of its reliance on one great machine, the railroad.

In his novel *The Octopus*, Frank Norris created one of the most evocative critical views of Western settlement. The primary emphasis of his account was the railroad in California. He wrote the following:

> But suddenly there was an interruption. Presley had climbed the fence at the limit of the Quien Sabe ranch. Beyond was Los Muertos, but between the two ran the railroad. He had only time to jump back upon the embankment when, with a quivering of all the earth, a locomotive, single, unattached, shot by him with a roar, filling the air with the reek of hot oil, vomiting smoke and sparks; its enormous eye, cyclopean, red, throwing a glare far in advance, shooting by in a sudden crash of confused thunder; filling the night with the terrific clamour of its iron hoofs....

> Before Presley could recover from the shock of the eruption while the earth was still vibrating, the rails still humming, the engine was far away, flinging the echo of its frantic gallop over all the valley. For a brief instant it roared with a hollow diapason on the Long Trestle over Broderson Creek, then plunged into a cutting farther on, the quivering glare of its fires losing itself in the night, its thunder abruptly diminishing to a subdued and distant humming. All at once this ceased. The engine was gone.

> But the moment the noise of the engine lapsed, Presley—about to start forward again—was conscious of a confusion of lamentable sounds that rose into the night from out the engine's wake. Prolonged cries of agony, sobbing wails of infinite pain, heart-rending, pitiful.

The noises came from a little distance. He ran down the track, crossing the culvert, over the irrigating ditch, and at the head of the long reach of track—between the culvert and the Long Trestle—paused abruptly, held immovable at the sight of the ground and rails all about him.

In some way, the herd of sheep—Vanamee's herd—had found a breach in the wire fence by the right of way and had wandered out upon the tracks. A band had been crossing just at the moment of the engine's passage. The pathos of it was beyond expression.

It was a slaughter, a massacre of innocents. The iron monster had charged full into the midst, merciless, inexorable. (Norris 1986, 12–15)

Most observers saw the railroad as an automatic community good (Norris 1986). As Norris described in *The Octopus*, although western farming expanded rapidly in the late 1800s, it remained entirely at the mercy of the railroads. When some western farmers became fed up with what they viewed as excessive rates being charged by the railroad to haul their crops, they banded together in an effort to exert collective action. Ultimately, farmers' groups helped to inspire a third-party political revolt of Populism in the 1890s.

Overall, farmers did not share in the general prosperity of the latter nineteenth century. In fact, many western farmers became convinced that the evolving industrial systems based in cities such as Chicago left those working the land a special victim. Beginning in the 1870s, western farmers attempted to mount an effective political campaign to rectify what they saw as the corruption of government and economic power, which they attributed to big businesses and railroads. In fact, much of the farmers' plight was caused by factors unrelated to industrialization, such as fluctuations in international markets for corn and wheat. However, perceptions are often more important than reality, and American farmers believed that the democratic system of their forebears was being subverted. In their reaction, they showed an amazing ability to mobilize as a force in national politics.

The most successful of the agrarian political movements was the People's Party, or the Populist Party, which, after the 1892 presidential campaign, appeared to have the strength to become a potent force in American politics. Its strength lay primarily in the southern and midwestern states, the agricultural heartland of the nation, although its leaders tried to reach out and attract eastern workers.

The People's Party platform of 1896 marked a number of important watersheds in American political history. First, it summed up two decades of resentment by farmers against a system that they believed ignored their needs and mercilessly exploited them, but it was not just big business to which they objected. The Populists worried that the alliance between business and government would destroy American democracy, and the various proposals they put forward had two aims. The goal was not just to relieve economic pressure on agriculture but also to restore democracy by eliminating what the Populists saw as the corrupt and corrupting alliance between business and government. The Populist Party disappeared after the election of 1896, absorbed for the most part into the Democratic Party.

It turns out that one of the other advantages of the frontier was the capacity of empowered groups of people to achieve early reform. The work of the Farmer's Alliance, Populism, and unions in the West and Midwest joined with strikes, unions, and other efforts in the

East to create a dramatic change in the ways that Americans viewed the role of humans and human labor.

The "Cross of Gold" Speech

Running as the Democratic and Populist candidate for president in 1896, William Jennings Bryan vocalized the feelings of the laborers and farmers of the West with his famous "Cross of Gold" speech. Led by candidate William McKinley, Republicans argued for the American economy to be based on a gold standard. Bryan and others who favored backing it with silver (now being found in abundance in Colorado and other western states) used this disagreement as an opportunity to discuss the severe economic class divisions of Gilded Age society:

> … Ah, my friends, we say not one word against those who live upon the Atlantic Coast, but the hardy pioneers who have braved all the dangers of the wilderness, who have made the desert to blossom as the rose—the pioneers away out there [pointing to the West], who rear their children near to Nature's heart, where they can mingle their voices with the voices of the birds—out there where they have erected schoolhouses for the education of their young, churches where they praise their creator, and cemeteries where rest the ashes of their dead—these people, we say, are as deserving of the consideration of our party as any people in this country. It is for these that we speak. We do not come as aggressors. Our war is not a war of conquest; we are fighting in the defence of our homes, our families, and posterity. We have petitioned, and our petitions have been scorned; we have entreated, and our entreaties have been disregarded; we have begged, and they have mocked when our calamity came. We beg no longer; we entreat no more; we petition no more. We defy them!

> … Having behind us the producing masses of this nation and the world, supported by the commercial interests, the laboring interests and the toilers everywhere, we will answer their demand for a gold standard by saying to them: You shall not press down upon the brow of labor this crown of thorns, you shall not crucify mankind upon a cross of gold.

Conclusion: Working-Class Politics

Populism brought new political organization to workers who had banded together in unions and other organizations. Although the political party did not survive, its priority of creating a voice for the working class did.

Sources and Further Reading: Norris, *The Octopus*; Trachtenberg, *The Incorporation of America*.

FRONTIER THESIS AND AMERICAN MEANING

Time Period: 1893 to the Present
In This Corner: Proponents of the heroic model of American settlement of the West
In the Other Corner: Proponents of the complex model of American settlement of the West
Other Interested Parties: American mythmakers, politicians, historians
General Environmental Issue(s): The West, settlement

What is the legacy of America's western settlement? Is this a past that we should be proud of? Or should it be a source of shame? This discussion has partly been a scholarly one. However, at times, the debate has taken shape in film and art as well.

The discussion began inadvertently when one scholar acknowledged the uniqueness of America's western experience. In 1891, historian Frederick Jackson Turner presented one of the best-known scholarly papers in history. In his address, Turner explored the demographic information he gathered throughout the American West. More important, however, Turner communicated his interpretation of the important role that the movement westward had played in American life and politics. In so doing, Turner became the first scholar to see that the effort to settle and make profitable the natural environment of the West had been a defining point for the American character.

Scholars differ over exactly how much weight to give Turner's ideas in the history of the nation, and certainly some of his points perpetuated the interest in the mythic West even to the exclusion of appreciation and understanding of the "real" West. However, Turner must be given credit for appreciating that the experience of managing and administering the shifting frontier provided a concrete expression of American principles and ideals. Often, as Norris noted, the outcome could be ruthless and ugly. However, the American character can be said to have defined itself in the American West. Turner was one of the first Americans to make this claim.

Turner urged Americans that "the American character did not spring full-blown from the Mayflower" but that "it came out of the forests and gained new strength each time it touched a frontier" (Turner, 1934). In his line of thought, the movement westward represented a defining moment of American heroism. Turner wrote, "What the Mediterranean Sea was to the Greeks, breaking the bond of custom, offering new experiences, calling out new institutions and activities, that, and more, the ever retreating frontier has been to the United States directly, and to the nations of Europe more remotely" (Turner, 1934).

In an interesting twist, Turner's overall celebration of American success was drawn to sound an alarm. As many pioneers proceeded westward with visions of a mythical garden of opportunity, Turner cautioned that the frontier was "closed": the West had been settled. Turner continued, "And now, four centuries from the discovery of America, at the end of a hundred years of life under the Constitution, the frontier has gone, and with its going has closed the first period of American history" (Turner, 1934). With this development, he felt the nation had lost a spiritual "safety valve" that would need to be filled with other attractions. Some thinkers have used Turner's argument to new frontiers—new challenges—ranging from raw wilderness to outer space (Smith 1978, 250–59).

Creating a New Western History

Clearly, many scholars believe the mythic West possessed a romance that endures. Advertisers and makers of popular culture continue to rely on images of cowboys and raw wilderness to suggest the West as the uncivilized "other" in American life. During an era of development and reliance on technology, enduring images of the romanticized West continue to attract Americans. The irony of the mythic West is that it falsely romanticizes aspects of a truly hard-earned life in a difficult environment. Historian Patricia Nelson Limerick wrote the following:

Many American people have held to a strong faith that humans can master the world—of nature and of humans—around them, and Western America put that faith to one of its most revealing tests. A belief in progress has been a driving force in the modern world; a depository of enormous hopes for progress, the American West may well be the best place in which to observe the complex and contradictory outcome of that faith. (Limerick 1987, 29–30)

Limerick has been one of the most visible scholars arguing for and inspiring what is now referred to as the "New Western History." Relative to this newer approach, previous models of history as well as the romanticized views of pulp fiction and western film have become known as the "Old Western History." Walter Nugent wrote the following:

By the early 1980s, western historians lamented that their field had become a backwater. Their conventions featured papers with such titles as "Is There Life after the Frontier?," "Will the West Survive as a Field in American History?," and "The West and Western History: Whither Goeth It?". Beneath this cloud of pessimism, a good many historians (including a young generation who were eagerly choosing careers in western history despite its tarnished reputation) were beavering away, producing articles and monographs of high quality on a widening range of subjects, including the Turner-less twentieth century (his famous "thesis" stops with what he took to be the "closing of the frontier" in 1890).

The seminal event in Nugent's view came in 1987, "when the stiff, fresh breeze of Patricia Nelson Limerick's *The Legacy of Conquest: The Unbroken Past of the American West* blew across the field," but her work clearly grew out of that of Yale's Howard Lamar, as did other revisionist works by historians Donald Worster, Richard White, and William Cronon.

Although this new approach took many forms, it generally sought to overcome the over-simplification inspired by Turner's thesis. The "new" approach included four basic elements. First, the West as a place, not the frontier as a process, should be our focus. Second, the unique convergence of diverse people in the West is one of its defining characteristics, and it should be used to frame the historical stories. Third, the processes have not subsided and continue today. Fourth, the western story is neither one of triumph over adversity, with the resulting ennoblement of the American character. The region's experience is fraught with moral ambiguity and should be re-explored anew at each juncture.

Because of the region's uniquely enduring reliance on its delicate ecology, "New Western History" has specifically sought to contextualize the history within these natural implications. Donald Worster's book *Dust Bowl*, for instance, studies the potential implications of irrigation and political power in the West. Worster, in particular, clarified the significance of western history as environmental history. In an era when many buffs, novelists, and film makers seek to exploit the region's stories, Worster urged historians to "help the American West to become a more thoughtful and self-aware community than it has been, a community that no longer believes in its special innocence, but accepts the fact that it is inextricably part of a flawed world" (Nugent 1994).

The new ideas were given a profound public unveiling when, on March 14, 1991, the exhibition "The West as America: Reinterpreting Images of the Frontier, 1820–1920" opened at the National Museum of American Art. The purpose of the exhibition and book, written by National Museum of American Art curator William Truettner and a team of seven scholars of western art, was to document how frontier images define our idea of the national past. The exhibit, particularly as it was produced with the help of federal funding, received both magnanimously impressed and rabidly appalled reactions.

Historian Elliott West, who has created many books putting the ideas of the "New Western History" to work, described the change in history this way: "The new history really is part of something larger. It is a maturing understanding of the West, a comprehension that takes into account the full length of its history, its severe limitations and continuing conflicts, its ambivalence, and its often bewildering diversity" (West 2006).

Nugent added, "Turner and other contributors to the western-frontier mythology of triumphal expansion, notably Buffalo Bill Cody, Owen Wister, Theodore Roosevelt, and John Wayne, glamorized the story, however subtly or unsubtly" (Nugent 1994).

Conclusion: Old West, New West

Historians will continue to debate the proper interpretation of the American settlement of the West. Without a doubt, however, in classrooms around the nation students are confronting a variety of perspectives on the settlement of the West as well as many other portions of American history.

Sources and Further Reading: Nugent, "Western History, New and Not So New"; Smith, *Virgin Land*; White, *It's Your Misfortune and None of My Own*.

THE PANAMA CANAL OPENS ROUTES OF TRADE AND ECOLOGICAL CHANGE

Time Period: 1900–2007
In This Corner: Trade and expansionists' interests, American government
In the Other Corner: Competing nations
Other Interested Parties: American manufacturers
General Environmental Issue(s): International trade, biodiversity

The convergence of new shipping technologies combined with political and economic desires in the early 1900s to create an American climate of outward expansion. The greatest symbol of this era was also one of the most controversial developments: the Panama Canal. Involving significant technological problems, the canal's functionality was only one of the significant hurdles in its development. Keeping workers healthy in an exotic, foreign environment turned into one of the most dramatic challenges of the project. Many Americans argued that such an effort was ill-conceived and not worth the expense and effort. However, they were arguing against one of the most powerful U.S. presidents in history: Theodore Roosevelt.

Roosevelt used a new Navy and new policies to redefine the U.S. role abroad. Referred to as "the Roosevelt Corollary," his policy shift notified neighboring nations in the Americas that the United States would consider the stability of this region as a part of its own security. Smooth and stable trade and diplomatic relations within this "sphere of influence" required that the United States be prepared to get involved in the affairs of nations in this area, whether through diplomacy or military. The Panama Canal was at once a reason for this new rationale and also a tool for its expansion.

One key to this changing role in the hemisphere concerned the management of topography in a profoundly new way. Roosevelt initiated plans for a canal through Panama that would greatly speed trade with Asia. The United States had been seriously interested in an isthmian canal since the Clayton-Bulwar treaty with Great Britain in 1850. Interest in constructing a canal in Panama actually dated back to 1524 when Charles V of Spain ordered the first survey of a proposed canal route through the isthmus of Panama. Even at this early date, people already realized the advantages and commercial value of a route that would avoid sailing the 10,000-mile journey around Cape Horn at the tip of South America.

However, more than three centuries passed before the first construction effort was attempted. In the interim, private interest groups tried to stimulate canal construction throughout Central America. Beginning in 1880, a private French Company (La Compagnie Universelle du Canal Interoceanique) pushed for construction of a canal at the isthmus of Panama. The second Walker Commission, the U.S. Isthmian Canal Commission of 1899–1902, ordered by President McKinley, favored a Nicaragua route, as did both popular and official U.S. support (McCullough 1977, 15–20).

Instead, the United States helped Panama gain its independence and then signed the Bunau-Varilla Treaty, which granted the United States the right to build, operate, and defend a canal through Panama.

Technological innovation did not only concern how to make a massive waterway function in the Panamanian jungle; first, an American workforce needed to be able to survive heat and illnesses associated with the region. Specifically, the disease malaria, which was transmitted by mosquitoes in the jungle, proved to be the most difficult part of the project. Without chemical pesticides, Americans chose to first eradicate the habitat that bred the pests.

Knowing that the Anopheles mosquito cannot fly far without landing for at least a moment on some sort of vegetation, scientists ordered workers to clear a 200-yard-wide swath around areas in which people lived and worked. Sanitation teams led other efforts to eradicate the areas in which the mosquitoes bred, draining more than 100 square miles of swamp. To better manage areas of standing water, the teams created nearly 1,000 miles of earthen ditching, some 300 miles of concrete ditch, 200 miles of rock-filled trench, and almost 200 miles of tile drain. In addition to cutting hundreds of acres of wild vegetation, scientists also ordered the teams to spray standing water with thousands of gallons of oil and hatched and released thousands of minnows to eat the Anopheles larvae and bred spiders, ants, and lizards to feed on adult insects (McNeill 2001, 197).

When vegetation prevented the oil from spreading so that it could smother the larvae, barrels of poison (a mixture of carbolic acid, resin, and caustic soda) were applied monthly around the edges of many water pools and streams. These efforts reduced the mosquitoes' numbers sufficiently so that malaria incidents were significantly reduced in populated areas. Two hundred eleven employees died of malaria during fiscal year 1906–1907, declining significantly from a peak of 7.45 per 1,000 in 1906 to 0.30 per 1,000 in 1913. This achievement greatly increased American chances of canal-building success and also provided important new understandings about disease control.

The opening of the Panama Canal to world commerce on August 15, 1914 represented the realization of a dream of more than 400 years. Construction by the United States of the fifty-mile waterway bisecting the Republic of Panama was among the great technological innovations of the early twentieth century (McCullough 1977, 30–35).

Contemporary Issues and the Canal

The unique status of the Panama Canal has given scientists an excellent opportunity for study over the last century. The Smithsonian Institution's 1910 Panama Biological Survey provided baseline data for Panama Canal construction, even studying details such as fish migration.

Interestingly, scientists used this data in 2002 to return and conduct a comparative study. All of the original species found in each stream in Meek and Hildebrand's initial 1916 survey were still there. Over this century, they found that three fish species had colonized the Chagres River and five had colonized the Rio Grande. Therefore, both sides of the isthmus became more species rich as a result of the canal connection.

During the early twenty-first century, ownership of the canal is gradually shifting to Panama. This has also brought a new analysis of some other related long-term considerations of the canal's management. For instance, the greatest threat to the stability of the canal's ecosystem is the depletion of the rainforests surrounding the canal. The canal is run on fresh water, which only comes from Lake Gatun in the middle of the canal. The rainforests supply a continuous source of fresh water to the lake. Since the canal was built, a ten-mile watershed buffer on each side of the canal has been protected from depletion by the United States military. Thus far, no pumps have been needed to manage the canal's artificial watershed; however, this could change without continued management.

As the canal reverts to Panamanian ownership, some observers question whether or not the government can be effective at preserving the areas from squatters and the basic neglect that it allows in the rainforest throughout the nation.

This debate is made more complicated by discussion of expanding the canal. Lake Gatun does not have sufficient water to supply an expanded canal as well as increasing urban demands. Therefore, in addition to the construction of a new set of giant-sized locks, canal expansion proposals also call for the construction of a new dam on the Indio River, the river closest to Lake Gatun, with a reservoir size of at least 4,500 hectares (45 kilometers) and the subsequent diversion of water to supplement the canal. It has been estimated that upward of 3,500 people living in the region, mostly poor campesino farmers, will have to be relocated because their lands will be flooded by the new dam. Canal expansion may also lead to deforestation, massive excavation of lands, and significant environmental impacts associated with dam construction.

Other difficulties with the expansion include cost. Cost projections have been made for the Panama Canal expansion project ranging from $2 billion to upward of $12 billion. Critics have begun to vocally call for the government to do other things with its funds.

A rural peasants' organization called La Coordinadora Campesina Contra las Embalses (CCCE) is opposing construction of a new dam in the canal watershed and any plans involving the inundation of their lands. Among other activities, the CCCE has organized in communities throughout the new canal watershed and has staged numerous protests in the cities of Colón and Panama City.

As with many mega-project development initiatives, the proposed expansion of the Panama Canal presents formidable challenges that, depending on how they are managed, could either impede or contribute to Panama's future development and environmental sustainability. Viewed strictly as a strategy for economic development, the project raises many serious questions in regard to whether it is the most appropriate course of action, questions that thus far the Panama Canal Authority has yet to address.

Sources and Further Reading: McCullough, *Path Between the Seas*; McNeill, *Something New Under the Sun: An Environmental History of the Twentieth-Century World*.

PRIVATE DEVELOPMENT AND THE WESTERN NATIONAL PARKS

Time Period: 1860s to the present
In This Corner: Railroad companies, tourist development in the West, boosters for western expansion, political leaders
In the Other Corner: Native Americans, parties resisting expansion
Other Interested Parties: Conservationists, environmentalists
General Environmental Issue(s): Preservation, railroads, the West

By the twentieth century, most Americans agreed that the preservation of natural areas was one of the greatest symbols of our nation's stature and stability. Although most Americans came to support preservation, the actual use of these spaces often aroused confusion and disagreement. From the establishment of the first parks in the American West, one of the most persistent issues was the role of private corporations, sponsors, or donors. If this was a publicly owned facility, argued some Americans, private corporations should have no role in them. As funding to support the national parks became harder to find at the end of the twentieth century, new precedents were considered. It turns out, however, that these "new" ideas had been around as long as the parks.

The role of private corporations in the western parks began with the involvement of the railroad companies that dominated all aspects of life in the American West. The legendary origins of the American parks idea linked the appreciation of romantic beauty with the expansion of Anglo settlers into the West. Although the fireside conversation over Yellowstone's fate very likely occurred, this legend did not properly represent the important forces that were at work to spread the idea of national parks throughout the American West. Although some preservationists genuinely wanted to protect certain areas, the movement for parks was much more the product of the efforts of tourist development. In fact, throughout the history of America's national parks, significant debate has occurred over just how much of a role private interests, including corporations, should have in America's jewels.

Many conservationists find it a less than pleasing legacy: that the first national parks have less to do with ideas of preserving natural resources and much more to do with the development of the western United States. Possibly the most obvious demonstration of this is the role of railroad companies and financiers. In the case of Yellowstone, the Philadelphia financier Jay Cooke played the most important role. Cooke was excited by the results of the 1870 Washburn survey, and he set out in 1871 to create interest in Yellowstone that would profit the railroad.

From the idea's inception then, the national parks served corporate profit motives. Historian Richard Sellars wrote:

> The Northern Pacific continued to influence the Yellowstone Park proposal, beginning even before the 1870 expedition that gave birth to the campfire tradition. With their land grants stretching across the continent, American railroads were already seeking to establish monopolistic trade corridors. By preventing private land claims and limiting competition for tourism in Yellowstone, the federal reservation of the area served, in effect, as a huge appendage to the Northern Pacific's anticipated monopoly across the southern part of the Montana Territory. (Sellars 1997, 9–11)

Initially, some of this publicity came from a well-known 1870 article in *Scribner's* magazine titled "The Wonders of the Yellowstone," which was illustrated by Thomas Moran. In fact, it was support from *Scribner's* and a loan from Cooke that enabled Moran to join the expedition. Although Moran's interest grew from the aesthetic wonder of the Yellowstone landscape, he was unknowingly offering his talents to the service of the railroad just as the Union Pacific relied on the work of Albert Bierstadt.

Just as Cooke expected, Hayden's survey demonstrated that the Yellowstone region would not support the typical model of economic development. From the developer's perspective, the land was "worthless": devoid of valuable minerals, with erratic stands of timber, and too steep for grazing. Cooke determined that the area's future lay in tourism. Cooke wrote, "… Let Congress pass a bill reserving the Great Geyser Basin as a public park forever—just as it has reserved that far inferior wonder the Yosemite Valley and big trees" (Runte 1997, 58). Cooke and Hayden agreed that Yellowstone's showcase of natural wonders merited federal protection.

The Northern Pacific Railroad adopted Yellowstone's cause with an eye toward establishing a resort in the region. Cooke's responsibility was to create the publicity that would make Americans want to visit. Stories about the region had circulated in the east for years. Now, when the park bill was signed, interest in Yellowstone as the first national park would increase as newspapers and magazines published the news of the bill's passage. Cooke planned a grand public showing of Moran's mammoth canvas for June 1872.

Beyond Yellowstone

By the late 1800s, private interests had taken a leading role in making many western parks tourist destinations. Some of the initial structures within parks preceded park status. When the U.S. Army began their occupation of Yellowstone in 1886, for instance, they constructed Fort Yellowstone with building plans and details similar to those in their other military facilities. Yosemite, conversely, contained scattered cabin villages and the well-known Wawona Hotel, which lay in the foothills to the southwest of the Yosemite Valley.

Picking up on the model of Yellowstone, however, most western parks were developed by railroads. Such transportation companies realized that moving raw materials and products would remain their bread and butter, but tourism could greatly enhance their business. Therefore, when considering such park areas, it was in their best interests to create the facilities that would attract resort-class tourism. Before such developments, many parks hosted accommodations such as tent cabins or hotels designed out of the mainstreams of American architecture.

The first model came in 1903 with the Northern Pacific Railroad's construction of the Old Faithful Inn at Yellowstone National Park. Built near the geyser that bore its name, the Old Faithful Inn was constructed with a soaring rustic lobby and gnarled log balconies. The design flair was obviously intended to convey a frontier feeling but also used the prevalent building supply available in the region, logs. In its use of materials and grand spacing, the inn sought to impress its visitors that it was worthy of its awesome natural setting.

Competitiveness between railroad companies perpetuated this tradition in other parks. The Atchison, Topeka, and Santa Fe Railway also realized the enormous potential for using architecture as a marketing strategy. The Santa Fe set its sights on the south rim of the Grand Canyon, which had not yet been declared a national park. For its attraction, it built the grand El Tovar hotel, which opened in 1905. It was designed to house tourists for lengthy stays and provided access to a replica of a Hopi pueblo. Entertainment included native dancers and craft demonstrations. Architect Mary Colter added nearby attractions, including the medieval-themed Hermit's Rest and Lookout in 1914, and she added the more appropriately themed Indian Watchtower at Desert View in 1931.

Similar to developing rides in a theme park, railroads hired architects such as Colter and Robert Reamer to create spaces that served the purposes of tourism but also helped to package the locale's image and sense of place. In Yosemite National Park, railroad sponsors built the Sierra Club that included both LeConte Memorial Lodge (1903, rebuilt 1919) and Parsons Memorial Lodge (1915). Using native materials and distinctive settings, the architects became extremely creative in drawing influences from romantic and nostalgic styles. With the backing of the railroads, the structures ultimately connected with the style from New York's Great Adirondack Camps. Known as "rustic," the designs looked toward nature and allowed the surrounding landscape to influence their designs. These designs within existing parks became quite popular in the 1920s and then spread significantly through the work-relief programs of the Depression.

Conclusion: Private Interest in National Parks Today

Was this appropriate for corporate entities to determine the contents and development of areas set aside as national parks? Although this point was little debated in the early 1900s, it remains a bone of contention today. In the twenty-first century, when private development is allowed to occur in national parks, many preservationists, environmentalists, and others cry foul.

For instance, recently, there was significant national debate about the role that private companies should be allowed in existing national parks. As funding for basic park functions has become increasingly uncertain, many activities such as administering concessions within parks have been opened to private contractors. Congress established a Concession Program in the National Park Service (NPS) through the passage of the 1965 Concession Policy Act and then updated the policy in 1998 with a compromise bill. Regulations stipulated that private contractors can be used for activities that are deemed necessary and appropriate for public use and enjoyment of the unit of the National Park System in which they are located. By extension, contractors and their activities are expected to remain consistent to the highest practicable degree with the preservation and conservation of the resources and values of the unit.

In the twenty-first century, the concession program of the NPS administers more than 600 contracts that gross more than $800 million annually. During peak park seasons, approximately 25,000 civilians are employed in the hospitality industry contained within national parks.

Sources and Further Reading: Nash, *Wilderness and the American Mind*; Runte, *National Parks: The American Experience*; Sellars, *Preserving Nature in the National Parks: A History*.

MOUNT RAINIER AS AN EXCEPTIONAL WESTERN PARK

Time Period: 1880s to 1920s
In This Corner: Tourist developers in the Northwest, railroads, preservationists
In the Other Corner: Prospective park interests in other regions
Other Interested Parties: Urban boosters, political leaders
General Environmental Issue(s): National parks, the West, preservation

In the state of Washington, railroads were instrumental in the state's economic development. Similar to other western states, railroads also were involved in the development of Washington's best-known national park, Mount Rainier. Unlike other western parks, however, railroad companies did not take over the planning of the park's attractions and features. They played a crucial role in the evolution of the park; however, railroads were not instrumental in its internal appearance and organization. This distinction demonstrates a striking difference between Washington and other western states holding national parks.

Railroads Reach into the Wilderness

When Washington was still a territory in 1853, its first governor, Isaac I. Stevens, also headed the survey of the northern route for a transcontinental railroad. His survey was not accepted until 1862 when Congress chartered the Union Pacific and the Central Pacific

Mount Rainier, located in the Pacific Northwest, developed much differently from national parks in other areas. Library of Congress.

Railroads that would follow the route that Stevens had surveyed. In addition, the Northern Pacific was added in 1864.

To create its railroad, the Northern Pacific was given the largest land grant of any of the western railroads. The grant consisted of a strip of land 200 feet wide as a right of way, plus a swath of alternate sections along the railroad's entire length, ten miles to either side of the railroad in the states and twenty miles in the territories. Included in this land grant was a swath through the Cascade Mountain Range that included the future site of Mount Rainier National Park.

The company had financial difficulty from the start. For this reason, Congress allowed the company to mortgage the land to settlers before the line was completed. This risky endeavor proved tragic when, in 1873, the Northern Pacific Railroad Company went bankrupt without finishing its line. Washington residents demanded that the grant be rescinded so that the lands would be open for other uses. With the help of additional financing, the Northern Pacific continued its construction. The railroad fueled the ongoing competition between the state's two primary cities, Seattle and Tacoma, by openly considering each as its terminus. The final selection went to Tacoma, largely because its use brought the company an additional two million acres of timber land along the Columbia and Cowlitz valleys.

The state's connection to the outside world would not occur for another decade when, in 1881, railroad financier Henry Villard brought all the existing railroads in Oregon together with the Northern Pacific and the Oregon Railway and Navigation Company and completed the transcontinental connection through Idaho and Montana two years later.

Completion of the Northern Pacific in 1883 was followed in short order by the addition of three more transcontinental lines over the next two and a half decades. In the mid-1880s, the Union Pacific constructed a branch from its main transcontinental line known as the Oregon Short Line. This line went from Ogden, Utah, through southern Idaho and eastern Oregon to the Columbia River, and made the Union Pacific the second transcontinental line to reach the Pacific Northwest. In 1889, railroad magnate James J. Hill added another transcontinental line. The Great Northern used many local feeder lines rather than federal land grants to help finance it. Together, these lines brought great economic development to the state.

These rail lines combined with branch railroads to link urban areas, particularly Tacoma and Seattle, to coal fields in the Cascade foothills and also stimulated agricultural development. It was this coal that was used by steamships to transport the timber, the extraction of which composed one of the state's primary industries. These industries combined to stimulate interest in areas around the Cascades as development neared Mount Rainier. Farther south and nearer to Mount Rainier, Northern Pacific surveyors discovered coal beds in the Carbon River drainage in 1875 while surveying the Northern Pacific's eventual route over the Cascades via Stampede Pass.

As the Northern Pacific began construction of its main line over the Cascades in 1884–1885, it stimulated interest in the timber resources on the plateau between the White and Puyallup rivers and the clearing of bottomlands for agriculture. Branch railroads sprung from the trunk lines primarily to exploit the timber and coal resources in the Cascade foothills rather than to profit from passenger traffic to Mount Rainier. Inadvertently, however, this economic development created the state culture that prioritized development of Mount Rainier as a tourist destination.

A National Park for the Northwest

Created in 1899, Mount Rainier National Park was the nation's fifth national park, following only Yellowstone in 1872 and Yosemite, Sequoia, and General Grant National Parks in 1890. The unique situation of Mount Rainier made it the primary example to distinguish idealistic purposes of national parks from the more utilitarian functions of national forests, or "forest reserves" as they were known at the time.

When the national forest system was created in the 1890s, many conservationists assumed that the parks would be subsumed by the conservation impulse. Instead, the establishment of Mount Rainier National Park reaffirmed the nation's intent to set aside other distinct areas for their outstanding scenic and scientific value for the enjoyment of present and future generations.

Partly because of the timing of its establishment, Mount Rainier offered a unique example of the preservation impulse, particularly as it related to the involvement of private interests such as railroad companies. Historian Theodore Catton wrote the following:

> The legislation which established the park was in some ways precedent-setting. Mount Rainier was the first national park to be created from lands that were already set aside as forest reserves, forming a precedent for numerous national parks established in the twentieth century. Lands within the park boundary which had been granted to the Northern Pacific Railroad Company were reclaimed under the act in order to make the national park whole. This insistence on federal ownership of the land became another hallmark of American national parks in the twentieth century. In other respects, the act which created Mount Rainier National Park followed the Yellowstone prototype and reinforced an emerging pattern of national park legislation. (Catton 1996)

Unlike the other western parks, wrote Catton, this park "was very much a local affair." The primary push to establish the park came from local mountaineering clubs and regional business proponents. After 1900, tourists from Seattle and Tacoma increasingly traveled to the park by automobile; this was fairly unique compared with the other more remote, western parks.

Defining Mount Rainier for All

Mount Rainier grew as a unique cooperative effort among local preservationists in the early 1900s. However, this process began much earlier with the efforts of one family: the Longmires. After his own first ascent of Mount Rainier in 1883, James Longmire discovered the mineral springs and natural clearing that would later bear his name. With the idea of a resort in mind, by 1885, he had cleared a trail accessing the site and built a cabin nearby. By 1889, Longmire had some adventurous visitors using his guest cabins and two bathhouses. His advertisements in Tacoma newspapers advertised "Longmire's Medical Springs." He added a small, two-story hotel in 1890. Catton wrote the following:

> Most visitors to Mount Rainier in the 1890s and early 1900s were not content to end their trip at Longmire Springs or even the Nisqually Glacier; they wanted to break out of the timber and experience the panoramic views and wildflower-strewn meadows

for which Paradise was renowned. From Longmire Springs they took the trail built by Leonard Longmire and Henry Carter in 1892—paying a small toll for the privilege— and proceeded upwards, usually making rest stops at Carter Falls and Narada Falls on the way. Some made the magnificent timberline park their destination; others passed through Paradise Park on their way to the mountain's summit.

The national park designation in 1899 added to the local interest in Mount Rainier's natural features, ranging from mountain peaks to springs as well as to high country meadows. Before the park had any regulations in place or a ranger staff, local business people established tent camping and horse rental near the area known as Paradise Park as well as Longmire's hotel accommodations. Although the park began constructing an access road to the area in 1904, local entrepreneurs were well ahead of them. The emphasis for local boosters became to package the attractions of the park to appeal to tourists.

Interest in camping had been combined with the natural attractions in the area of Paradise Hotel since approximately 1895 when Charlie Comstock opened a coffee shop, which he called the Paradise Hotel, and Captain James Skinner established a tent camp nearby. In 1898, John L. Reese combined these businesses into one venture that he named Camp of the Clouds. Catton observed that the steady growth of Reese's camp provides a rough index to the increasing popularity of Paradise Park. In 1903, Reese had seven tents and a cook tent. Camp of the Clouds increased to thirty tents in 1906, forty tents in 1909, sixty tents in 1911, and seventy tents in 1914. By then, Reese also had moved the kitchen and dining room into two wood-frame buildings (Catton 1996).

Even after the establishment of the national park, Camp of the Clouds received very little supervision from the Department of the Interior. Beginning in 1902, Reese needed to obtain annual permits for the camp. Catton notes that park officials all held Reese in high regard and considered his camp to be an asset not only from the visitors' standpoint but also from the standpoint of protecting the resources. Relations with the national park officials did not go so well for others.

National Park Forms New Relationship with Railroads

The hotel built by Longmire, which became known as the Longmire Springs Hotel, was the most obvious holdover from previous tourism when Mount Rainier became a national park. In 1899, drawn-out negotiations began with the Longmire family to purchase their land and its attractions. Although these negotiations were problematic from the start, they took a more severe turn in 1905 when the government provided the Tacoma and Eastern Railroad Company with a five-year lease of two acres immediately south of the Longmire claim. Although the family worried about the hotel's impact on their business, they soon discovered that the increasing tourist travel to the park was more than even two hotels could satisfy.

Mount Rainier National Park's second hotel, called the National Park Inn, opened for business on July 1, 1906. Similar to arrangements in other western parks, this inn was built and operated by the Tacoma and Eastern Railroad Company. Unlike the Longmire hotel, the National Park Inn sought to appeal to upper-class travelers by providing elegant meals supplied by the commissary of the Chicago, Milwaukee, and Puget Sound Railway Company in Tacoma. One writer compared the two as follows:

There are two hotels here, one catering to plain and simple abundance, the other boasting a French chef. With a big bonfire on the grounds one appeals to the love of out of doors; while the other entertains the guest through the evening with music before the open fire-place in the social hall. There are the usual hotel accommodations, and also the well patronized tents in connection. Across from the hotels are the springs, iron and sulphur bubbling side by side. There are, of course, bath houses, and this is a resort to which many come for a day, or a week, or a month, and some never go further. It satisfies them. (Catton 1996)

With this unique arrangement, Mount Rainier offered a pattern of development different from most other Western parks. Unlike the Longmire Springs Hotel, the National Park Inn occupied leased ground and operated under contract with the government. The contract was the responsibility of the secretary of the interior. Although the railroads played a role in the park's development, it was obviously a lesser one than in many other sites.

Conclusion

Catton reports that the total number of visitors climbed from 1,786 in 1906 (the first year that the park staff kept an official count) to 7,754 in 1910, to 15,038 in 1914, and, finally, leaping to 34,814 in 1915. Mount Rainier's remarkable growth came with the overall economic assistance of railroads. However, the park had considerable popularity before the railroad's involvement in hotel and attraction management.

In short, Mount Rainier National Park appears not to have needed the benefaction of the railroads as did more remote parks. Mount Rainier, conversely, was located within a small number of miles of cities such as Seattle and Tacoma. These two cities would prove to be the real driving forces of Mount Rainier National Park's development, providing local initiative for road and hotel construction and a concentrated population of park users with an effective political voice.

The best-known tourist attraction at Mount Rainier, Paradise Inn, was built in 1916 without assistance from railroads. Financed by a group of local businessmen who lacked the seemingly unlimited funds of the powerful railroads, this inn was constructed on a smaller scale than most of the others but with a subtle yet powerful architectural presence that fulfilled its purpose as the hub of alpine climbing and one of the earliest ski resorts in the country.

Therefore, unlike other national park areas in the west, Mount Rainier did not grow from the desire of railroads to exploit tourist opportunity. The opportunity existed without the rail companies. Instead, the impact of the railroads was to bring settlers to Washington state by stimulating economic development. In Washington, the railroads' primary significance was in binding the Pacific Northwest more closely to the national economy. They not only brought a flood of new settlers to Washington and carried Washington's products to eastern markets, but their advent encouraged an influx of investment capital into the Pacific Northwest as well.

Sources and Further Reading: Catton, *Wonderland*, 1996; Runte, *National Parks: The American Experience*; Sellars, *Preserving Nature in the National Parks: A History*; National Park Service, *Mount Rainier*, http://www.nps.gov/archive/mora/adhi/adhit.htm.

MUCKRAKERS SET THE TONE FOR NATIONAL REFORM

Time Period: 1890s to 1920s
In This Corner: Social reformers, journalists
In the Other Corner: Business and government leaders
Other Interested Parties: American public, Progressive leaders
General Environmental Issue(s): Social reform

The Gilded Age of the late 1800s concentrated wealth with a small number of the super rich. Few Americans questioned the ethics with which wealthy industrialists managed their corporations or acquired their wealth. In many factories, workers had little if any say in their work environment and its safety. Rivers, forests, wildlife, and land resources were used without restraint or limits. In short, few checks and balances acted to limit development of any type. Around the turn of the century, a few brave journalists demanded that the nation reevaluate its view of progress.

This new breed of writer-activists were known as muckrakers. Muckrakers demanded action to counter the social ills that they explored in their writing. Thanks to the interest of a supportive American public that had become more interested in reform, the muckrakers helped to change the nature of American ethics regarding business, the environment, and human workers.

In at least two cases, the writers received swift and dramatic action. Ida Tarbell was born in the oil regions of northwestern Pennsylvania. She watched as her father's oil tank business failed as a result of the unfair practices of one company in particular, the Standard Oil Trust. To Tarbell, the efforts of John D. Rockefeller to place his company in control of the nation's energy, in the early 1870s, formed a turning point for the nation. In the episode known as "the Oil War," Rockefeller sought to dominate petroleum markets worldwide.

Initially, small producers banded together and defeated his efforts, but Rockefeller built his Standard Oil Trust from the ashes of this initial setback. By the end of the 1870s, Standard controlled approximately 80 percent of the world's oil supply. By dominating transportation and refining, Rockefeller dominated the market.

An alarming picture of "Big Oil" took shape as Standard used ruthless business practices, including rollbacks and insider pricing, to squeeze out its competitors. Tarbell, who had become a successful journalist and editor, went to work to show Americans what Rockefeller was really doing. Across the nation, readers awaited each installment of the story serialized in nineteen installments by *McClure's* between 1902 and 1904. Although Tarbell teemed with bitterness for what Rockefeller had done to the private businessmen of Pennsylvania's oil fields, she appreciated the value of allowing the details of the story to express her point of view. The articles were compiled into the *History of the Standard Oil Company* (1904), and President Theodore Roosevelt, who succeeded McKinley in 1901, used the public furor to pursue a federal investigation. Tarbell's investigation inspired new efforts to enforce antitrust laws. In 1911, the Supreme Court ordered the dissolution of Standard Oil.

Upton Sinclair also achieved significant changes through his writing. Although Sinclair's most famous book, *The Jungle*, was published after Tarbell's, the book's topic gained almost immediate action. *The Jungle* described the meat-packing industry of Chicago, Illinois. Sinclair made the industry the setting for the tale of a Lithuanian immigrant, Jurgis Rudkus.

Although Sinclair made the difficult experience of new immigrants the primary theme of the narrative, readers were drawn to the squalor-filled setting in which their meat and other food was prepared. In the end, *The Jungle* exposed to a wide audience the horrors of the Chicago meat-packing plants and the immigrants who were worked to death in them.

The meatpackers protested Sinclair's characterization: "We regret that if you feel confident the report of your commissioners is true, you did not make the investigation more thorough, so that the American public and the world at large might know that there are packers and that if some are unworthy of public confidence, there are others whose methods are above board and whose goods are of such high quality as to be a credit to the American nation." However, the public reacted so strongly that the government launched an investigation of the meatpacking plants of Chicago. Based on their findings, Roosevelt pushed through the 1906 Pure Food and Drug legislation that placed regulatory authority with the federal government. It was one of the first moments in which progressivism placed the responsibility for protecting citizens' everyday lives on the federal government.

Throughout the twentieth century, this expectation consistently increased. As president, Roosevelt held the ideal that the government should be the great arbiter of the conflicting economic forces in the nation, especially between capital and labor, guaranteeing justice to each and dispensing favors to none. A change emerged in very basic ways that many Americans perceived the nature of economic relationships in American life. This renewed interest in social equality, of course, would grow further with the emergence of the American middle class during the twentieth century.

Sources and Further Reading: Chernow, *Titan: The Life of John D. Rockefeller*; Tarbell, *All in the Day's Work: An Autobiography*.

ALICE HAMILTON CONNECTS SOCIAL REFORM WITH HUMAN HEALTH

Time Period: 1910s to 1930s
In This Corner: Alice Hamilton and social reformers
In the Other Corner: Business leaders, government leaders
Other Interested Parties: Reformers, American public
General Environmental Issue(s): Public health, social reform

The critiques of social reformers in the early 1900s often contained much more passion than substance. Their criticism of environmental degradation most often concerned the appearance or other physical manifestations of industrial pollution. By the 1910s, physicians and scientists began to connect such critiques with verifiable medical findings. Very quickly it became clear to many Americans that industrial development claimed a serious impact on human health. It is remarkable to learn that such a dramatic alteration to American ideas emanated from the actions and efforts of very few individuals. In fact, our expectations of health protection can be traced to one individual: Alice Hamilton.

Although President Theodore Roosevelt and his progressive allies worked to change laws to help Americans in general, other progressive reformers used medical skills at the grassroots level to help urban Americans. Hamilton is one of the best known. She established the field of occupational medicine, served as the first woman professor at Harvard Medical

School, and was the first woman to receive the Lasker Award in public health (Opie 1998, 286).

After taking her first academic appointment in 1897, Hamilton was appointed professor of pathology at the Women's Medical School of Northwestern University in Evanston, Illinois, and, in 1902, she accepted a position as a bacteriologist at the Memorial Institute for Infectious Diseases in Chicago. Dr. Hamilton became familiar with Jane Addams's Hull House, where she began to apply her medical knowledge to the needs of the urban poor. During her stay at the Hull House, she established medical education classes and a well-baby clinic.

During the typhoid fever epidemic in Chicago in 1902, it was Hamilton who connected improper sewage disposal and the role of flies in transmitting the disease. Based on her findings, the Chicago Health Department was entirely reorganized. She then noted that the health problems of many of the immigrant poor were attributable to unsafe conditions and noxious chemicals, especially lead dust, to which they were being exposed in the course of their employment. In 1910, the state of Illinois created the world's first Commission on Occupational Disease, with Hamilton as its director. The commission led to new laws and regulations for Illinois and contributed to the workers' rights movement in the United States. They introduced a new notion that workers were entitled to compensation for health impairment and injuries sustained on the job.

Because of her work in Illinois, Hamilton was asked by the U.S. commissioner of labor to replicate her research on a national level. She noted hazards from exposure to lead, arsenic, mercury, and organic solvents, as well as to radium, which was used to manufacture watch dials. Although many reformers argued that the industrial environment had become unsafe, Hamilton was one of the first scientists to prove it. Industrialists could do little but agree with her scientific findings.

Source: Opie, *Nature's Nation.*

FEDERAL EFFORTS TO REGULATE HUMAN HEALTH

Time Period: 1880s to the present
In This Corner: Reformers, proponents of federal involvement in American life, members of scientific community
In the Other Corner: Interests resistant to federal involvement, manufacturers
Other Interested Parties: American consumers
General Environmental Issue(s): Health, diet, disease

It was one thing for Americans to slowly decide that their federal government needed to be more involved in stimulating development by financing certain large-scale transportation projects; it was entirely another case for the federal government to actively oversee public health. As new scientific understanding connected human health to outside stimulants, including diet, information and action were slow to come to the public.

Reformers began using the federal government to get the word out on public health. It would, however, require generations of change.

Federalizing Health and Diet

The connection between science and the health of the American public evolved late in the nineteenth century. The first federal efforts at monitoring health focused on one of the most important sectors of the nation's population in the late 1700s: seamen. From 1798 to 1902, when national security and global trade relied on the stability of this segment of the population, the Marine Hospital Service (MHS)—and from 1902 to 1912, the Public Health and Marine Hospital Service—made sure sailors had the best healthcare the nation could muster. Members of the maritime industries, of course, are not government employees; however, their well-being had a direct impact on the nation's economy. Seamen traveled widely and often became sick at sea or in foreign nations. Therefore, their healthcare became a national problem.

In 1798, Congress established a network of marine hospitals in port cities around the world. Here, doctors cared for these sick and disabled seamen in facilities financed by taxing American seamen 20¢ per month. This payment was one of the nation's first direct taxes, as well as its first medical insurance program.

The Progressive Era of the late 1800s and early 1900s brought new calls from the public for the federal government to help improve the American standard of living. Of primary concern were well-known epidemics of contagious diseases, such as smallpox, yellow fever, and cholera that had caused many deaths worldwide. Congress's interest in enacting laws to stop the importation and spread of such diseases resulted in a significant expansion of the responsibilities of the MHS.

The increase in passenger travel by steamship, for instance, meant that the MHS was responsible for supervising national quarantine, including ship inspection and disinfection, the medical inspection of immigrants, the prevention of interstate spread of disease, and general investigations in the field of public health, including yellow fever epidemics.

The effort to diagnose and treat infectious disease required the application of new science. To help inspect and diagnose passengers of incoming ships, the MHS established a small bacteriology laboratory in 1887. The Hygienic Laboratory was first located at the marine hospital on Staten Island, New York, and then later moved to Washington, DC, where it became ultimately the National Institutes of Health.

In 1902, Congress passed an act to expand the scientific research work at the Hygienic Laboratory and to provide it with reliable funding. In an effort to spread the impact of good health practices, the bill also required the surgeon general to organize annual conferences of local and national health officials. To reflect these new responsibilities, the name of the MHS became the Public Health and Marine Health Service. Finally, the Public Health Service (PHS) was established in 1912, just in time to confront one of the nation's most serious health issues (U.S. Department of Health and Human Service, *Health*).

Influenza Pandemic Mobilizes Federal Action

Increasing interactions because of global trade increased the occurrence and awareness of disease. Although World War I was a global tragedy, it also contributed to one of the most significant pandemics in world history. The influenza (or the flu) pandemic of 1918–1919 killed between twenty and forty million people. It has been cited as the most devastating epidemic in recorded world history. More people died of influenza in a single year than in four years of the black death bubonic plague from 1347 to 1351. Known as "Spanish Flu" or "La Grippe," the influenza of 1918–1919 was a global health disaster (Crosby 1990, 23–29).

Ultimately, influenza in the fall of 1918 infected approximately one-fifth of the world's population. The flu usually most affects the elderly and the young, but this strand hit hardest on people aged twenty to forty. It infected 28 percent of all Americans, killing an estimated 675,000. The flu's connection to World War I was clear: of the U.S. soldiers who died in Europe, half of them fell to the influenza virus, an estimated 43,000 servicemen. The *Journal of the American Medical Association* wrote in 1918, "The effect of the influenza epidemic was so severe that the average life span in the U.S. was depressed by 10 years. The death rate for 15 to 34-year-olds of influenza and pneumonia was 20 times higher in 1918 than in previous years" (Stanford University, *The Influenza Pandemic of 1918*).

As the influenza pandemic circled the globe, there were very few regions that did not feel its effects. Primarily, the spread of the flu followed the path of its human carriers along trade routes and shipping lines. Outbreaks swept through North America, Europe, Asia, Africa, Brazil, and the South Pacific in 1919. In India, the mortality rate was extremely high at around fifty deaths from influenza per 1,000 people.

Ports such as Boston, where materials were shipped out to the battlefront, were the most heavily affected American cities. The flu first arrived in Boston in September 1918. Then, however, the disease became more diffuse when soldiers brought the virus with them to those they contacted. In October 1918 alone, the virus killed almost 200,000. As people celebrated the end of the war on Armistice Day with parades and large parties, many U.S. cities suffered from public health emergencies. Throughout the winter, millions became infected and thousands died.

In one effort to stall the spread of the disease, public health departments distributed gauze masks to be worn in public. Basic parts of everyday life changed: places of business closed, funerals were limited to fifteen minutes to fit in more services, and some towns and railroads required a signed certificate to enter. Bodies piled up throughout the nation. Besides the lack of healthcare workers and medical supplies, there was a shortage of coffins, morticians, and gravediggers. The public emergency very closely resembled the black death of the Middle Ages (Crosby 1990, 15–25).

Philadelphia, the hardest hit of all U.S. cities, was struck in October 1918. By the end of the first week, 700 residents were dead; 2,600 died by October 12, and the death toll continued to rise. Although no single group or neighborhood was entirely spared, immigrant neighborhoods, where basic sanitation and overall health were poorest, were the hardest hit. By November 2, the death toll in Philadelphia from the flu reached a staggering 12,162 people.

Conclusion: Creating a New Model for Public Health

Through this public health disaster, Americans learned a valuable lesson: the federal government needed to be used proactively to assist in preventing such outbreaks and help ensure and educate Americans on better health practices.

Sources and Further Reading: Crosby, *America's Forgotten Pandemic: The Influenza of 1918*; Opie, *Nature's Nation*; Stanford University, *The Influenza Pandemic of 1918*, http://virus.stanford.edu/uda; U.S. Department of Health and Human Services, *Health*.

SPINDLETOP: PETROLEUM SUPPLY CREATES NEW OPPORTUNITIES

Time Period: 1901–1920
In This Corner: Petroleum developers searching for new supplies
In the Other Corner: Alternative energy sources
Other Interested Parties: American consumers
General Environmental Issue(s): Energy management, petroleum

The age of petroleum required a massive change in the supply of oil. Although new drilling technologies helped to increase supply, entire new regions were required to be developed. By 1900, companies such as Standard Oil sought to develop new fields all over the world. In terms of the domestic supply of crude, however, the most significant breakthrough came in Texas. With one 1901 strike, the limited supply of crude oil became a thing of America's past. It is no coincidence, then, that the century that followed was powered by petroleum.

This important moment came in East Texas where, without warning, the level plains near Beaumont abruptly give way to a lone, rounded hill before returning to flatness. Geologists call these abrupt rises in the land "domes" because hollow caverns lie beneath. Over time, layers of rock rise to a common apex and create a spacious reservoir underneath. Often, salt forms in these empty, geological bubbles, creating a salt dome. Over millions of years, water or other material might fill the reservoir. At least, that was Patillo Higgins's idea in eastern Texas during the 1890s.

Higgins and very few others imagined such caverns as natural treasure houses. Higgins's intrigue grew with one dome-shaped hill in southeast Texas. Known as Spindletop, this salt dome, with Higgins's help, would change human existence.

Texas had not yet been identified as an oil producer. Well-known oil country lay in the eastern United States, particularly western Pennsylvania. Titusville, Pennsylvania introduced Americans to massive amounts of crude oil for the first time in 1859. By the 1890s, petroleum-derived kerosene had become the world's most popular fuel for lighting. Thomas Edison's experiments with electric lighting placed petroleum's future in doubt; however, petroleum still stimulated boom wherever it was found. But in Texas? Every geologist who inspected the "Big Hill" at Spindletop told Higgins that he was a fool.

With growing frustration, Higgins placed a magazine advertisement requesting someone to drill on the Big Hill. The only response came from Captain Anthony F. Lucas, who had prospected domes in Texas for salt and sulfur. On January 10, 1901, Lucas's drilling crew, known as "roughnecks" for the hard physical labor of drilling pipe deep into the earth, found mud bubbling in their drill hole. The sound of a cannon turned to a roar, and suddenly oil spurted out of the hole. The Lucas geyser, found at a depth of 1,139 feet, blew a stream of oil over one hundred feet high until it was capped nine days later. During this period, the well flowed an estimated 100,000 barrels a day, well beyond any flows witnessed previously. Lucas finally gained control of the geyser on January 19. By this point, a huge pool of oil surrounded it. Throngs of oilmen, speculators, and onlookers came and transformed the city of Beaumont into Texas's first oil boomtown.

The flow from this well, named Lucas 1, was unlike anything witnessed before in the petroleum industry: 75,000 barrels per day. As news of the gusher reached around the world, the Texas oil boom was on. Land sold for wildly erratic prices. After a few months, more than 200 wells had been sunk on the Big Hill. By the end of 1901, an estimated $235 million had been invested in oil in Texas. This was the new frontier of oil; however, the industry's scale had changed completely at Spindletop. Unimaginable amounts of petroleum, and the raw energy that it contained, were now available at a low enough price to become part of every American's life.

It was the businessmen who then took over after Higgins and other petroleum wildcatters. Rockefeller's Standard Oil and other oil executives managed to export petroleum technology and exploited supplies worldwide. The modern-day oil company became a version of the joint-stock companies that had been created by European royalty to explore the world during the period of mercantilism of the 1600s. Now, however, behemoth oil companies were transnational corporations, largely unregulated and seeking one thing: crude oil. Wherever "black gold" was found, oil tycoons set the wheels of development in motion. Boomtowns modeled after those in the Pennsylvania oil fields could suddenly pop up in Azerbaijan, Borneo, or Sumatra (Yergin 1993, 117–19).

In the United States, the supply of oil was a far cry from the domestic supplies of a century later. As East Texas gushers created uncontrollable lakes of crude, no one considered the idea of shortage or conservation. Even the idea of importing oil was a foreign concept. California and Texas flooded the market with more than enough crude oil, and then, from nearly nowhere, Oklahoma emerged in 1905 to become the nation's greatest oil producer.

Texas recaptured the top spot in 1928 and has been the nation's largest producer of petroleum ever since, and its image has been inextricably linked with oil in popular culture thanks to films such as *Giant* (1960) and *Boomtown* (1949), television programs such as *Dallas*, and oil executives-turned-politicians such as George Bush, George W. Bush, and Dick Cheney.

Sources and Further Reading: Black, *Petrolia: The Landscape of America's First Oil Boom*; Olien, *Oil and Ideology: The American Oil Industry, 1859–1945*; Yergin, *The Prize: The Epic Quest for Oil, Money & Power.*

THE NEW NIAGARA AND AMERICA'S NEW PRESERVATION ETHIC

Time Period: 1890–1920
In This Corner: Preservationists
In the Other Corner: Industrial developers
Other Interested Parties: Park officials, regional residents
General Environmental Issue(s): Preservation, parks, industrialization

America learned its "preservation ethic" in pieces at a variety of locales. Overall, the role of nature at the close of the nineteenth century was a fascinatingly complex cacophony of mixed impulses. A single site could force Americans to measure and define their commitment to viewing nature as an instrument for their use and economic growth or, on the contrary, as an exceptional place to be restricted from change and development. In the 1890s, idealism slowly gained ground on the practicality with which many Americans viewed their natural surroundings.

The shift to considering nature on any level other than economic value marked a profound shift in human society, particularly one constructed on capitalism. However, the stark inconsistency of the American view of nature at times presented a bizarre dichotomy. Possibly no site better exemplifies this complexity than Niagara. During the 1800s, the great waterfall served as a demonstration of America's changing sensibilities.

Niagara Falls began the century as the young nation's primary tourist attraction, a dramatic example of the natural sublime, but the relentless spirit of industrialization had also permeated Niagara. By the 1880s, the motive of power was not only attracting milling interests; in 1889, the Cataract Construction Corporation announced plans to make the falls the dynamo for an entire industrial region.

Involving sponsorship from the Edison Electric Company and Westinghouse (among others), the company in 1893 committed its resources to making the falls the hub for alternating current electricity in the region. As construction moved forward, Niagara's tourists failed to see the irony. The role of "electric mecca" made Niagara Falls even more attractive as a tourist destination. As historian William Irwin wrote, "Confident that the New Niagara did not usurp nature, tourist promoters and power developers alike championed the engineer's additions to the Niagara landscape" (Irwin 1996, 114).

The main emphasis of this interest was an enormous power tunnel that was 6,700 feet long and 21 feet high. The tunnel rerouted a portion of the water through turbines and dropped it out just below the base of the falls. The tunnel's outlet soon became a stop on the tours carried by the *Maid of the Mist* tour boats. At the other end of the tunnel, of course, was the powerhouse. Built to be a comparable monument to the falls, the powerhouse was designed and built by the famous architectural firm McKim, Mead, and White. Inside, the powerhouse held the state-of-the-art technology of the age: the electric dynamo.

A symbol of technological success in the popular imagination, the "New Niagara" became the backdrop for futuristic waxing, including H. G. Wells' science fiction, Buck Rogers' adventures, and many others. The electricity, of course, also attracted cutting-edge manufacturing, including the model factory of the Shredded Wheat Company. Referred to as the Natural Food Company or the Natural Food Conservatory, the cereal factory attracted more than 100,000 visitors per year. The use of the term "natural" was meant to denote the healthfulness of the product. However, it certainly contributed to this image that the power for the factory came from the mighty, natural forces of Niagara.

With such rapid development at Niagara, it is not surprising that the contrary impulse was not far behind the industrialists. In 1879, Frederick Law Olmsted, the nation's leading landscape architect and preservationist, was commissioned to prepare a special report for the New York State Survey of the area around Niagara Falls. At that point, tourism had not been prioritized by developers and planners. In fact, only a small portion of the falls were visible to the tourist. Olmsted felt that the spectacle of the falls—and its soothing power—needed to be made accessible to more visitors.

Although it is most appropriately grouped under the preservation impulse, Olmsted's plan made tourist satisfaction a priority. With that priority in place, industrial blight could not be tolerated in specific scenic areas. Olmsted's plan focused on Goat Island, an island that separated the Canadian and U.S. Falls. When Olmsted and Vaux received the commission to work on the project, they set out to purchase Goat Island as well as neighboring Bath Island, which had a small factory on it. They returned the islands to a natural state and added footpaths for visitors. In addition, they took out an amusement park on the mainland and replaced it with a reception hall and picnic grounds.

Following Olmsted's efforts, Niagara Falls became the focus of J. Horace McFarland, a leader of the American horticulture and preservation movements in the early 1900s. Eventually, his efforts grew into early efforts at city planning and suburbanization that scholars refer to as the City Beautiful movement. McFarland served as a vigorous leader in formulating and disseminating the ideal of preserving, not merely conserving, natural resources, particularly in the northeastern United States.

As a specialist on horticulture and the propagation of roses, McFarland became editor of the "Beautiful America" column in the *Ladies' Home Journal* in 1904. In this same year, the American League for Civic Improvement merged with the American Park and Outdoor Society, and McFarland was elected its first president. Immediately, he set out to use his new-found influence to preserve the Niagara Falls from further industrial exploitation. After Olmsted's work, developers set out to construct more dams along the river but were thwarted by preservationists who protected the falls proper as a state park in 1885. The developers retaliated with plans to construct huge conduits that would capture the river and divert its flow around the cataract to powerhouses set away from the Niagara Gorge.

McFarland argued publicly against such ideas. He pointed out that the cataract would be ruined and that previous efforts at preservation would be rendered meaningless if the grand power development moved forward. Harnessing the forces of the new organization as well as a general upper-class interest in nature preservation, McFarland pressed the Roosevelt administration to ensure permanent protection for Niagara Falls.

McFarland's primary role was to ensure that the controversy over diverting the river gained national attention. From his office in Harrisburg, Pennsylvania, McFarland alerted scores of government, civic, and business leaders to the impending tragedy of a waterless Niagara. In October 1906, he wrote *Shall We Make a Coal-Pile of Niagara?* In this essay, McFarland urged that "Every American—nay, every world citizen, should see Niagara many times, for the welfare of his soul and the perpetual memory of a great work of God." He continued, "The engineers calmly agree that Niagara Falls will, in a very few years, be but a memory. A memory of what? Of grandeur, beauty and natural majesty unexcelled anywhere on earth, sacrificed unnecessarily for the gain of a few!" The article included illustrations of before and after, and McFarland led the reader to comprehend the outcome: "The words might well be emblazoned in letters of fire across the shamelessly-uncovered bluff of the American Fall: 'The Monument of America's Shame and Greed'" (Runte 1997, 86–87).

McFarland's perspective was unique because of his understanding of horticulture and aesthetic beauty. However, he also was a successful and realistic businessman. He was not above appealing to America's pocketbook to rationalize preservation. Based on tourism alone, he wrote, the destruction of Niagara was truly "folly unbounded. To the railroads of the country and to the town of Niagara Falls visitors from all over the world pay upward of $20 million each year—a sure annual dividend upon Nature's freely-bestowed capital of wonders." Extended to the nation at large, the figure "would thus stand at over three hundred million dollars," he estimated, but at Niagara "all this will be wiped out, for who will care to see a bare cliff and a mass of factories, a maze of wires and tunnels and wheels and generators?" (Runte 1997, 87–89).

Ultimately, McFarland's successful campaign culminated in the signing of a treaty with Great Britain in 1909 safeguarding the falls, whereby the control of Niagara Falls was taken away from the state of New York and the Province of Ontario and placed under the joint authority of the International Niagara Falls Control Board.

However, the questions brought up at Niagara were not just about the waterfall. They were about the American view of nature and which ethic would be given priority. After a nineteenth century dominated by an exploitative view of natural resources, Americans showed themselves to be receptive to the preservation ethic seen at Niagara in the twentieth century. The efforts of conservationists and preservationists formed an important part of the foundation for the environmental movement of the twentieth century. Historian Sam Hays describes these social changes in this manner:

> The broader significance of the conservation movement stemmed from the role it played in the transformation of a decentralized, nontechnical, loosely organized society, where waste and inefficiency ran rampant, into a highly organized, technical and centrally planned and directed social organization which could meet a complex world with efficiency and purpose. (Hays 1999, 265)

Sources and Further Reading: Hays, *Conservation and the Gospel of Efficiency*; Irwin, *The New Niagara*; Nash, *Wilderness and the American Mind*; Runte, *National Parks: The American Experience*.

THE LACEY ACT CREATES THE FIRST FEDERAL LAW FOR WILDLIFE CONSERVATION

Time Period: 1890s to 1910
In This Corner: Commercial hunters, settlers
In the Other Corner: Sport hunters, upper-class female reformers
Other Interested Parties: Progressive politicians
General Environmental Issue(s): Species management, hunting

One of the first examples of the efforts to apply the emerging "conservation culture" of the 1890s was, quite literally, for the birds. In 1894, George "Bird" Grinnell used the pages of *Forest and Stream* to announce that it was time to "de-commercialize" the killing of wild game. His call for the conservation of bird populations evolved into the Federal Migratory Bird Act and Treaty, otherwise known as the Lacey Bill. This prohibition on the interstate shipment of wild species killed in violation of state laws marked the first federal conservation measure and signaled a significant shift in American environmentalism.

The call for such legislative action came from the extensive impact of American hunters. By the late 1800s, commercial hunting for fowl to adorn the menus of elegant restaurants and wealthy homes or to provide plumage—feathers—to adorn fashionable women's hats of the era had severely impacted the population of many species. When hunters took primarily for their own use, the impact was much less significant. When they began selling in global markets, however, there was no limit to what consumers would buy. Therefore, commercial hunters knew no restraint.

The best known of these species, of course, are the passenger pigeons. Commercial hunters throughout the Midwest sought to catch vast portions of the immense flocks of these birds that frequented the region. Using nets, commercial hunters often caught entire flocks as if they were fish in the ocean. By the close of the century, passenger pigeons neared extinction. Sought for plumage, water fowl also saw a severe drop in their numbers. For instance, populations of the Eskimo curlew and the snowy egret had been reduced to mere remnants of their historical populations.

Popular support for regulation came from the upper-class women who often had purchased the goods made from the birds. Grinnell and Audubon clubs helped to alert upper-class women of the dramatic consequences of their decadences. Particularly because of their influence in society, these female activists were able to bring about reform on behalf of the nation's bird populations.

Introduced by John Lacey, a congressman from Iowa and also a member of the Boone and Crockett Club, the Lacey Bill limited the trade of illegal game. Passed into law in May 1900, the Lacey Bill also contained wording that incorporated Yellowstone Park into the judicial district of Wyoming, thereby applying its laws to the game within the park. Although the bill emphasized enforcement, there was little manpower in place to carry out its initiatives. However, the bill marked a new understanding of natural resources and the important role that the federal government could have in helping to conserve them. In particular, the Lacey Act began an effort that continues today to regulate and monitor migratory populations—in other words, populations that cross international borders and, therefore, require international cooperation.

Ultimately, the Lacey Act proved ineffective at stopping these interstate shipments. In an effort to improve the regulation's ability to be enforced, Congress passed the Weeks-McLean Law in 1913. This dynamic area of law continued to shift; however, it ultimately evolved to include the "Hunting Stamp Act," which used registration fees by hunters to help finance efforts at species regulation and monitoring.

Sources and Further Reading: Hays, *Conservation and the Gospel of Efficiency*; Nash, *Wilderness and the American Mind*; Price, *Flight Maps*; Reiger, *American Sportsmen and the Origins of Conservation*; Steinberg, *Down to Earth*; National Conservation Training Center, *Origins of the U.S. Fish and Wildlife Service*, http://training.fws.gov/history/origins.html; U.S. Fish and Wildlife Service, *A Guide to the Laws and Treaties of the United States for Protecting Migratory Birds*, http://www.fws.gov/migratorybirds/intrnltr/treatlaw.html.

BOONE AND CROCKETT CLUB USES VIRILITY TO ATTRACT ENVIRONMENTAL SUPPORT

Time Period: 1890s to 1920
In This Corner: Upper-class boys, wealthy conservationists, progressive reformers
In the Other Corner: Supporters of traditional ways of training children for work
Other Interested Parties: Outdoors enthusiasts
General Environmental Issue(s): Early environmentalism, conservation, outdoors

Growing up in nineteenth-century America generally involved training for a specific occupation from a very young age. Apprenticeships were often used for training youths to be tradesmen, and many occupations involved highly specialized skills. In addition, most young boys spent a great deal of time learning about the outdoors from elders who supported themselves by hunting, fishing, and interacting with natural resources.

Industrialization concentrated many children in cities. By the late 1800s, many wealthy urban families sought to manufacture methods for artificially providing their children, particularly boys, with these practical skills and outdoor experience. During the 1890s, this cultural interest galvanized in the form of a few organizations for boys that were particularly connected with hunting.

Sport hunters began to seek ways to differentiate themselves from their commercial brethren in the late 1800s. Particularly in upper-class circles, sport hunting became linked with other examples of desirable masculinity—a forerunner of contemporary ideas of machismo—by the 1890s. Eventually, this connection became particularly important for the evolution of American ideas of conservation. To do so, however, this version of masculinity needed to be disseminated through cultural means. One of the most important of these was an organization called the Boone and Crockett Club, which was a forerunner of the Boy Scouts of America (BSA).

Once again, a primary arbiter of this new sensibility was the journalist George Bird Grinnell, who was instrumental in founding the Audubon Society. Grinnell became close friends with Theodore Roosevelt during the 1880s, and together they would begin a group that marked an entirely new idea in Americans' relationship with nature. In *Forest and Stream* editorials, Grinnell wrote that he desired "a live [national] association of men bound together by their interest in game and fish, to take active charge of all matters pertaining to the enactment and carrying out of laws on the subject" (Reiger 1988, 118). Roosevelt shared this opinion and invited many of his sportsmen friends to a meeting in Manhattan at which he suggested the formation of the group.

As the group took shape, it was designed to influence boys regarding the priorities to which Roosevelt and others were committed. Often, meetings brought men and boys together in active recreation. The Boone and Crockett Club had five organizing principles: to promote "manly" sport with the rifle; to promote travel to wild portions of the country; to work for preservation of large game; to stimulate interest in natural history and animal habits; and to stimulate the exchange of ideas among members. As a forerunner of the BSA, the Boone and Crockett Club, obviously, placed a specific stress on hunting.

Extending from these elite New York high-society roots, the Boone and Crockett Club spread its influence nationally, although almost always limited to upper-class portions of society. In the late 1890s, the organization began publishing books on conservation and hunting. The books, of course, also disseminated the sportsmen's code at the core of the organization. Most important, the code emphasized the need to use whatever was killed (Nash 1982, 152–53).

The Boone and Crockett Club's ethical foundations were exhibited in the early 1890s when the group became involved with Yellowstone National Park. Critical of the mismanagement of game at the site, Boone and Crockett Club members considered taking over management of the park themselves (Reiger 1988, 125). The core of the issue was the effort to build a railroad through the park. Boone and Crockett members feared that this might increase the wholesale slaughter of game animals in the pristine area.

This was just one example of the crossover between hunting and conservation. The efforts of hunters contributed to some of the earliest efforts at game management and land preservation. It should be pointed out, however, that the Boone and Crockett Club was part of an unapologetically sexist effort to instill virility and outdoor activity in males, whereas females were normally assumed to be uninterested in such activities.

Sources and Further Reading: Hays, *Conservation and the Gospel of Efficiency*; Nash, *Wilderness and the American Mind*; Price, *Flight Maps*; Reiger, *American Sportsmen and the Origins of Conservation*; Steinberg, *Down to Earth*.

THE BOY SCOUTS OF AMERICA INVOLVE YOUNG MEN IN OUTDOORS

Time Period: 1890s to the present
In This Corner: Upper-class boys, wealthy conservationists, progressive reformers
In the Other Corner: Supporters of traditional ways of training children for work
Other Interested Parties: Outdoors enthusiasts
General Environmental Issue(s): Early environmentalism, conservation, outdoors

The Gilded Age of the nineteenth century brought to wealthy Americans a genuine interest in rustic living and the outdoors. This was clearly demonstrated by the public's interest in places such as the Adirondacks and the far West. Additionally, the popularity of the Boone and Crockett Club demonstrated that many Americans considered the influence of the outdoors to be an essential part of young people's lives. Rapidly, many Americans were changing their ideas about what constituted the proper preparation for their children, particularly for boys. Cultural and educational mechanisms took shape to create new opportunities for children in hopes that they could meet expectations.

Reaching much more broadly across society by the early 1900s, the priority of involving young people in the outdoors fueled many wealthy urbanites to send their children to summer camps that could provide their children with a connection to the culture of outdoors that city living lacked. By the end of the first decade of the twentieth century, this national interest in the outdoors resulted in the formation of the American tradition of scouting.

Among other influential Americans, President Theodore Roosevelt spurred a national interest in virility during the Progressive Era. He was concerned that young people growing up in a highly mechanized society would become "soft" and out of touch with genuine, hard work. To keep this from happening, he worked with others to initiate various clubs and organizations, including the Boone and Crockett or Izaak Walton Clubs. Each group had an offspring for younger male members, with Sons of Daniel Boone proving the most popular. Neither, however, truly sought to reach young men of all economic classes. Ernest Thompson Seton, artist and wildlife expert, founded the Woodcraft Indians in 1902 to serve this purpose (Nash 1982, 147–48).

Seton chose to unveil the group through articles in *Ladies' Home Journal*. Shortly afterward, Seton helped Robert Stephenson Smyth Baden-Powell to import the British model of the Boy Scouts to the United States. Seton became the first chief scout of the BSA. The BSA that Seton created prioritized the "whole child" and encouraged the need for boys to be well-rounded. For instance, the first Boy Scout manual, *Scouting for Boys*, contained chapters titled Scoutcraft, Campaigning, Camp Life, Tracking, Woodcraft, Endurance for Scouts, Chivalry, Saving Lives, and Our Duties as Citizens. In thirty years, the handbook sold an alleged seven million copies in the United States, second only to the Bible.

Officially, the BSA incorporated in February 8, 1910, and was granted a federal charter by Congress on June 15, 1916. As its purpose, the BSA strove to provide boys and young adults with an effective educational program designed to build desirable qualities of character, to train them in the responsibility of participatory citizenship, and to develop in them physical and mental fitness.

Working in cooperation with the Young Men's Christian Association, the BSA was popular from its outset. The BSA network spread throughout the nation and, in 1912, included

These Boy Scouts cooking over an open fire at Camp Ranachqua, New York, c. 1919, were part of the most popular American tradition for young men and boys. Prior to organized sports and other activities, boys became involved with scouting in hopes of offsetting the influences of their "overcivilized" life. Library of Congress.

Boys' Life, which would grow into the nation's largest youth magazine. Most educators and parents welcomed scouting as a wholesome influence on youth. Scores of articles proclaimed such status in periodicals such as *Harper's Weekly*, *Outlook*, *Good Housekeeping*, and *Century*. Attitudes toward the outdoors were actively changing during the 1910s as Americans grew more and more aware of the benefits of contact with the natural environment.

Early scouting undoubtedly fostered male aggression; however, such feelings were to be channeled and applied to "wilderness" activities. Many scholars see such an impulse as a reaction to Frederick Jackson Turner's 1893 pronouncement that the frontier had "closed." Turner and many Americans wondered how the nation could continue to foster the aggressive, expansionist perspective that had contributed so much to its identity and success. The first BSA handbook explained that in the previous century, all boys lived "close to nature," but since then, the country had undergone an "unfortunate change" marked by industrialization and the "growth of immense cities." The resulting "degeneracy," instructed the handbook, could be altered by BSA leading boys back to nature (Fox 1981, 347).

Roosevelt's personality guided many Americans to seek adventure in the outdoors and the military. BSA sought to acculturate young men into this culture with an unabashed connection to the military. Weapons and their careful use, as well as survival skills, constructed the basis for a great many of the activities and exercises conducted by Baden-Powell, a major-general in the British army. The original Boy Scout guidebook was partly based on the army manual that Baden-Powell had written for young recruits. World War I would only intensify youth involvement in scouting.

Throughout the rest of the twentieth century, scouting became an important component of the preparation of young boys and girls. Particularly as suburbanization continued to alienate young Americans from the outdoors, outlets such as scouting provided at least a glimpse of "basic" ways of living and an essential connection between humans and the natural environment.

Sources and Further Reading: Fox, *The American Conservation Movement*; Hays, *Conservation and the Gospel of Efficiency*; Nash, *Wilderness and the American Mind*; Price, *Flight Maps*; Reiger, *American Sportsmen and the Origins of Conservation*; Steinberg, *Down to Earth*.

PROGRESSIVES DEMAND FEDERAL ACTION ON CONSERVATION

Time Period: 1890s to 1910s
In This Corner: Social reformers, progressives, engineers
In the Other Corner: Industrial interests, big business, government officials
Other Interested Parties: Voting public
General Environmental Issue(s): Conservation, resource management

Many of these articles have discussed separate cases of a shifting sensibility in the 1890s as Americans began to consider a new ethic of resources use that is best called "conservation." These separate cases and their supporting cultural elements coalesced in the early twentieth century to form a revolutionary new mandate for federal activity in regard to natural resources use. The Progressive Era energized many Americans to identify social ills and to use the government to correct them. The impulse to discontinue waste of resources and the pollution, physical and spiritual, of American communities rapidly became an expression of Americans' unique connection to the land.

The earliest interest in environmental policy grew out of wealthy urbanites of the Gilded Age who often combined an interest in hunting and fishing with efforts to maintain recreational sites. This also fueled efforts by women's groups to limit styles of fashion, which included the use of exotic birds and feathers as hat decorations. Efforts to manage the use of certain species and eliminate the practice altogether took root in the 1890s and extended into the early 1900s. Magazines on topics ranging from gardening to hunting fed Americans' interest. Magazines such as *The Horticulturalist*, *Field and Stream* (then known as *Forest and Stream*), *Godey's Lady's Book*, and *Better Homes and Gardens* helped to merge the women's magazine with practical publications specifically concerned with home design. These popular interests, however, required a leader who could guide them toward concrete expression and policy initiatives.

The leadership of President Theodore Roosevelt and his Chief of Forestry Gifford Pinchot galvanized the upper-class interest with national policies. The aesthetic appreciation of wealthy urbanites grew into progressive initiatives to create national forests and national parks with a unifying philosophy for each. These policies would develop in two directions, preservation and conservation. Roosevelt greatly admired the national parks as places where "bits of the old wilderness scenery and the old wilderness life are to be kept unspoiled for the benefit of our children's children." With his spiritual support, preservationists argued that a society that could exhibit the restraint to cordon off entire sections of itself had ascended to the level of great civilizations in world history (Fox 1981, 19–25).

Although Roosevelt possessed preservationist convictions, his main advisor on land management, Pinchot, argued otherwise for the good of the nation. Conservationists, such as Pinchot, sought to qualify the preservationist impulse with a dose of utilitarian reality. The mark of an ascendant society, they argued, was the awareness of limits and the use of the government to manage resources in danger of exhaustion. Forest resources would be primary to Pinchot's concern. The first practicing American forester, Pinchot urged Americans to manage forests differently than had Europe. Together, Roosevelt and Pinchot (as head of the National Forests) fueled the popular interest in nature. For instance, Pinchot had a national mailing list for the Forest Service of more than 100,000 private citizens. He also made a specific effort to talk about forest issues in frequent public appearances, and he penned articles for popular magazines.

The conflict between preservation and conservation revolved around each side's emerging definitions of themselves. Preservationists such as J. Horace McFarland urged that sites such as Niagara Falls required a hands-off policy that would maintain the natural aesthetic that they found so appealing. Roosevelt and others took a more practical approach that ultimately became known as conservation. Conservationists were buoyed by Roosevelt's vociferous and active ideas. In 1908, he stated some of these points in the nation's first Conference of Governors for Conservation:

> The wise use of all of our natural resources, which are our national resources as well, is the great material question of today. I have asked you to come together now because the enormous consumption of these resources, and the threat of imminent exhaustion of some of them, due to reckless and wasteful use … calls for common effort, common action. (Nash)

The image of Roosevelt as the active conservationist also contributed to the growing appreciation of "rustic" ways of life. His interest in a "vigorous life" and outdoor activity fed the development of organizations such as the BSA and the Boone and Crockett Club (Cutright 1985, 191–95).

During the early 1900s, each of these cultural and political details became part of a land-use ethic termed "conservation." As a method of applying practical management principles to natural resources, conservation would influence American use of forests, rivers, wildlife, and fish by the end of the first decade of the 1900s.

Sources and Further Reading: Hays, *Conservation and the Gospel of Efficiency*; Nash, *Wilderness and the American Mind*; Pinchot, *Breaking New Ground*; Price, *Flight Maps*; Reiger, *American Sportsmen and the Origins of Conservation*; Steinberg, *Down to Earth*.

PINCHOT ARGUES FOR CONSERVATION AS A DEVELOPMENT STRATEGY

Time Period: 1890s to 1910s
In This Corner: Social reformers, progressives, engineers
In the Other Corner: Industrial interests, big business, government officials
Other Interested Parties: Voting public
General Environmental Issue(s): Conservation, resource management

Gifford Pinchot, who became one of the national politicians most identified with the Progressive Era and served as governor of Pennsylvania, is generally regarded first and foremost as the "father" of American conservation. Largely through his efforts, Americans gained a clearer understanding of the finite supply of many natural resources and the capacity of Americans to better manage their usage. In his writings, Pinchot argued for an ethic of managed use that became identified with the term "conservation."

Although this terminology and ethic could be applied to any resource, Pinchot began his work on trees. His unrelenting concern for the protection of the American forests resulted in the formation of the Society of American Foresters, which first met at his Washington home in November 1900. Through his work, however, Americans learned a new way of applying scientific knowledge to the natural world.

Born on August 11, 1865, in Simsbury, Connecticut, Pinchot grew up in an influential family of upper-class merchants, politicians, and land owners. He traveled widely and attended the nation's best schools. Then, in 1885, when he set off for Yale University, his father gave him the idea to study forestry. No American had yet taken up forestry as a profession. As Pinchot later recalled, "I had no more conception of what it meant to be a forester than the man in the moon.... But at least a forester worked in the woods and with the woods—and I loved the woods and everything about them.... My Father's suggestion settled the question in favor of forestry."

Pursuing a graduate degree in forestry required that Pinchot study in Europe. After one year of school in France, he returned to work as the nation's first professional forester. He worked as a resident forester for Vanderbilt's Biltmore Forest Estate for three years. In 1898, Pinchot was named chief of the U.S. Division of Forestry. Shortly thereafter, his close friend Theodore Roosevelt replaced the assassinated President William McKinley. The stage was set for the most formative period in American conservation history.

In 1905, Roosevelt restructured the federal government so that the responsibility for managing the forest reserves was transferred from the Department of the Interior to Agriculture and the new Forest Service. With Pinchot in charge of the new Forest Service, he professionalized the management of the national forests and increased their size and number. In 1905, the forest reserves numbered sixty units covering fifty-six million acres; in 1910, there were 150 national forests covering 172 million acres.

Pinchot's administration of the forests became one of the chief national models of natural resource management. He called for effective organization and management "conservation" and "wise use." In his public appearances and mailings from the Forest Service, Pinchot presented Americans with the utilitarian philosophy of the "greatest good for the greatest number in the long run."

Eventually, Pinchot wrote about the conservation movement and helped to shape its legacy, as well as his own. In 1910, he wrote the book *The Fight for Conservation* in which he attempted to call the nation to action on behalf of conservation:

> The most prosperous nation of to-day is the United States. Our unexampled wealth and well-being are directly due to the superb natural resources of our country, and to the use which has been made of them by our citizens, both in the present and in the past. We are prosperous because our forefathers bequeathed to us a land of marvelous resources still unexhausted. Shall we conserve those resources, and in our turn transmit

them, still unexhausted, to our descendants? Unless we do, those who come after us will have to pay the price of misery, degradation, and failure for the progress and prosperity of our day. When the natural resources of any nation become exhausted, disaster and decay in every department of national life follow as a matter of course. Therefore the conservation of natural resources is the basis, and the only permanent basis, of national success. There are other conditions, but this one lies at the foundation.

Perhaps the most striking characteristic of the American people is their superb practical optimism; that marvelous hopefulness which keeps the individual efficiently at work. This hopefulness of the American is, however, as short-sighted as it is intense. As a rule, it does not look ahead beyond the next decade or score of years, and fails wholly to reckon with the real future of the Nation. I do not think I have often heard a forecast of the growth of our population that extended beyond a total of two hundred millions, and that only as a distant and shadowy goal. The point of view which this fact illustrates is neither true nor far-sighted. We shall reach a population of two hundred millions in the very near future, as time is counted in the lives of nations, and there is nothing more certain than that this country of ours will some day support double or triple or five times that number of prosperous people if only we can bring ourselves so to handle our natural resources in the present as not to lay an embargo on the prosperous growth of the future....

We are in the habit of speaking of the solid earth and the eternal hills as though they, at least, were free from the vicissitudes of time and certain to furnish perpetual support for prosperous human life. This conclusion is as false as the term "inexhaustible" applied to other natural resources. (Pinchot 1998)

Some critics would argue that Pinchot's land-use decisions did not cohere with his philosophical writings. However, in these writings and in the example of forest management, he presented Americans with a new vision of "conservation."

Sources and Further Reading: Hays, *Conservation and the Gospel of Efficiency*; Miller, *Gifford Pinchot and the Making of Modern Environmentalism*; Nash, *Wilderness and the American Mind*; Pinchot, *Breaking New Ground*; Price, *Flight Maps*; Reiger, *American Sportsmen and the Origins of Conservation*; Steinberg, *Down to Earth*.

MUIR ARGUES FOR THE SOUL OF WILDERNESS

Time Period: 1890s to 1910s
In This Corner: Social reformers, progressives, preservationists
In the Other Corner: Conservationists, industrial interests, big business, government officials
Other Interested Parties: Voting public
General Environmental Issue(s): Conservation, resource management

Attitudes toward nature are at least partly shaped by cultural tastes and preferences. When many Americans became more concerned and interested in the natural environment at the end of the nineteenth century, one of the most popular voices came from the nature writer John Muir. He described exotic, far-off places in the context of their natural systems and geography. For other readers, however, Muir represented the voice of an out-of-control extremist. Even some conservationists found his unchecked passion for nature at times irresponsible. Nevertheless, Muir's consistent writing on behalf of the natural environment made him an important shaper of Americans' environmental attitudes just emerging in the late 1800s.

A naturalist and conservationist, he is renowned for his exciting adventures in California's Sierra Nevada, among Alaska's glaciers, and worldwide travels in search of nature's beauty. His interest in nature, however, had come second in his life. First, he had lived a life of industry and mechanics. Born in Dunbar, East Lothian, Scotland, in 1838, Muir emigrated to the United States in 1849, when his family started a farm in rural Marquette County, Wisconsin. An inventor and tinkerer, Muir temporarily lost his eyesight in an industrial accident in 1867. During his struggle to regain his eyesight and health, Muir vowed that, when he could see again, he would only be concerned with nature's beauty. When his eyesight returned, Muir kept his promise. He set out on a walking trek throughout the United States.

When he arrived in San Francisco in March 1868, Muir ventured to the Yosemite Valley for the first time. Admired by many Americans, Yosemite had been set aside as a state park. Its beauty so overwhelmed Muir that he vowed to dedicate himself to knowing it as deeply as possible. Working various jobs, Muir spent as much time as possible learning about the region's geology and landforms. In an era of few trained scientists, Muir's powers of observation soon made him an authority. Arguing against the accepted science of the time, Muir hypothesized that glaciers had sculpted many of the features of the valley. To back up his observation, Muir in 1871 discovered an active alpine glacier below Merced Peak. He soon began to write about his findings and observations for a national audience.

By 1880, Muir had married and taken up permanent residence on a ranch in the area. Although the Yosemite area had been designated a state park in 1864, Muir began to actively fight for its status to be the same as Yellowstone National Park, which had been created in 1872. In September 1890, Muir's articles "Treasures of the Yosemite" and "Features of the Proposed Yosemite National Park" were published in *Century* magazine. Shortly thereafter, Congress passed a bill creating Yosemite National Park.

Muir began to travel widely and to write about the natural beauty that he found. With his new-found popularity, Muir joined with a few friends on May 28, 1892 and founded the Sierra Club, one of the nation's first environmental organizations. Muir was elected president, and the club began its efforts on behalf of the preservation of wild nature. Guided by Muir's passion, preservationists went beyond the rational view of many conservationists and pressed for preserving nature for nature's sake.

During his lifetime, Muir published sixteen books that educated Americans on the environment. His description of Yosemite stands as a representation of his writings:

> The most famous and accessible of these cañon valleys, and also the one that presents their most striking and sublime features on the grandest scale, is the Yosemite, situated in the basin of the Merced River at an elevation of 4000 feet above the level of the sea. It is about seven miles long, half a mile to a mile wide, and nearly a mile deep in the solid granite flank of the range. The walls are made up of rocks, mountains in size, partly separated from each other by side cañons, and they are so sheer in front, and so compactly and harmoniously arranged on a level floor, that the Valley, comprehensively seen, looks like an immense hall or temple lighted from above.

But no temple made with hands can compare with Yosemite. Every rock in its walls seems to glow with life. Some lean back in majestic repose; others, absolutely sheer or nearly so for thousands of feet, advance beyond their companions in thoughtful attitudes, giving welcome to storms and calms alike, seemingly aware, yet heedless, of everything going on about them. Awful in stern, immovable majesty, how softly these rocks are adorned, and how fine and reassuring the company they keep: their feet among beautiful groves and meadows, their brows in the sky, a thousand flowers leaning confidingly against their feet, bathed in floods of water, floods of light, while the snow and waterfalls, the winds and avalanches and clouds shine and sing and wreathe about them as the years go by, and myriads of small winged creatures birds, bees, butterflies—give glad animation and help to make all the air into music. Down through the middle of the Valley flows the crystal Merced, River of Mercy, peacefully quiet, reflecting lilies and trees and the onlooking rocks; things frail and fleeting and types of endurance meeting here and blending in countless forms, as if into this one mountain mansion Nature had gathered her choicest treasures, to draw her lovers into close and confiding communion with her. (Muir, *Travels in Alaska*)

As a writer, Muir taught Americans of the 1890s about the importance of experiencing and protecting our natural heritage. Ultimately, his writings contributed greatly to the creation of Yosemite, Sequoia, Mount Rainier, Petrified Forest, and Grand Canyon National Parks. In addition, however, his writings taught an ethic, now called preservation, that became a core part of the entire concept of American national parks.

After reading Muir's book *Our National Parks*, President Theodore Roosevelt arranged a trip to California in 1901. Muir and Roosevelt hiked and camped throughout the park discussing preservation the entire time. Muir obviously made an impact on Roosevelt. During his presidency, Roosevelt created 235 million acres of protected land, which increased the National Forest System by 400 percent. He also created twenty-three national monuments and five new national parks, including Grand Canyon.

Theodore Roosevelt and John Muir on Glacier Point, Yosemite Valley, California, 1906. These men formed a historic friendship that helped to shape the American ideas of conservation and preservation that would grow into the modern environmental movement. Library of Congress.

Muir sought to use all of his political influence in the early 1900s when planners voiced their desire to build a dam in Yosemite and to flood the Hetch Hetchy Valley. The dam was officially approved in 1913; however, the years of public debate taught Americans about what was necessary to fully preserve national parks. Muir died the next year in 1914.

Sources and Further Reading: Fox, *American Conservation Movement*; Hays, *Conservation and the Gospel of Efficiency*; Miller, *Gifford Pinchot and the Making of Modern Environmentalism*; Muir, *Travels in Alaska*; Nash, *Wilderness and the American Mind*; Pinchot, *Breaking New Ground*; Price, *Flight Maps*; Reiger, *American Sportsmen and the Origins of Conservation*; Steinberg, *Down to Earth*.

PINCHOT, MUIR, AND THE CONSERVATION MOVEMENT MEET AT HETCH HETCHY

Time Period: 1890s to 1910s
In This Corner: Social reformers, progressives, engineers
In the Other Corner: Industrial interests, big business, government officials
Other Interested Parties: Voting public
General Environmental Issue(s): Conservation, resource management

The distinction between two well-intentioned approaches to land use, conservation and preservation, were destined for conflict in the early 1900s. With well-known spokespeople and legions of followers, each viewpoint could call itself part of an emerging environmental consciousness. Despite their commonalities, however, solutions to a single issue could appear vastly different to preservationists and to conservationists. This debate first openly clashed in the early 1910s over a river valley known as Hetch Hetchy.

No simple river valley in the middle of nowhere, Hetch Hetchy was near enough to San Francisco that, after the fire in the early 1900s, the valley might be considered as a possible supplier of water in case of another fire. To do so, however, the valley would need to be transformed into a reservoir, flooded by damming the Tuolumne River, which ran through it. To conservationists, this plan represented a sensible use of natural resources. To preservationists, however, the most important information about Hetch Hetchy was that the valley fell within the borders of Yosemite National Park in northern California. Yellowstone and Yosemite had been designated the nation's first national parks in the 1870s. Still, thirty years later and with other parks in existence, no one had yet specifically determined what that designation actually constituted.

Because of the mixed motives and lack of organization in the creation of early parks, it was inevitable that interested parties would soon clash over one site. The parks were vulnerable to competing interests, including industry and development. More surprising, however, early conservationists often squared off against preservationists on what the term "national park" should mean. Utilitarian conservationists favoring regulated use rather than strict preservation of natural resources often advocated the construction of dams by public authorities for water supply, power, and irrigation purposes.

Although the difference between preservation and conservation may not have been clear to Americans at the beginning of the twentieth century, popular culture and the writings of "muckraking" journalists clearly reflected a time of changing sensibilities. When the mayor of San Francisco moved forward the plan to flood Hetch Hetchy, the stage was set for one of the nation's first full-blown environmental battles. Preservationists, rallied by popular magazine articles by naturalist John Muir, boisterously refused to compromise the authenticity of a National Park's natural environment.

Reviving romantic notions and even transcendental philosophies, Muir used this pulpit to urge that "Thousands of tired, nerve-shaken, over-civilized people are beginning to find out that going to the mountains is going home; that wildness is a necessity; and that mountain parks and reservations are useful not only as fountains of timber and irrigating rivers, but as fountains of life." He called those wanting to develop the site "temple destroyers." He went on to say the following:

> … Damming and submerging it 175 feet deep would enhance its beauty by forming a crystal-clear lake. Landscape gardens, places of recreation and worship, are never made beautiful by destroying and burying them. The beautiful sham lake, forsooth, should be only an eyesore, a dismal blot on the landscape, like many others to be seen in the Sierra. For, instead of keeping it at the same level all the year, allowing Nature centuries of time to make new shores, it would, of course, be full only a month or two in the spring, when the snow is melting fast; then it would be gradually drained, exposing the slimy sides of the basin and shallower parts of the bottom, with the gathered drift and waste, death and decay of the upper basins, caught here instead of being swept on to decent natural burial along the banks of the river or in the sea. Thus the Hetch Hetchy dam-lake would be only a rough imitation of a natural lake for a few of the spring months, an open sepulcher for the others.
>
> Hetch Hetchy water is the purest of all to be found in the Sierra, unpolluted, and forever unpollutable. On the contrary, excepting that of the Merced below Yosemite, it is less pure than that of most of the other Sierra streams, because of the sewerage of camp grounds draining into it, especially of the Big Tuolumne Meadows camp ground, occupied by hundreds of tourists and mountaineers, with their animals, for months every summer, soon to be followed by thousands from all the world.
>
> These temple destroyers, devotees of ravaging commercialism, seem to have a perfect contempt for Nature, and, instead of lifting their eyes to the God of the mountains, lift them to the Almighty Dollar.
>
> Dam Hetch Hetchy! As well dam for water-tanks the people's cathedrals and churches, for no holier temple has ever been consecrated by the heart of man. (Muir, *Travels in Alaska*)

In response, Pinchot defined the conservationist philosophy by claiming that such a reservoir represented the "greatest good for the greatest number" of people and therefore should be the nation's priority. In his testimony to Congress, Pinchot said the following:

> … we come now face to face with the perfectly clear question of what is the best use to which this water that flows out of the Sierras can be put. As we all know, there is no use of water that is higher than the domestic use. Then, if there is, as the engineers tell us, no other source of supply that is anything like so reasonably available as this

one; if this is the best, and, within reasonable limits of cost, the only means of supplying San Francisco with water, we come straight to the question of whether the advantage of leaving this valley in a state of nature is greater than the advantage of using it for the benefit of the city of San Francisco.

Now, the fundamental principle of the whole conservation policy is that of use, to take every part of the land and its resources and put it to that use in which it will best serve the most people, and I think there can be no question at all but that in this case we have an instance in which all weighty considerations demand the passage of the bill.

… I think that the men who assert that it is better to leave a piece of natural scenery in its natural condition have rather the better of the argument, and I believe if we had nothing else to consider than the delight of the few men and women who would yearly go into the Hetch Hetchy Valley, then it should be left in its natural condition. But the considerations on the other side of the question to my mind are simply overwhelming, and so much so that I have never been able to see that there was any reasonable argument against the use of this water supply by the city of San Francisco. (Nash)

Although preservationists had great passion in this instance, the logic of Pinchot's arguments was undeniable to the American public. The dam and reservoir would be approved in 1913, but the battle had fueled the emergence of the modern environmental movement (Fox 1981, 147–49). Considered a disaster by many environmentalists, this episode forced preservationists to denote more specifically what it meant and should mean to call a place a "national park."

Sources and Further Reading: Fox, *The American Conservation Movement*; Hays, *Conservation and the Gospel of Efficiency*; Miller, *Gifford Pinchot and the Making of Modern Environmentalism*; Muir, *Travels in Alaska*; Nash, *Wilderness and the American Mind*; Pinchot, *Breaking New Ground*; Reiger, *American Sportsmen and the Origins of Conservation*; Steinberg, *Down to Earth*.

LONG-TERM IMPLICATIONS OF SODBUSTING AND CONVERSION OF THE AMERICAN WEST FOR AGRICULTURE

Time Period: 1800s to the present
In This Corner: American settlers, farming interests, federal, state, and local governments
In the Other Corner: Scientists, long-term health of the region
Other Interested Parties: Native peoples
General Environmental Issue(s): Agriculture, aridity

Jefferson's call to fill the "empty" lands of the West found many interested settlers. Many settlers found settlement made much easier by the Midwest's convenient lack of trees. Not only did the trees not need to be felled, but fields could be sown more quickly without the need to remove roots and stumps. Conversion of the grasslands into arable land became a defining portion of the pioneer experience. However, in hindsight, some contemporary ecologists and historians have led us to more clearly comprehend the implications of pulling up the Plains' grasses. Although this should not necessarily influence our view of the strong individuals who settled the Great Plains, it may be important information to consider in light of contemporary interest in more sodbusting.

Most settlers to the Midwest were surprised by the landscape that confronted them. Willa Cather wrote about this experience in *My Antonia*: "There seemed to be nothing to see; no fences, no creeks or trees, no hills or fields. There was nothing but land: not a country at all, but the material out of which countries are made." The land of this region, described as blank, empty, and worse by most settlers, required adjustment. Improving the land, of course, revolved around clearing the grasses that dominated the prairie. One Nebraska settler described the process in this manner:

I commenced my sod mansion last Monday and took some of the material on the ground such as brush & poles tuesday it misted rain some so I could not work at it, Wednesday I broke sod & commenced laying the walls, hauling the sod about 80 rods off of R.R. land. Why dont you get it on your own you ask, well we are not going to use our own soil to build with when the R.R. owns every other section around here. I am building my walls 2 1/2 ft thick and have got them 3 1/2 ft high, but the weather today and tonight looks as though I would not do much at it tomorrow this morning when we got up it was misting rain and this afternoon it snowed some and tonight (for I am writing by candle light) the wind is blowing from the north and looks as though it would freeze some before morning. well if it will give us a good rain it would do us good for it is dry here so that the dust blows my eyes full and makes my face all dirty. now I expect you think being so dry the sod would all break to pieces. not a bit of it you can grab a piece 10 ft long and start off with it and wear it out dragging and not tear it. I cut my sod 2 1/2 ft long and plowed about 4 1/2 in' deep & 10 in' wide and it is pretty heavy work to handle them. (Oblinger 1873)

However, a lack of trees also meant no readily available supply of building materials. The only cheap material on hand was the earth itself. The supply of sod was immediately available. Plowing the grass produced a mat of grass roots and earth up to four inches thick and eighteen inches wide. This was cut by ax into manageable lengths of sod and used in the same way as bricks to build a house. So that the wall of earth would not be too high and unstable, settlers sometimes dug into the earth and laid the floor four feet below the ground level. Even if a settler did build a house of logs or boards, very often their animals would live in one made of earth. Often, the first boards were used for the animal's shelter and only later did the home change forms.

The effort to make settlement of the midwestern prairie was considered to be of national significance. For this reason, once the Civil War had begun and cleared the slavery issue that was impeding expansion westward, federal laws placed the government directly behind efforts to stimulate the use and alteration of the prairie. Sodbusting, in short, became a federally supported undertaking.

Creating a Policy Framework to Stimulate Settlement

From the inception of the United States, many constituents had argued for liberal policies to aid in the disposition of these lands. From 1830 onward, groups called for free distribution of such lands. This became a demand of the Free-Soil party, which saw such distribution as a means of stopping the spread of slavery into the territories.

Policy efforts of the 1840s emphasized two primary concepts: "pre-emption" and "graduation." Historian Donald Stevens wrote the following:

> Pre-emption referred to giving squatters the privilege to purchase land at the minimum price before the holding of a public auction. Congress passed a permanent pre-emption law in 1841; it allowed such sales, however, only after the survey. Proponents of graduation, who Senator Thomas Hart Benton of Missouri led in Congress, wanted to reduce the price of land left on the market after the auctions (the established policy allowed the purchase of land not sold at the auction for the minimum price). The Graduation Act of 1854 authorized the federal government to sell public land at a reduced price based on its length of time on the market. (White)

After the passage of the Graduation Act, more than 25.7 million acres sold at the reduced rates. Sales in the southwest (primarily in Missouri, Alabama, and Arkansas) alone involved 18.3 million acres. The Graduation Act was also a boon for speculators who collected many additional large tracts at the low prices and then sold them later.

The stage was set, however, for the land dispersal policies to also contend with the issue of slavery. This occurred in 1854 with the Kansas-Nebraska Act, otherwise known as "An Act to Organize the Territories of Nebraska and Kansas." This act repealed the Missouri Compromise, which had outlawed slavery above the 36° 30′ latitude in the Louisiana territories and, thereby, reopened the national struggle over slavery in the western territories.

In 1854, Senator Stephen Douglas introduced a bill arguing for dividing Missouri into two territories, Kansas and Nebraska. He argued for allowing voters to determine whether or not the new territories would allow slavery. Here is some of the text of the act:

> *Be it enacted by the Senate and House of Representatives of the United States of America in Congress assembled,* That all that part of the territory of the United States included within the following limits, except such portions thereof as are hereinafter expressly exempted from the operations of this act, to wit: beginning at a point in the Missouri River where the fortieth parallel of north latitude crosses the same; then west on said parallel to the east boundary of the Territory of Utah, the summit of the Rocky Mountains; thence on said summit northwest to the forty-ninth parallel of north latitude; thence east on said parallel to the western boundary of the territory of Minnesota; thence southward on said boundary to the Missouri River; thence down the main channel of said river to the place of beginning, be, and the same is hereby, created into a temporary government by the name of the Territory Nebraska; and when admitted as a State or States, the said Territory or any portion of the same, shall be received into the Union with or without slavery, as their constitution may prescribe at the time of the admission: Provided, That nothing in this act contained shall be construed to inhibit the government of the United States from dividing said Territory into two or more Territories, in such manner and at such times as Congress shall deem convenient and proper, or from attaching a portion of said Territory to any other State or Territory of the United States: *Provided further*, That nothing in this

act contained shall construed to impair the rights of person or property now pertaining the Indians in said Territory so long as such rights shall remain unextinguished by treaty between the United States and such Indians, or include any territory which, by treaty with any Indian tribe, is not, without the consent of said tribe, to be included within the territorial line or jurisdiction of any State or Territory; but all such territory shall excepted out of the boundaries, and constitute no part of the Territory of Nebraska, until said tribe shall signify their assent to the President of the United States to be included within the said Territory of Nebraska or to affect the authority of the government of the United States make any regulations respecting such Indians, their lands, property, or other rights, by treaty, law, or otherwise, which it would have been competent to the government to make if this act had never passed. [from http:// www.ourdocuments.gov/print_friendly.php?flash=true&page=transcript&doc=28&title= Transcript+of+Kansas-Nebraska+Act+(1854)]

After months of debate, the Kansas-Nebraska Act passed on May 30, 1854. Proslavery and anti-slavery settlers rushed to Kansas, each side hoping to determine the results of the first election held after the law went into effect. Opponents of the Kansas-Nebraska Act helped found the Republican Party, which opposed the spread of slavery into the territories. As a result of the Kansas-Nebraska Act, the United States moved closer to Civil War.

Homestead Act

In its 1860 election platform, the Republican party adopted this policy as one of its own. Of course, the southern states had been the most outspoken opponents of the policy. When these states seceded to become the Confederacy, the path was clear for adoption of the Homestead Act.

The Homestead Act, May 20, 1862

AN ACT to secure homesteads to actual settlers on the public domain.

Be it enacted, That any person who is the head of a family, or who has arrived at the age of twenty-one years, and is a citizen of the United States, or who shall have filed his declaration of intention to become such, as required by the naturalization laws of the United States, and who has never borne arms against the United States Government or given aid and comfort to its enemies, shall, from and after the first of January, eighteen hundred and sixty-three, be entitled to enter one quarter-section or a less quantity of unappropriated public lands, upon which said person may have filed a pre-emption claim, or which may, at the time the application is made, be subject to pre-emption at one dollar and twenty-five cents, or less, per acre; or eighty acres or less of such unappropriated lands, at two dollars and fifty cents per acre, to be located in a body, in conformity to the legal subdivisions of the public lands, and after the same shall have been surveyed: Provided, That any person owning or residing on land may, under the provisions of this act, enter other land lying contiguous to his or her said land, which shall not, with the land so already owned and occupied, exceed in the aggregate one hundred and sixty acres.

Although a few tracts of land were deemed necessary to hold resources with known, appreciable value, the vast majority of the western landscape was open range land and remained uninhabited by 1860 while the federal government determined the best method for "disposing" of the land and convincing people to move westward. The passage of the Homestead Act by Congress in 1862 was the culmination of more than seventy years of controversy over the disposition of these lands.

The act allowed anyone to file for a quarter section of free land (160 acres). If, after five years, the owner had built a house on it, dug a well, or broken (plowed) ten acres, his ownership became permanent. After the first five years, many landowners fenced a specified amount and finally took up permanent residence there. Additional land could also be acquired if one claimed a quarter section of land for "timber culture" (commonly called a "tree claim"). This required the planting and successful cultivation of ten acres of timber on unowned property.

To spur settlement, the federal government kept the process for staking a claim very simple. Settlers filed their intentions at the nearest Land Office, where plots of land were checked for any previous ownership claims. If the land was unowned, the interested settler could temporary claim it by paying a filing fee of $10 and a $2 commission to the land agent. At this point, the homesteader could proceed to his plot and initiate the process of building a home and farming the land. The addition of such improvements needed to occur within five years for the owner to have "proved" his land. The settler needed to have two neighbors serve as witness to the improvements. With their signed verification, the homesteader could take legal possession of the land with the payment of a $6 fee. Once they received their patent for the land, most homesteaders proudly displayed this document inside their home.

Using the Prairie Land

The supply of sod was immediately available. Plowing the grass produced a mat of grass roots and earth up to four inches thick and eighteen inches wide. This was cut by axe into manageable lengths of sod and used in the same way as bricks to build a house. So that the wall of earth would not be too high and unstable, settlers sometimes dug into the earth and laid the floor four feet below the ground level. The pioneers who worked the expansive western landscape conformed their lives and, eventually, their expectations to the western environment.

Of course, stripping the grass was only one part of the effort to transform the Plains' ecology. Some settlers saw trees as essential for civilizing the prairies. Often, their ideas were substantiated by science in the nineteenth century. Nathaniel Egleston, chief forester for the USDA in the late nineteenth century, for example, wrote that "tree planting is almost the first necessity of life" on the Great Plains and suggested that, without trees, "barbarism" would be "the inevitable result."

Geographer Blake Gumprecht wrote that proponents of tree planting "warned that the removal of forests in other parts of the world had led to the downfall of once-great Societies." He quoted C. S. Harrison, founder of a Nebraska town, who said, "The curse of God falls heavily on the people who ignore His grand designs and rob the lands of forests." When Harrison founded a town, he optimistically named it Arborville. Names and trees in general, it was believed, had symbolic value.

Gumprecht argued that, to the typical European-American sensibility, trees represented fertility. He wrote the following:

Promoters, in particular, believed that the planting of trees was essential for proving to Easterners that the entire Western region was capable of producing lush vegetation and, therefore, inhabitable. The record of Europeans adding trees to the Midwestern landscape, in fact, pre-dated settlement in Kansas, where a missionary added fruit trees in 1838. By the 1850s, Nebraska had become known as "The Tree Planters State." (Gumprecht)

Gumprecht reported that, a year prior to the Nebraska territory being opened for settlement, a steamboat delivered 55,000 trees and shrubs to squatters there. In Nebraska, the tree-planting obsession grew in the 1870s when cities and towns taxed residents for the planting of trees along streets. It was also made a crime to injure or destroy a tree.

This cultural interest grew into a full-blown movement in 1872 when Nebraska newspaperman J. Sterling Morton proposed a holiday known as Arbor Day.

In the end, however, trees were a way of making the odd prairie look more familiar. Most settlers arrived in the American West from European roots, and many were born or grew up in more wooded parts of the United States. The planting of a few seedlings around a homestead was simply a way to recreate a more familiar setting. Arbor Day founder Morton, a native of upstate New York, once remarked that "there is a comfort in a good orchard, in that it makes the new home more like the old home in the East." In a study of ecological adaptations in a Kansas county, James Malin noted that settlers from more humid states "missed trees possibly more than anything else."

Congress next got into the act with the Timber Culture Act, which offered an additional 160 acres to homesteaders who planted a specific percentage of trees on their plot. Eventually, this initiative contributed to the popular sentiment that "rain followed the plow." Based

Tree planting on Arbor Day in a New York public school demonstrates one example of this long-standing American tradition. Library of Congress.

on loose scientific understandings, this belief argued that planting additional trees increased precipitation.

There could not, however, be enough trees added to offset the massive impact that the plow had on the prairie. As mechanized agriculture intensified the plowing further, the alterations to the Plains ecology, of course, contributed to or caused the event known as the Dust Bowl.

Desertification and the Dust Bowl Years

Discussed in a separate essay, the event of the 1930s that is known as the Dust Bowl occurred at least partly because of American sodbusting during the earlier eras of settlement. As a result, Paul Sears introduced many people to the concept of desertification in his book *Deserts on the March*. Readers were shocked to learn from Sears that deserts were neither permanent nor dead. In addition, he presented the ecological shifting that took place when land became arid and eventually desert. Still, many observers doubted his scientific understanding.

Interestingly, this disaster on the Plains helped to usher in the next phase of soil use in the Midwest as planners and ecologists applied the new science of soil conservation to the effort to make the region inhabitable. From the 1930s, the region was a hotbed for new ideas about how to manage the ecology of the region. In general, the Great Plains helped to instruct the entire nation on soil conservation policies. In general, the Depression awoke the nation to the interrelated problems of poverty and poor land use. The public glimpsed some of this suffering in the South in the photographs of the Farm Security Administration and those in Walker Evans and James Agee's *Let Us Now Praise Famous Men*, which told a tale of poor land and poor people complicated by tenancy and racism, but it was the experience of the Great Plains that captured most of the national attention.

Newspaper accounts of dust storms, the government-sponsored documentary classic *The Plow That Broke the Plains*, and John Steinbeck's novel *Grapes of Wrath* placed evocative images in the public's mind. Clearly, these images functioned as an antithetical one to "manifest destiny" of the 1800s. Americans pushed the ecology of the region too hard and had been too reluctant to listen to the warnings of scientists. As a reaction, New Deal efforts would be drawn directly from the ideas of many of the scientist-critics.

On the policy front, the New Deal first sponsored a major study to establish what had gone wrong. In his report, Henry Wallace stipulated the following:

> Particular attention has been given to the suggestion in your letter of July 22nd, as follows: We have supposed that the modes of settlement and of development which have been prevalent represented the ordinary course of civilization. But perhaps in this area of relatively little rain, practices brought from the more humid part of the country are not most suitable under the prevailing natural conditions.

> A trip through the drought area, supplementing data already on record, makes it evident that we are not confronted merely with a short term problem of relief, already being dealt with by several agencies of the Federal Government, but with a long term problem of readjustment and reorganization.

The agricultural economy of the Great Plains will become increasingly unstable and unsafe, in view of the impossibility of permanent increase in the amount of rainfall, unless over cropping, over grazing and improper farm methods are prevented. There is no reason to believe that the primary factors of climate temperature, precipitation and winds in the Great Plains region have undergone any fundamental change. The future of the region must depend, therefore, on the degree to which farming practices conform to natural conditions. Because the situation has now passed out of the individual farmer's control, the reorganization of farming practices demands the cooperation of many agencies, including the local, State and Federal governments.

We wish to make it plain that nothing we here propose is expected or intended to impair the independence of the individual farmer in the Great Plains area. Our proposals will look toward the greatest possible degree of stabilization of the region's economy, a higher and more secure income for each family, the spreading of the shock of inevitable droughts so that they will not be crushing in their effects, the conservation of land and water, a steadily diminishing dependence on public grants and subsidies, the restoration of the credit of individuals and of local and State governments, and a thorough going consideration of how great a population, and in what areas, the Great Plains can support. (Wallace, *Report of the Great Plains Drought Area*)

The findings of Wallace's report contributed to new efforts to conserve and manage the soils of the Midwest.

Formalizing Soil Conservation

The establishment of the Soil Conservation Service (SCS) was one of the most important portions of this movement to change agriculture because it prioritized delivering new scientific information directly to the farmers who were using the land. The SCS at first worked through demonstration projects and the labor of members of the New Deal's Civilian Conservation Corps.

In 1937, FDR encouraged states to pass a standard soil conservation districts act, which would organize areas into agricultural groupings. Through these districts, the USDA established a presence in the countryside working directly with farmers and ranchers in a relationship that focused researchers on common problems and provided the means to work on transferring new ideas and technology to the land. Demonstration projects drew together the engineers, agronomists, and range management specialists.

Although the SCS influenced agriculture nationwide, it seemed particularly appropriate in the Plains where farmers had struggled with wind erosion and devised a number of methods to combat it. State agricultural experiment stations and later USDA stations specializing in soil erosion provided answers. To provide vegetative cover, the SCS advocated water conservation through detention, diversion, and water spreading structures and by contour cultivation of fields and contour furrows on range land. SCS representatives also advocated stripcropping and the creation of borders of grass, crops, shrubs, or trees to serve as wind barriers. For critically erodible land, however, the SCS advocated abandoning agriculture all together.

Speaking to the Unique History of the Prairie

Inspired by new scientific information regarding the unique ecology of the American grassland, one historian in particular began to rethink the American experience with the arid West. James Malin, who taught at the University of Kansas, wrote the following:

> The forms of life of each natural geographical region (vegetation, animals, and microorganisms) are primarily characteristic of the particular region or type of region. They are closely interrelated in an unstable equilibrium with each other and with climate, soil materials, physiography and time. Such a description of the natural environment of man is the product of twentieth century science crystallized out of scattered nineteenth century beginnings.
>
> The heart of North America is a Grassland. This region is being given a new significance by reason of the Air Age and its reorientation of the world outlook in terms of circumpolar land-masses—just a century after the Anglo-American forest man met for the first time the problems of living in a grassland. The eastern extremity of the grass country is a triangular prairie peninsula between the Great Lakes and the Ohio river, the point extending east as far as Indiana, with detached outlying openings farther east in Ohio and Pennsylvania, south in Kentucky and Tennessee, and north in Michigan. From this eastern forest boundary, the grassland extends westward through passes in the Rocky Mountains across the continental divide to the southwestern deserts, and farther north, to the famous prairie of the inland empire of the Pacific northwest. To the northward, the grassland extends through the prairie provinces of Canada to the northern forest boundary line, and to the southward into the Mexican desert plains.
>
> The region is occupied by a wide variety of grasses with their associated plants, and of animals; forms of life mostly characteristic of, and peculiar to, the environment. There is present also, a minority representation of forms of life typical of other regions, especially along the boundary transitions. The grassland is complete in itself, a relatively stabilized product of nature, the outgrowth of climate, soil, vegetation, animals, and microorganisms, all interacting together. This general kind of region is not unique to North America but is a characteristic formation, of greater or lesser extent, occupying parts of all continents, unique only in respect to the particular species of biological forms that enter into the combinations. (Malin)

Malin was well ahead of his time; however, his writing combined with the findings of scientists to begin to correct the historical record by the close of the twentieth century.

Conclusion: Sodbusting Resurgence

Despite the important evolution of ecological understanding of the Plains, recent years have brought a new interest in sodbusting. Thanks to mechanical technology, more than 6.4 million acres of marginal grasslands in Montana and Colorado and an estimated twenty-seven million acres scattered across the rest of the United States have been transformed in the past decade. Most often, the land used previously for grazing is being converted into wheat fields by large-scale agricultural companies.

The contemporary sodbusters are either big operators who buy land and plow on a major scale or small ranchers who break their own land for a quick cash fix. Sodbusters buy range land at prices that are relatively low because of today's depressed livestock industry, plow and plant the acreage in wheat, and then sell the cultivated land, sometimes to buyers unfamiliar with the region and the fragility of the range's topsoil.

Sources and Further Reading: Cooke et al., *Report of the Great Plains Drought Area Committee*, http://newdeal.feri.org/hopkins/hop27.htm; Gumprecht, "Transforming the Prairie," 2001; Helms, *Conserving the Plains: The Soil Conservation Service in the Great Plains*; Malin, *The Grassland of North America*, http://www.kancoll.org/books/malin/mgchap02.htm; Oblinger, *Letter from Uriah W. Oblinger to Mattie V. Oblinger and Ella Oblinger*; Opie, *Nature's Nation*; Stevens, *A Homeland and a Hinterland*, http://www.cr.nps.gov/history/online_books/ozar/hrs4.htm; Wallace, *Report of the Great Plains Drought Area*, http://newdeal.feri.org/hopkins/hop27.htm; White, *It's Your Misfortune and None of My Own*; Worster, *Dust Bowl: The Southern Plains of the 1930s*.

USING CHICAGO TO MAKE THE GREAT LAKES THE NATION'S FIFTH COAST

Time Period: 1880–1920
In This Corner: Chicago bosses, federal government, western farmers
In the Other Corner: Other cities with development aspirations
Other Interested Parties: American consumers
General Environmental Issue(s): Land use, agriculture, aridity

The effort to develop seaports was one of the defining characteristics of the United States. With ports on the Atlantic, Pacific, and Gulf, settlement moved inland. The primary exception was the Mississippi River, which allowed access to the interior. Many developers, however, were not satisfied with this river's access to the heartland. As government efforts for westward settlement progressed and the railroad and other technology synched the nation together, businessmen and developers became enraptured with the idea of manufacturing a port—and an entire coast—for the nation's new region, known as the West. Although it may seem that other port cities would have fought this development, in fact, the effort to make the Great Lakes the nation's fifth coast benefited trade networks throughout the country. The only naysayers would be those who knew such an effort had never before succeeded on such a grand scale.

In short, it was the popularity of western commodities that fostered the growth of Great Lakes trade. Among a handful of other cities, the Great Lakes development relied on the structuring of Chicago, Illinois, as a trading hub. First, however, Chicago grew on the supplies of lumber harvested from the Great Lakes region. The small town of Chicago was transformed into a booming trade town. The upper Mississippi River and the Illinois-Michigan Canal, which was competed in 1848, were the avenues on which to move the rafts of logs and lumber in the era before railroads.

In the canal, Chicago wholesalers could sell Michigan and Canadian lumber to buyers in the farmlands of the Midwest. Thanks to the Chicago markets, lumber prices fell by approximately 50 percent by 1850. Historian William Cronon wrote that the late 1800s economic boom in the Midwest grew less from human labor and more from "autonomous ecological

This photo from 1910-20 shows the Santa Fe elevator at the head of the canal, Chicago, Illinois. The ability to store and systematize grain supplies in such devices revolutionized midwestern farming and also helped to incorporate the railroad for widespread use. Library of Congress.

processes that people exploited on behalf of the human realm.... Although people might use it, redefine it, or even build a city from it, they did not produce it" (Cronon 1991, 149–50). In the case of the great forests of the Midwest, this process began when settlers saw the trees not as forest but as a product: "timber."

Native American residents had been removed from the northern Great Lakes by treaties in the 1830s. Buying and selling these lands became big business, particularly for those in the lumber industry. By the 1860s, Frederick Weyerhaeuser and other timbermen made fortunes by purchasing the vacated lands, deforesting them, selling the lumber to farm regions, and then selling the deforested lands to settlers before any taxes needed to be paid. Ultimately, the ability to transform midwestern and western natural resources into profitable business opportunity depended on transportation access.

Creating the Tools for Settlement

Chicago's abilities also helped spur agriculture throughout the American West, symbolically and literally, by the plow. On the typical pioneer farm, there was no end to the work that needed to be done. In addition to farming, the household needed to shoe and feed horses and oxen and repair the plows and other equipment. As the number of farmers in the Midwest grew, businessmen found that there was significant opportunity available to service the equipment of the rapidly increasing number of agriculturalists. One of these entrepreneurs was John Deere.

While servicing equipment, Deere learned of the most serious problem that pioneer farmers encountered in trying to farm the midwest soil: the cast-iron plows brought from the east were designed for the light, sandy New England soil; the rich midwest soil was much thicker. The dirt clung to the plow bottoms, clogging their forward motion. Plowmen were forced to slow their progress by stopping repeatedly to scrape the sticky soil from the plow. The plowing that was necessary to pull up the deeply rooted grasses became one of the settlers' most dreaded and laborious tasks. The difficult plowing made many homesteaders consider moving on or returning to the eastern United States before they had improved their land.

Deere's experiments left him convinced that a highly polished and properly shaped moldboard plow could function much better in the western United States by scouring itself as it turned the furrow slice. In 1837, he modeled his first plow along these specifications out of a broken saw blade. The plow worked as he had hoped, and Deere immediately began to sell them to farmers in the Midwest. Ultimately, Deere's steel plow almost single-handedly allowed settlers to break the soils found west of the Mississippi River and to prepare them for farming. However, Deere's efforts to sell his product are what truly changed agriculture in the western United States. Additionally, in a market in which most plows were made to order, Deere produced a surplus. Once he had produced a full railcar load, Deere took the plows to rural areas and offered his "self-polishers" for sale. For the first time, marketing treated rural communities like any other.

Of course, once the vast fields of grain had been plowed and grown, they needed to be picked or cut. The McCormick Reaper, which was perfected in Virginia in the 1830s, began large-scale use in the West after Cyrus McCormick moved his factory to Chicago in 1848 (Opie 1998, 237–39).

Describing his innovation, one historian wrote, "McCormick used seven fundamental principles in the reaper he invented. A divider separated the wheat so it could be cut. The reel pushed the wheat toward the knife and then pushed the cut wheat onto the platform. A straight reciprocating knife cut the wheat. Fingers held the wheat while it was cut. The platform held the cut wheat. The main wheel and gears drove all the moving parts of the reaper. The front-side draft traction provided the reaper a firm grip on the ground" (Wong 1992).

By the time that his Chicago factory opened, McCormick was estimated to have manufactured about 1,300 reapers. Using his new factory and marketing efforts, between 1848 and 1850, he manufactured and sold 4,000 reapers. The influence of the plow and the reaper could be seen throughout the western United States.

Making a Commodity of Meat

Chicago developed many new resources, but none was so specifically identified with as the West's greatest space optimizer: cattle. Once again, it was the application of the railroad as infrastructure that allowed the city to define itself as the primary cog for an industry that would spur development in faraway areas such as Texas and Idaho.

Conceived by George Henry Hammond and Gustavus F. Swift, the refrigerated train car had widespread implications for the nature of American eating. Up to this point, meat had been available from local butchers or private livestock supply. Preservation techniques such as salting, pickling, or smoking were considered technological advances earlier in the 1800s. The first refrigerated railroad car was used in 1867. When Swift began the widespread use of the refrigerated cars in the 1880s, however, he enabled meat from the Chicago stockyards to reach markets throughout the nation.

Chicago's Union Stockyards had been constructed in the 1860s. By the late 1880s, Chicago's meatpackers, particularly Swift and Philip D. Armour, controlled the nation's market. Their specialty, however, was not fresh meat; instead, they popularized processed meat. By mixing gelatin or other materials with chopped meats, these entrepreneurs made them capable of being packaged into cans. For the first time, meat could be sold in a form that would not spoil.

Ironically, the use of the refrigerated car in the late 1800s also made fresh meat able to be more easily transported and sold throughout the nation. With the use of the refrigerated car, however, the Union Stockyards by 1883 processed 1.9 million cattle and more than 5.6 million pigs annually.

Together, these innovations meant that there were new large-scale markets for meat that entirely changed the history of the American West. The stockyards, disassembly lines, railroads, cattle drives, and cowboys became vital cogs in feeding the nation. Each acre of range land in the West now could be profitable if it hosted or fed animals for market. Cattle, but especially pigs, became the most profitable way for farmers to transform fields of corn into cash.

Industrial Agriculture Redefines the West

Farm crops also expanded in scale. Liberated from the limits of self-sufficiency and connected to larger markets by the railroad, western farmers began to use the vast open spaces to create massive farming establishments. Usually referred to as bonanza farms, these farms were most popular in states such as Minnesota in which the Northern Pacific Railroad sold off huge acreage.

Historian Lauren McCroskey of the NPS described the process for marketing this massive style of agriculture when she wrote the following:

> George W. Cass, president of the Northern Pacific, and George Cheney, a board member, needed to prove to potential buyers that the land was fertile and productive. The bigger the farm, the more convincing their case—so the first bonanza farms were composed of massive amounts of acreage and typically grew only wheat. The Northern Pacific investors hired wheat expert Oliver Dalrymple to develop and manage their demonstration farm. The Cass-Cheney farm (which was also referred to as the Dalrymple farm) was located about 20 miles west of Fargo, North Dakota. It grew from 5,000 acres in 1877 to 32,000 acres by 1885, and yielded as much as 600,000 bushels of wheat per year. The bonanza farm needed 600 men to plant seed and 800 to harvest. The scale of the mechanical enterprise was similarly massive, including 200 plows, 200 self-binding reapers, 30 steam threshers, and 400 teams of horses or mules. One manager typically oversaw a single 2,500-acre tract. (McCroskey)

Other huge farms, ranging from 3,000 to more than 30,000 acres, soon appeared in the Red River Valley. The soil, topography, and climate in the upper Midwest and West were ideal for large-scale farming. The necessary workforce was drawn from new immigrants from northern Europe as well as from off-season loggers from Minnesota's timber industry. Typically, workers were hired on a limited-term contract. Although most owners of these vast farms did not live in the area, the absentee landowners hired local managers to run the farms (Opie 1998, 234–39).

Through the creation of bonanza farms, Minnesota and North Dakota, and specifically the Red River Valley, became one of the country's largest wheat-producing areas. Between 1875 and 1890, the bonanza farms modeled after Dalrymple's became highly profitable, but the scale of such enterprise could not be sustained. Eventually this intense use of the soil resulted in exhaustion of the necessary minerals. Many of the farms were no longer profitable by the 1890s. However, regional industry helped keep this style of agriculture going into the 1900s.

These great farms helped support a new midwestern grain industry that possessed awe-inspiring scale. Minneapolis was one of the many cities to prosper in milling. By the 1880s, wrote McCroskey, "27 mills produced more than two million barrels of flour annually, making Minneapolis the nation's largest flour-manufacturing center." This primarily occurred through corporate consolidation. "In 1876, 17 firms operated 20 mills; in 1890, 4 large corporations produced almost all of the flour made in Minnesota" (McCroskey).

Leading these corporate entities, continued McCroskey, "the Pillsbury Company completed its gigantic A Mill in 1880. Containing two identical units, it had a capacity of 4,000 barrels of flour per day when it opened. By 1905 the mill had tripled its output. Its owners claimed that it was the largest flour mill in the world" (McCroskey).

By the early twentieth century, the day of the bonanza farm had passed. Run as a large-scale business, the farms operated on a very narrow profit margin. The price of land began to rise after 1900, as did corporate taxes. As the costs of labor and equipment rose, wheat prices fell. Many of the huge farms were gradually broken up, but the geographical feature that grew from grain storage remained a prominent part of the landscape. In the 1850s, many midwestern cities began utilizing a device found on many farms that grew grain: the silo. The urban model, however, was much larger and became known as an *elevator*. Intimately linked to the system of growing grain and transporting it by railroad, the elevator became a revolutionary device when used to store grain for market-driven reasons. Historian William Cronon wrote, "Chicagoans began to discover that a grain elevator had much in common with a bank—albeit a bank that paid no interest to its depositors" (Cronon 1991, 120). Primarily, the elevator stored the grain and helped growers maintain a supply throughout the year. It functioned as a kind of holding tank.

As if carrying a precious mineral, farmers brought their wheat or corn to the elevator operator. The operator gave the grower a receipt that could be redeemed for grain when the original grower wanted to make a withdrawal. Early on, grain was measured, traded, and sold in sacks. To simplify its trading, grain operators adopted a liquid form of measuring that liberated the grain from the sack. This was the change that allowed the elevator to take over the landscape of each midwestern city.

In cities such as Chicago, the supply of grain seemed endless. Supplying much of the nation's needs, the Chicago elevator operators were grain brokers. Much like a bank, the brokers bought and sold grain. The elevator, however, allowed this financial market to become a futures market. Cronon wrote, "Grain elevators and grading systems had helped transmute wheat and corn into monetary abstractions, but the futures contract extended the abstraction by liberating the grain trade itself from the very process which once defined it: the exchange of physical grain" (Cronon 1991, 126).

Conclusion: Chicago as a Primary City

Finally, the great manufacturing center of the West also had internal problems with which to contend. Infrastructural engineering was evident in the responses now given to age-old problems. In early Chicago, for instance, the sewage system was primitive, with gutters serving as drains in many streets. Although underground pipes were added later, they typically discharged either directly into Lake Michigan or into the river that flowed into the lake. The water cribs were being pushed farther out into the lake to escape the wastes, but the effort was not successful. The main impact on city residents was disease.

In 1854, a cholera epidemic took the lives of 5.5 percent of the population. Deaths from typhoid fever between 1860 and 1900 averaged sixty-five per 100,000 persons a year. The worst year was 1891, when the typhoid death rate was 174 per 100,000 persons. Disease resulting from water polluted by human waste brought about a state of emergency.

In 1887, Rudolph Hering, chief engineer of the drainage and water supply commission, observed that the Great Lakes drainage system was separated from the Mississippi River drainage system by a ridge approximately eight feet high, which was located approximately twelve miles west of the lakeshore. With this knowledge, Hering devised a plan to alleviate the city's sewage problem by reversing the flow of the Chicago River and thereby carrying all city wastes away from the lake. In 1889, the Metropolitan Sanitary District of Greater Chicago was created to make a canal through the ridge and carry wastes away from the lake and down to the Mississippi River. The twenty-eight-mile canal, which was referred to as the "Sanitary and Ship Canal" or "main channel," was completed in 1900. Chicago had, essentially, built a river to carry its waste away (Hays 1999, 96–99).

With its centrality, Chicago had also gained the problems that accompanied industrialization. Being the regional center of manufacturing and trade also brought with it many challenges during the twentieth century.

Sources and Further Reading: Cronon, *Nature's Metropolis*; Labaree, *America and the Sea*; Opie, *Nature's Nation*; White, *It's Your Misfortune and None of My Own*; Wong, "McCormick's Revolutionary Reaper"; Worster, *Dust Bowl: The Southern Plains of the 1930s*.

POWERING INDIVIDUAL TRANSPORTATION

Time Period: 1900–1920
In This Corner: Auto manufacturers
In the Other Corner: Electric manufacturers, mass transportation interests
Other Interested Parties: American consumers and travelers
General Environmental Issue(s): Transportation

How could a single resource become so intertwined with everyday American life? The single greatest development lay in the transportation sector. This sector of Americans' lives changed radically during the early twentieth century. A steady supply of cheap crude oil became the necessary resource to support everyday American life after 1900, particularly after Americans had begun their love affair with the automobile. However, this affair almost never began. Significant debate met each developer of methods to power the first automobiles. Ironically, many of these debates have returned today as Americans consider the next generation of energy for transport.

The Fateful Path Away from Electrics

When Oliver Evans built the first motor vehicle in the United States in 1805, his prime mover of choice was steam. A combination dredge and flatboat, it operated on land and water. Richard Dudgeon's road engine of 1867, which resembled a farm tractor, could carry ten passengers. By the late 1890s, nearly 100 manufacturers were marketing steam-driven automobiles. Twin brothers Francis E. and Freelan O. Stanley of the United States received the greatest fame of the steamer era with their 1897 creation of the "Stanley Steamer." Most of the models of steam cars burned kerosene to heat water in a tank that was contained on the car. The pressure of escaping steam activated the car's driving mechanism, which moved the vehicle. The popularity of the steam car declined at about the time of World War I, and production came to an end in 1929. This was not, however, caused by a decline in interest in automobility. Instead, powerful interests had swayed Americans toward a new model of vehicle construction.

Some Americans adopted electrically powered automobiles that were built in Europe by the 1880s. In the United States, William Morrison is credited with the first "electric" in 1891. Other manufacturers followed quickly, however, and between 1896 and 1915, fifty-four American manufacturers turned out almost 35,000 vehicles. Many people thought the great models of the era, including, the Columbia, the Baker, and the Riker, were just the beginning; however, the first decade of the twentieth century turned out to be the high point of the American manufacture of electrics. Most early electrics, however, did not run efficiently at speeds of more than twenty miles per hour. In addition, most were difficult to charge. Early electrics were assumed to be limited to city use. The true revolution in automobility would first require a new source of energy.

This moment of historical convergence brings together the timing of the strike at Spindletop with the public's growing interest in the speed and independence of the automobile. The massive quantities of petroleum drove its price downward. Edison's experiments with electric lighting forced those in the industry, particularly Rockefeller of Standard Oil, to search for new uses for petroleum. When Henry Ford and other entrepreneurs began manufacturing autos in the United States, they followed the urging of Rockefeller and others and opted for a design powered by an internal-combustion engine that used gasoline, derived from petroleum. Patillo Higgins's dream at Spindletop helped to define human life and world power in the twentieth century by revolutionizing transportation in everyday American life (Yergin 1993, 40–43).

Henry Ford Democratizes the Automobile

American inventors vied to be the first with a workable automobile model before 1900. J. Frank and Charles E. Duryea are credited with manufacturing the first working gasoline-powered automobile in the United States in 1893. The year 1896 brought the first commercial production of the Duryea car. However, in the same year, the automobile revolutionary, Henry

Ford, also successfully operated his first auto in Detroit. How would such inventions get to the consumer? By 1899, Percy Owen opened the first automobile salesroom in New York City. Auto shows soon followed. They helped to get information to the public about American manufacturers' interest in making the United States the world's first automobile nation.

The manufacture of automobiles quickly shifted from the efforts of independent inventors to become the world's largest industry. Manufacturers of many items experimented with mass production; however, automobiles became the first new product to be entirely defined by this new manufacturing process. Although Ford is often credited with first mass producing automobiles, Ransom E. Olds in 1901 became the first auto manufacturer to use mass-production methods. In its first year, his company manufactured a shocking 400 vehicles, each selling for approximately $650. Unlike other manufacturing processes, mass production was a relatively vague term. It was impossible to patent. Therefore, other manufacturers, including Henry M. Leland and Ford, also experimented with using mass-production processes in the manufacturing of automobiles. However, other portions of early automobiles proved to be extremely contentious.

The gas-powered engine, for instance, had been patented by George B. Selden, an American attorney, in 1879. The bulk of American manufacturers formed an association in 1903 that would recognize the Selden patent and pay its holder a royalty on each car that came off of their assembly lines. Ford, however, demonstrated his rebelliousness by refusing to go along with the other manufacturers. In fact, he filed suit to break the hold of the patent on the manufacture of gas-powered engines. It was this independent sense of business that helped to make Ford a revolutionary figure in American industry. In 1911, Ford was vindicated when a court ruled that Selden's patent only applied to one engine design. Manufacturers were free to use other designs without paying Selden royalties (Kay 1997, 154–58). In the interim, Ford had set the table for his great success.

Using mass production, Ford's assembly line produced the first of his famous Model T autos in 1908. During the next twenty years, Ford's plants manufactured more than fifteen million Model Ts. Although it only came in one color and lacked pizzazz, the Model T, nicknamed the "flivver" and the "tin lizzie," revolutionized the automotive history. Although auto manufacturing came to a standstill during World War I, the nation emerged from the war more committed to the auto than ever before (Kay 1997, 184–88).

No longer an extravagant novelty, the motorcar had become a necessity. By the early 1920s, most of the basic mechanical problems of automotive engineering had been solved. Manufacturers set out to create a new variety of styles that would appeal to a variety of consumer needs. By the mid-1920s, Henry Ford had decided to abandon the difficult-to-drive Model T and to replace it with the Model A, which was equipped with a more conventional gearshift. New consumers flocked to the new variation. Other manufacturers followed in the late 1920s, including Chrysler, which began production of the Plymouth in 1928. From Ford's uniform "tin lizzie," autos would now become a way for consumers to express who they were and who they wished to be.

Mass production ensured that, by the 1920s, the car had become no longer a luxury but a necessity of American middle-class life. The landscape, however, had been designed around other modes of transport, including an urban scene dependent on foot travel. Cars enabled an independence never before possible, if they were supported with the necessary service structure. Massive architectural shifts were necessary to make way for the auto.

Sources and Further Reading: Brinkley, *Wheels for the World: Henry Ford, His Company and a Century of Progress*; Kay, *Asphalt Nation*; Yergin, *The Prize: The Epic Quest for Oil, Money & Power*.

ESTABLISHING THE INFRASTRUCTURE FOR COORDINATED ADMINISTRATION OVER AMERICA'S NATIONAL PARKS

Time Period: Early 1900s
In This Corner: Preservationists
In the Other Corner: Tourist developers, non-environmentalists
Other Interested Parties: Conservationists
General Environmental Issue(s): Parks, preservation

Even when the first American national parks had been created, they possessed no definition or system for maintenance. Valued most for their natural oddity, these parks and monuments lacked any ability to be preserved once faced with competition or challenge. A historic failure made this park system definition possible and necessary.

Discussed in other essays, the episode at Hetch Hetchy demonstrated the intellectual weakness of the American parks movement in relation to the idea of preservation. Grinnell, Pinchot, and Roosevelt had led the utilitarian perspective on conservation to a broad audience. Agencies including the USGS and the Forest and Reclamation Services gave the federal government a significant involvement in conservation. However, no comparable bureau in Washington spoke for preservation. A wealthy Chicago businessman named Stephen T. Mather recognized this shortcoming. When he brought it to the attention of Franklin K. Lane, the secretary of the interior, in 1915 Lane invited him to join him as his assistant and advisor.

As is often the case, the battle over Hetch Hetchy marked a loss for preservationists while also clarifying the route for future policy initiatives. The shortcomings of the environmental movement that were shown by the Hetch Hetchy episode focused future efforts of environmentalists to clarify and sharpen the definition and meaning of national park.

Crusading for an independent federal bureau to oversee the U.S. national parks, Mather and Conrad Albright effectively blurred the distinction between utilitarian conservation and preservation by emphasizing the economic value of parks as tourist attractions that might spur economic development. Mather initiated a vigorous public relations campaign that led to widespread media coverage in popular magazines. Mather also used funds from western railroads to produce *The National Parks Portfolio*, a lavishly illustrated publication that he sent to congressmen and other influential citizens (Sellars 1997, 48–50).

Although the Interior Department was responsible for fourteen national parks and twenty-one national monuments by 1916, it had no organization to manage them. Typically, parks were staffed by the army, not for security purposes but because they were the only available federal employees. In Yellowstone and in the California parks, military staff developed park roads and buildings and enforced regulations against hunting, grazing, timber cutting, and vandalism. They had little or no ability to educate or advise visitors; therefore, national parks included little of what today is referred to as "interpretation."

Congress quickly responded to Mather's publicity campaign. President Woodrow Wilson signed legislation on August 25, 1916 that created the NPS and placed it within the Interior

Department. The act made the bureau responsible for the Interior's national parks and monuments, as well as Hot Springs Reservation in Arkansas (made a national park in 1921) and other national parks and reservations "of like character" that would subsequently be created by Congress. The act also stipulated how the Park Service was to carry out site management. The agency's efforts, read the act, "[will include] … to conserve the scenery and the natural and historic objects and the wild life therein and to provide for the enjoyment of the same in such manner and by such means as will leave them unimpaired for the enjoyment of future generations."

The Secretary was specifically given the right to

> … sell or dispose of timber in those cases where in his judgment the cutting of such timber is required to control the attacks of insects or diseases or otherwise conserve the scenery or the natural or historic objects in any such park, monument, or reservation. He may also provide in his discretion for the destruction of such animals and of such plant life as may be detrimental to the use of any of said parks, monuments, or reservations. He may also grant privileges, leases, and permits for the use of land for the accommodation of visitors in the various parks, monuments, or other reservations herein provided for, but for periods not exceeding thirty years; and no natural curiosities, wonders, or objects of interest shall be leased, rented, or granted to anyone on such terms as to interfere with free access to them by the public. (Sellars 1997)

Mather was chosen to be the first director of the Park Service. His assistant, Horace Albright, became the assistant director (and would eventually become director). Early initiatives clearly laid out a dual mission for the bureau: conserving park resources and providing for their enjoyment. Although the bureau's existence emphasized the idea of preservation, the rationale for setting sites aside was for their future use and enjoyment by the general American public. For instance, automobiles, which were not permitted in Yellowstone, would now be allowed throughout the system. Roads would be built to assist tourists in accessing sites by vehicle (Sellars 1997, 55–61).

With this unified mission, the national parks changed significantly during ensuing decades. One primary change, however, was in the accessibility of new parks. For a variety of reasons, before 1916 the national parks were primarily in the western United States Although the existence of such parks demonstrated an ethical change in the American relationship with nature, it did not necessarily change the everyday lives of the majority of Americans. During the 1920s, however, the national park system added Shenandoah, Great Smoky Mountains, and Mammoth Cave National Parks in the Appalachian region but required that their lands be donated. Private donors, including John D. Rockefeller, Jr. and other philanthropists, joined with the involved states to slowly acquire and turn over to the federal government the bulk of the lands that would combine new parks that were created during the next decade (Sellers 1997, 108).

The Park Service also took on another eastern U.S.–based mandate by accepting oversight of many historic sites during the 1930s. Starting in 1890, Congress had directed the War Department to manage historic battlefields, forts, and memorials as national military parks and monuments. These included sites such as Gettysburg, Chickamauga, and Chattanooga. After succeeding Mather as director in 1929, Albright succeeded in fulfilling his predecessor's desire to have Congress move military parks to the jurisdiction of the NPS. The popular, largely eastern sites increased the Park Service's annual visitation. Over time, however, they also complicated what the service intended by carrying out "preservation." The efforts to maintain certain aspects of the natural surroundings in Yellowstone, for instance, were a very different mission from facilitating visitors' interpretation of Pickett's Charge at the Gettysburg Battlefield.

Sources and Further Reading: Carr, *Wilderness by Design*; Runte, *National Parks: The American Experience*; Sellars, *Preserving Nature in the National Parks: A History*.

ATOMIC TECHNOLOGY DEFINES THE AMERICAN CENTURY, COLD WAR, AND BEYOND

Time Period: 1945 to the Present
In This Corner: U.S. military, political leaders
In the Other Corner: Soviet Union military, political leaders
Other Interested Parties: Atomic scientists, world public
General Environmental Issue(s): Nuclear technology, Cold War

The mushroom cloud startled each of the handful of scientists, researchers, and military advisors who looked on. The explosion over the Nevada Desert in 1945 during the first successful test of an atomic bomb was much more than any witness had expected, and they were only recognizing the powerful explosion, not the radioactive debris and residue that rained down afterward.

This was the first time that the power of atomic technology stunned its human creators, but it would not be the last. In fact, the significance of this innovation forces a basic question: can a single technology alter a century's worth of diplomacy and world affairs? In the case of the atomic technology, the answer appears to be affirmative.

A New World View

By 1944, World War II had wrought a terrible price on the world. The European theater would soon close with Germany's surrender. Although Germany's pursuit of atomic weapons technology had fueled the efforts of American scientists, the surrender did not end the project. The Pacific front remained active, and Japan did not accept offers to surrender. "Project Trinity" moved forward, and it would involve the Japanese cities Hiroshima and Nagasaki as the test laboratories of initial atomic bomb explosions. *Enola Gay* released a uranium bomb on the city of Hiroshima on August 6, and *Bock's Car* released a plutonium bomb on Nagasaki on August 9. Death tolls vary between 300,000 and 500,000, and most were Japanese civilians. The atomic age, and life with the bomb, had begun

The war ended with the emperor's announcement to the Japanese people on August 15, 1945; however, bomb production did not cease with the cease-fire. The nuclear age had begun, and this technology would help to define the rest of the twentieth century. Although nuclear weapons and technology would play a defining role in making the twentieth century the "American century," the symbol of American success would become the standard of living with which we lived: a middle class with technologies not even available to the wealthiest members of many other societies.

When the first atomic weapons exploded over Japan in 1945, observers from all over the world knew that human life had changed in an instant. In the years since, proponents of nuclear technology have struggled to overcome its dangerous potential and to have it be identified as a public good. Although technology rapidly became a tool for environmental action, it also presented a downside that inadvertently helped to propel environmental concern. Nothing else embodies this development like atomic technology, including the bombs developed to end World War II. However, this era of nuclear successes created an exuberance for many other technologies as well. Together, the technological developments of the post–World War II era brought the United States a standard of living unrivaled in the world. Atomic technology serves as a symbol of the emergence of the United States as a world power based on its ingenuity and technological development. Although these strengths translated into economic progress, they also fueled military dominance as well.

Although nuclear technology began under the control of the military, its implications influence every American. The dangerous implications of nuclear reactions have made it an important part of domestic politics since the 1960s. All of this together illustrates that twentieth-century life has obviously been significantly influenced by "the bomb," although it has been used sparingly—nearly not at all. A broader legacy of atomic technology can be seen on the landscape, from Chernobyl to the Bikini Atoll or from Hiroshima to Hanford, Washington. With this broader legacy in mind, nuclear technology may have impacted everyday nature more than any other innovation. In fact, between its explosive ability and its toxic products, atomic technology changed the order of the natural world, clearly placing with humans the ability to destroy everything.

During the Cold War era, Americans focused on stabilizing everyday life and improving the quality of living for the growing middle class. Nuclear weapons were an important part of this stability. Atomic bombs, which were being tested in the American West, were not known to be a threat to the environment.

Living with the Bomb

Although the nuclear bomb had shortened the war for Americans, it was a horrific weapon that seemed to contradict nearly everything for which the United States stood. One historian describes the announcement of the bomb's use in Japan as a "psychic event of almost unprecedented proportions" (Boyer 1994, 22). Eventually, the Soviet nuclear threat created additional anxiety for Americans. In response, some Americans constructed bomb shelters in backyards. Nearly every community confronted its mortality by creating Civil Defense plans and community bomb shelters. In schools, along with fire drills, young children learned, inaccurately, to duck under their desks for safety in the event of a nuclear attack. Although it brought an end to World War II, nuclear weapons created a kind of insecurity never before seen in human history.

From duck-and-cover drills to science fiction films, Americans learned to live with the bomb that they had created. American anxiety came from both the Soviet threat and the awareness that they had introduced the world to the potential of nuclear destruction. The awe that many felt on witnessing nuclear detonations can be seen in artistic creations such as Andy Warhol's silkscreen *Atomic Bomb* (1965) and James Rosenquist's *F-111* (1964–1965). Feature films used nuclear war or the threat of it to a variety of ends. Most often, films touched the anxiety many Americans felt about their role in developing this deadly technology.

Dr. Strangelove or How I Learned to Stop Worrying and Love the Bomb (1964), *On The Beach* (1959), *The Day After* (1983), *The War Game* (1966), *Threads* (1985), and *War Games* (1983) wove a narrative around some aspect of nuclear weapons. Another entire science fiction genre evolved out of the unknown outcomes of radiation and nuclear testing, including the Godzilla films and 1950s B films such as *Them* (1954), which was one of many that based its narrative on the possible outcome of radiation on insects, vegetables, and humans.

On an individual basis, humans had lived before in a tenuous balance with survival as they struggled for food supplies with little technology; however, never before had such a tenuous balance derived only from man's own technological innovation. Everyday human life changed significantly with the realization that extinction could arrive at any moment. Some Americans applied the lesson by striving to live within limits of technology and resource use. Anti-nuclear activists composed some of the earliest portions of the 1960s counterculture and the modern environmental movement, including "radical" organizations such as the Sea Shepherds and Greenpeace, each of which began by protesting nuclear tests. More mainstream Americans would also eventually question the use of such devices.

Making War on the Brink

Although historians speculate that U.S. President Harry Truman elected to use the atomic bomb to win World War II, he also clearly wanted to set the stage for the next conflict. In short, Truman wanted to impress the Soviet Union, which, even as World War II ended, appeared to be moving independently in contrast to many of the Allies. It is now known that the Soviet Union was able to develop nuclear weapons in a relatively short time because of its espionage activities in the United States and the United Kingdom. On August 9, 1949, the Soviet Union detonated its first atomic device, ending the United States monopoly on having nuclear bombs. Prime Minister Winston Churchill announced on February 26, 1952 that the United Kingdom also had an atomic bomb. France and China also demonstrated nuclear capability, and these five countries were considered to be the Nuclear Powers throughout the Cold War period. The technology had clearly altered global affairs.

Nuclear technology had also altered the nature of warfare. In the first major conflict of the atomic era, America led United Nations (U.N.) forces pressing North Koreans out of South Korea and all the way through the Korean peninsula to the border with China at the Yalu River. Led by Douglas MacArthur, the American forces were poised to move even farther. MacArthur waited for the order from President Truman; however, it never came. Truman realized that further escalation would mean a larger conflict with the enemies possessing nuclear weapons. Much to MacArthur's chagrin, atomic weapons had ushered in the need for limited warfare. Truman would send the troops no further.

Truman's successor, General Dwight D. "Ike" Eisenhower, moved the nuclear era into its next phase by attempting to demilitarize the technology. In 1953, Eisenhower appeared before the U.N. and presented a speech that has become known as the "Atoms for Peace" speech, in which he clearly instructed the world on the technological stand-off that confronted it. The "two atomic colossi," he forecasted, could continue to "eye each other indefinitely across a trembling world," but eventually their failure to find peace would result in war and "the probability of civilization destroyed," forcing "mankind to begin all over again the age-old struggle upward from savagery toward decency and right, and justice."

To Eisenhower, "no sane member of the human race" could want this. In his estimation, the only way out was discourse and understanding. With exactly these battle lines, a war—referred to as cold, because it never escalated (heated) to direct conflict—unfolded over the coming decades. With ideology—communism vs. capitalism—as its point of difference, the conflict was fought through economics, diplomacy, and the stockpiling of a military arsenal. With each side possessing a weapon that could annihilate not just the opponent but the entire world, the bomb defined a new philosophy of warfare referred to as a Cold War (Boyer 1994, 22).

The Cold War was predicated on a balanced threat referred to as the doctrine of Mutually Assured Destruction (or, appropriately, MAD for short). So important was this balance to international political stability that a treaty, the Antiballistic Missile Treaty was signed by the United States and the United Soviet Socialist Republic in 1972 to curtail the development of defenses against nuclear weapons and the ballistic missiles that carry them. It was believed that such a system would upset the balance by offering one nation protection from the other.

In the United States, citizens learned to live with the threat of a nuclear attack from the Soviet Union. This threat, of course, was made possible by each nation's very advanced delivery system. Early delivery systems for nuclear devices were primarily bombers such as the American B-29 Superfortress. Later, ballistic missile systems, based on earlier designs used by Germany, were developed by both American and Soviet scientists. Similar rocket systems were also used for nonmilitary purposes, including launching satellites, and space travel. Ultimately, these rocket systems were attached to nuclear missiles referred to as Intercontinental Ballistic Missiles with which nuclear powers could deliver that destructive force anywhere on the globe.

Delivery technologies represented an important facet of the nuclear arms race. Each new development marked an important new chapter in the Cold War. Ultimately, when negotiations and treaties were used to defuse the Cold War, weapons systems were typically used as bargaining chips. This strange war in history, however, was fought on many fronts. Some of these fronts involved no weapons at all.

Fighting the Cold War through Lifestyle

Particularly in the United States, everyday life became a device to serve as a model of the potential of democratic success. The Truman administration construed of an era in which the entire world seemed a playing field that could either "go red" and become Communist or shift to the American model of democratic capitalism. Images of Americans enjoying the globe's highest standard of living became an important component of the Truman world view. Afterward, documents such as NSC-68 have been used to reconstruct the emerging logic of the day:

NSC-68

A. Nature of Conflict

The Kremlin regards the United States as the only major threat to the conflict between the idea of slavery under the grim oligarchy of the Kremlin, which has come to a crisis with the polarization of power described in Section I, and the exclusive possession of atomic weapons by the two protagonists. The idea of freedom, moreover, is peculiarly and intolerably subversive of the idea of slavery. But the converse is not true. The implacable purpose of the slave state to eliminate the challenge of freedom has placed the two great powers at opposite poles. It is this fact which gives the present polarization of power the quality of crisis....

The objectives of a free society are determined by its fundamental values and by the necessity for maintaining the material environment in which they flourish. Logically and in fact, therefore, the Kremlin's challenge to the United States is directed not only to our values but to our physical capacity to protect our environment. It is a challenge which encompasses both peace and war and our objectives in peace and war must take account of it.

1. Thus we must make ourselves strong, both in the way in which we affirm our values in the conduct of our national life, and in the development of our military and economic strength.
2. We must lead in building a successfully functioning political and economic system in the free world. It is only by practical affirmation, abroad as well as at home, of our essential values, that we can preserve our own integrity, in which lies the real frustration of the Kremlin design.
3. But beyond thus affirming our values our policy and actions must be such as to foster a fundamental change in the nature of the Soviet system, a change toward which the frustration of the design is the first and perhaps the most important step. Clearly it will not only be less costly but more effective if this change occurs to a maximum extent as a result of internal forces in Soviet society.

In a shrinking world, which now faces the threat of atomic warfare, it is not an adequate objective merely to seek to check the Kremlin design, for the absence of order among nations is becoming less and less tolerable. This fact imposes on us, in our own interests, the responsibility of world leadership. It demands that we make the attempt, and accept the risks inherent in it, to bring about order and justice by means consistent with the principles of freedom and democracy. We should limit our requirement of the Soviet Union to its participation with other nations on the basis of equality and respect for the rights of others. Subject to this requirement, we must with our allies and the former subject peoples seek to create a world society based on the principle of consent. Its framework cannot be inflexible. It will consist of many national communities of great and varying abilities and resources, and hence of war potential. The seeds of conflicts will inevitably exist or will come into being. To acknowledge this is only to acknowledge the impossibility of a final solution. Not to acknowledge it can be fatally dangerous in a world in which there are no final solutions.

All these objectives of a free society are equally valid and necessary in peace and war. But every consideration of devotion to our fundamental values and to our national security demands that we seek to achieve them by the strategy of the cold war. It is only by developing the moral and material strength of the free world that the Soviet regime will become convinced of the falsity of its assumptions and that the pre-conditions for workable agreements can be created. By practically demonstrating the integrity and vitality of our system the free world widens the area of possible agreement and thus can hope gradually to bring about a Soviet acknowledgement of realities which in sum will eventually constitute a frustration of the Soviet design. Short of this, however, it might be possible to create a situation which will induce the Soviet Union to accommodate itself, with or without the conscious abandonment of its design, to coexistence on tolerable terms with the non-Soviet world. Such a development would be a triumph for the idea of freedom and democracy. It must be an immediate objective of United States policy. (Federation of American Scientists, *NSC-68*)

Conclusion: Axis of Evil and Efforts to Develop Nuclear Technology

The importance of nuclear weapons to international relations did not cease with the end of the Cold War. The topic of proliferation, a rogue, independent nation gaining nuclear weapons and forcing the hands of the super powers, had gained credibility during the last decades of the Cold War. In the post–Cold War world, developing nuclear weapons has become a way of forcing the United States and other developed countries to listen to smaller nations.

For instance, both North Korea and Iran, nations labeled by President George W. Bush in 2001 as being part of the "Axis of Evil," face scrutiny for their alleged nuclear weapons programs. In each case, the effort to develop weapons programs changes daily.

In recent years, North Korea has backed off its initial refusal to negotiate with the United States. Over the past few years, North Korea has set off international condemnation by testing its first nuclear bomb. Then the nation agreed to begin a series of talks that are intended to eventually cease the rogue nation's weapons programs in exchange for the ending of sanctions, financial aid, and an eventual inclusion in the international community.

Iran, conversely, seems particularly convinced that the development of nuclear weapons is its only way to receive fair treatment from the developed nations, and it will not back down. Its uranium enrichment program refutes a U.N. Security Council deadline that called for the program to cease.

Although these nations would require decades to develop an arsenal that would endanger the United States, the effort to limit the number of nuclear nations has become a driving force behind American foreign diplomacy. Many experts speculate that it may also drive future military conflict.

Sources and Further Reading: Boyer, *By the Bomb's Early Light*; Garwin, *Megawatts and Megatons: A Turning Point in the Nuclear Age.*

PUBLIC LEARNS GARDENING AND RATIONING TO SUPPORT THE CAUSE

Time Period: 1940s
In This Corner: American public, particularly women
In the Other Corner: Axis powers, shortage
Other Interested Parties: Federal government
General Environmental Issue(s): Conservation, recycling, gardening, World War II

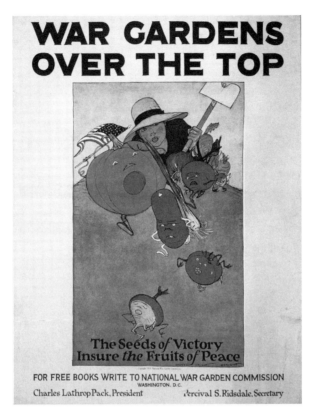

This poster is part of the National War Garden Commission's campaign to encourage Americans to raise more food and free resources for the U.S. military needs in World War I. Similar efforts were made during World War II to encourage Americans to use gardens to increase their self-sufficiency. Library of Congress.

World War II presented a time of global panic possibly unlike any seen before or since. As powerful, imperialist nations sought to dominate the world, much of the free world, including the United States, lived in fear of attack. Once the attack on Pearl Harbor pulled the United States into the fighting, civilian Americans rallied to support the war effort like never before. Severe rationing was imposed to conserve materials needed for the war effort. Rationing meant sacrifices for all. Even Americans who would never see the fighting became integral parts of the war efforts. Although many had doubts about whether or not consumer-oriented Americans could rally in such a fashion, the effectiveness of rationing meant not a single person was left unaffected by World War II. At a time when women were not allowed to serve in the armed forces, rationing served as a way that they served the patriotic cause.

In the spring of 1942, the Food Rationing Program was established at the federal level so that the government could control supply and demand. Rationing was introduced to avoid public anger with shortages. Although industry and commerce were affected, individuals felt the effects more intensely. Throughout the war, Americans gave up many material goods, but there also was an increase in employment. Individual efforts evolved into clubs and organizations coming to terms with the immediate circumstances, including scrap drives, taking factory jobs, goods donations, and other similar projects to assist those on the front. Government-sponsored ads, radio shows, posters, and pamphlet campaigns urged the American people to comply. With a sense of urgency, the campaigns used propaganda to reach the masses and encourage them to contribute by whatever means they had.

Although life during the war meant daily sacrifice, few complained because they knew it was the men and women in uniform who were making the greater sacrifice. A poster released by the Office of War Information stated simply, "Do with less so they'll have enough." Yet another pleaded, "Be patriotic, sign your country's pledge to save the food." On the whole, the American people united in their efforts.

Training sessions were held to teach women to shop wisely, conserve food and plan nutritious meals, as well as teach them how to can food items. The homemaker planned family meals within the set limits. The government also printed a monthly meal-planning guide with recipes and a daily menu. *Good Housekeeping* magazine printed a special section for rationed foods in its 1943 cookbook. Numerous national publications also featured articles explaining what rationing meant to America. Overall, rationing most likely helped Americans to eat more healthfully by reducing the amount of fats and red meats being eaten.

Although food manufacturers participated in the rationing effort, some companies took advantage of the wartime shortages to flaunt their patriotism to generate increased profit. The familiar blue box of Kraft Macaroni and Cheese Dinner gained great popularity as a substitute for meat and dairy products. Two boxes required only one rationing coupon, which resulted in eighty million boxes sold in 1943. Food substitutes created in laboratories became much more widely used. The best known example was real butter being replaced with oleo margarine. In addition, cottage cheese became a popular substitute for meat, with sales exploding from 110 million pounds in 1930 to 500 million pounds in 1944.

Organizing such efforts followed a basic structure. For instance, sugar rationing took effect in May 1943 with the distribution of "Sugar Buying Cards." Registration usually took place in local schools. Each family was asked to send only one member for registration and be prepared to describe all other family members. Coupons were distributed based on family size, and the coupon book allowed the holder to buy a specified amount; however, possession of a coupon book did not guarantee that sugar would be available.

Rationing forced Americans to view food very differently. Whereas some food items were scarce, others did not require rationing, and Americans adjusted accordingly. "Red Stamp" rationing covered all meats, butter, fat, and oils, and, with some exceptions, cheese. Each American was allowed a certain amount of points weekly with expiration dates to consider. "Blue Stamp" rationing covered canned, bottled, and frozen fruits and vegetables, plus juices and dry beans, and such processed foods as soups, baby food, and catsup. Ration stamps became a kind of currency, with each family being issued a "War Ration Book." Each stamp authorized a purchase of rationed goods in the quantity and time designated, and the book guaranteed each family its fair share of goods made scarce thanks to the war.

Many Americans living in urban and suburban areas took a page from the nation's agrarian past and grew much of their own produce in gardens. These came to be called "Victory Gardens."

In 1941, the Agricultural Department startled most Americans by suggesting that, if they wanted fresh vegetables on their dinner tables, they should plant a Victory Garden. Many urban Americans had little experience with growing a garden. Many novice gardeners experimented with hoe and shovel to do their part in the war effort. Armed with seeds and shovels, millions of small town backyard and city rooftop gardens suddenly sprouted across the country. The government reported that, by 1943, these "Sunday Farmers" had planted more than twenty million Victory Gardens. Their gardens were estimated to produce eight million tons of food! During the war, this production accounted for more than 50 percent of all the fresh vegetables consumed in the United States.

Other efforts on the home front included increased manufacturing and the emergence of women as a force in factory labor. Many of these factories, particularly in the Pacific Northwest and in California, created new models of mass production, especially of aircraft. In other areas, however, America's strength in industrial manufacturing combined with cutting-edge theoretical thought to make a weapon not just capable of fighting the Axis powers but also of defeating them.

For nonfood rationing, the government often used a point system. In addition to food, rationing encompassed clothing, shoes, coffee, gasoline, tires, and fuel oil. With each coupon book came specifications and deadlines. Rationing locations were posted in public view. Rationing of gas and tires highly depended on the distance to one's job. If one was fortunate enough to own an automobile and drive at the then specified speed of thirty-five miles per hours, one might have a small amount of gas remaining at the end of the month to visit nearby relatives.

The rationing effort also marked the first organized effort at recycling metals and other materials that could be reused. Saving aluminum cans meant more ammunition for the soldiers. Economizing initiatives seemed endless as Americans were urged to conserve and recycle metal, paper, and rubber. War bonds and stamps were sold to provide war funds, and the American people also united through volunteerism. Communities joined together to hold scrap iron drives, and schoolchildren pasted saving stamps in bond books.

All Americans did not react perfectly to rationing. There was a black market, where people could buy rationed items on the sly but at higher prices. The practice provoked mixed emotions from those who banded together to conserve as instructed, as opposed to those who fed the black market's subversion and profiteering. For the most part, black marketers dealt in clothing and liquor in Britain and meat, sugar, and gasoline in the United States.

The rationing effort lasted into 1946. Life resumed as normal, and the consumption of meat, butter, and sugar inevitably rose. Although Americans still live with some of the results of World War II, rationing has not returned. The active role taken by women, however, was a harbinger of important changes in American ideas of gender. In particular, the involvement of women in industrial production—work in which they actively earned a wage and achieved independence—presented women with choices that had not previously been available to them. Part of the reason for this was changes in their expectations; however, their hard work and dedication had also altered the ideas of American society.

Sources and Further Reading: Opie, *Nature's Nation*; Rupp, *Mobilizing Women for War*; Steinberg, *Down to Earth*; For women working in production in World War II, see Restone Arsenal, *Women and War*, http://www.redstone.army.mil/history/women/welcome.html; Teacher Oz's Kingdom of History, *Women and the Home Front during World War II*, http://www.teacheroz.com/WWIIHomefront.htm.

CITY BEAUTIFUL, URBAN RENEWAL, AND THE EFFORT TO REFORM THE MODERN CITY

Time Period: 1900 to the present
In This Corner: Urban planners, sociologists
In the Other Corner: City officials, business owners
Other Interested Parties: Local, state, and federal officials
General Environmental Issue(s): Planning, cities, urban

In the early 1900s, Americans developed new expectations for their own health and safety. For many decades, however, these expectations remained limited by economic class: wealthier Americans demanded and received reform, whereas many working-class communities remained unsafe through much of the twentieth century. Other essays have discussed initial attempts to reform the cities in which many working-class Americans lived. The revolutionary ideas of Jacob Riis and others, however, took physical form in efforts at large-scale planning that took on the ills of urban life and attempted to create solutions. This impulse has gone by a variety of names, including the City Beautiful movement and urban renewal.

Horace McFarland and City Beautiful

Cities, which had long been centers of economic concentration and production, began to be understood more as residential environments by the early 1900s. Although this same era saw a rise in suburbanization, noticeable improvements occurred within American cities as well. From 1860 to 1910, the number of Americans living in cities increased by 46 percent (Hines 1991, 81). With population centering in urban areas, Americans began to consider issues of crime, health, and poverty with added urgency.

The connection between nature and health was most obvious in the effort to plan natural areas into many cities. The American Park and Outdoor Art Association, begun in 1897, championed outdoor art and the cultivation of beautiful landscapes in great city parks. It consisted of about 230 landscape architects, park superintendents, commissioners, and laypeople who were inspired by Frederick Law Olmsted's park designs. They met annually and encouraged "proper" principles of park development, landscaping of factory grounds, school yards, railroad-station sites, and city streets. They railed against billboards and pleaded for state parks and forest preservation. They emphasized piecemeal, practical projects. One of the best known spokesmen for this movement was J. Horace McFarland, who was discussed in a previous essay. He would become known as the father of the American City Beautiful Movement.

Between 1900 and 1904, a social reform movement took shape that became known as the City Beautiful movement. Planners and reformers designed ideal landscapes, much like that of the Columbian Exposition, within cities in an effort to incorporate municipal art, civic improvement, and outdoor art associations into every city. These efforts joined in the American Civic Association to promote elaborate "model cities," including the design of the 1904 St. Louis World's Fair. Soon, however, the priorities of the City Beautiful effort could be seen in every American city as well as in the development of new suburbs and roadways to connect them. The common theme was planning and design for human living with an aesthetic priority on natural forms.

Expanding Scale of Urban Renewal

Merging the ideas of the City Beautiful movement with the planning and design of modernism, urban renewal after 1950 introduced projects of a massive scale. Government sponsorship enabled developers to flatten entire urban interiors that were not "functioning" well economically or sociologically and to allow planners to consider urban life from a nearly blank slate. Emphasizing skyscrapers, asphalt and concrete, urban renewal is often criticized today for prioritizing grand plans ands strategies over the lives of everyday people.

When planning and new, modern forms of architecture became popular during the New Deal, attention again focused on urban areas. The Housing Act of 1934, which established the Federal Housing Authority (FHA) and the Department of Housing and Urban Development, searched for methods of improving urban housing standards. This search led to redlining, a practice viewed as racist today because, by creating target-specific areas, the practice made it more difficult for ethnic minorities to secure mortgages in certain areas. This resulted in residential racial segregation in many areas of the United States. The Housing Act was altered in 1937 to create the nation's first public housing program, which created the large public housing projects that would become popular during urban renewal. Local governments could fund new housing, but first they needed to demolish slum housing.

Robert Moses, a planner in New York City from the 1930s into the 1970s, helped to create the larger-scale plans. In New York, Moses initiated new bridges, highways, housing projects, and public parks. Moses used new technologies and designs to remake the city's infrastructure, which was then tied into urban housing. In New York, Moses' emphasis on highways also empowered suburban growth, which would see explosive growth throughout the United States after World War II.

Whereas Arthur Levitt and others perfected the prefabricated suburb, the Housing Act of 1949 stimulated the wholesale demolition of urban slums. Entire neighborhoods were torn down and replaced by highways. Unabashedly structured around redlining, discrimination became the norm, and housing values declined rapidly in minority neighborhoods. The stage was set for a publicly funded era of urban renewal.

The main initiative for urban renewal came in Pittsburgh, Pennsylvania, where multimillionaire R. K. Mellon helped to stimulate a modern urban-renewal program in May 1950. A massive section of the downtown famous for its dirt and pollution was demolished and renamed the Golden Triangle. In its place emerged a futuristic landscape of parks, office buildings, and a sports arena. Soon other cities set out to create their own renewal zones.

The role of state and federal government in these new developments was ensured with the passage in 1956 of the Interstate Highway Act. State and federal governments gained complete control over new highways. The priority of the program was to bring traffic more easily in and out of cities' central cores, even if it meant dislodging or isolating urban neighborhoods. Very often, the redevelopment most impacted African-American communities located in urban areas. Some critics began to claim that "Urban renewal is Negro removal." Indeed, many African Americans and Latinos were left with no other option than moving into the public housing that was being constructed. Many white urban residents, conversely, continued the trend of moving to suburbs.

In Boston, one of the country's oldest cities, almost one-third of the old city was demolished, including the historic West End, to make way for a new highway, low- and moderate-income high-rises (which eventually became luxury housing), and new government and commercial buildings. Later, this would be seen as a tragedy by many residents and urban planners, and one of the centerpieces of the redevelopment, Boston City Hall, is still considered an example of the excesses of urban renewal.

Fairly rapidly, critics began to realize what was being lost with urban renewal. In 1961, Jane Jacobs published *The Death and Life of Great American Cities*, which criticized the "soulless" nature of urban renewal. Although criticism intensified, architects and designers did not create a rebuttal for decades. In the 1990s, the criticism of Jacobs, James Kunstler, and others helped to inspire the design field of New Urbanism.

Regarding the racism that was inherent in much of urban renewal, the Civil Rights Act of 1964 removed racial deed restrictions on housing. This led to the beginnings of desegregation of residential neighborhoods, but redlining continued to mean that real estate agents could steer ethnic minorities to certain areas.

By the 1970s, many major cities developed opposition to the sweeping urban-renewal plans for their cities. In Boston, community activists halted construction of the proposed Southwest Expressway. San Francisco's mayor, Joseph Alioto, became the first mayor to publicly denounce urban renewal. He forced California to discontinue highway construction through the heart of the city.

Policies initiated under President Lyndon Johnson's "War on Poverty" also eventually helped to slow the push for urban renewal. The Housing and Urban Development Act and The New Communities Act of 1968 guaranteed private financing for private entrepreneurs to plan and develop new communities. However, they were updated in 1974 to establish the Community Development Block Grant program. As its name suggests, this program was more redevelopment than renewal. Prioritizing existing neighborhoods and properties rather than demolitioning substandard housing and economically depressed areas, the block grant program provided assistance to many urban communities.

Conclusion: Creating Livable Cities

Since the 1970s, renovation has mixed with selective demolition in efforts to revitalize urban neighborhoods. New Urbanism has been embraced by many planners.

Urban renewal was one of the most utopian applications of modernist planning. It never lived up to these grand hopes. For some cities, it brought economic and cultural development. Almost always, however, the new vitality came with the loss of cultural heritage

Sources and Further Reading: Jacobs, *The Death and Life of Great American Cities*; Kunstler, *Geography of Nowhere: The Rise and Decline of America's Man-Made Landscape*.

SELLING AMERICA'S FUTURE TO THE AUTOMOBILE

Time Period: 1910s
In This Corner: Auto manufacturers, engineers working with internal-combustion engines
In the Other Corner: Gasoline producers, mass transit interests
Other Interested Parties: Consumers, political officials
General Environmental Issue(s): Automobile, transportation, energy, planning

New technologies allowed twentieth-century Americans to redefine everyday human life in a number of different ways. Flexible, individual transportation became the norm after the 1920s, and, by the 1950s, this priority had overwhelmed the American landscape. To Americans in general, the automobile became a symbol of the U.S. unrivaled standard of living. Today, however, our automobility has begun to look like our society's Achilles heel. Growing out of the same interest in creating a modern world that integrated new technologies into everyday life, automobility—the reliance on automobiles—presented one of the most ubiquitous technologies in American history. With each passing year in the twentieth century, the American landscape seemed to transform itself in various ways that would better accommodate the automobile and the living patterns that were connected with it.

The American Home and Automobility

With the national future clearly tied to cars, planners began perfecting ways of further integrating the car into American domestic life. Initially, these tactics were quite literal. In the early twentieth century, many homes of wealthy Americans soon required the ability to store vehicles. Most often these homes had carriage houses or stables that could be converted. Soon, of course, architects devised an appendage to the home and gave it the French name garage. From this early point, housing in the United States closely followed the integration of the auto and roads into American life.

Upper- and middle-class Americans began moving to suburban areas in the late 1800s. The first suburban developments, such as Llewellyn Park, New Jersey (1856), followed train lines or the corridors of other early mass transit. The automobile allowed access to vast areas between and beyond these corridors. Suddenly, the suburban hinterland around every city compounded. As early as 1940, about thirteen million people lived in communities beyond the reach of public transportation. As new construction subsequently began, more recent ideas and designs (such as the ranch house) remade the American suburb (Jackson 1985, 102).

Planners used home styles such as these to develop one site after another, with the automobile linking each one to the outside world. The ticky-tacky world of Levittown (the first of which was constructed in 1947) involved a complete dependence on automobile travel. This shift to suburban living became the hallmark of the late twentieth century, with more than half of the nation residing in suburbs by the 1990s. The planning system that supported this residential world, however, involved much more than roads. The services necessary to support outlying, suburban communities also needed to be integrated by planners.

Instead of the Main Street prototype for obtaining consumer goods, the auto suburbs demanded a new form. Initially, planners such as Jesse Clyde Nichols devised shopping areas, such as Kansas City's Country Club Plaza, that appeared as a hybrid of previous forms. Soon, however, the "strip" had evolved as the commercial corridor of the future. These sites quickly became part of suburban development to provide basic services close to home. A shopper rarely arrived without an automobile; therefore, the car needed to be part of the design program. The most obvious architectural development for speed was signage: integrated into the overall site plan would be towering neon signs that identified services. Also, parking lots and drive-thru windows suggest the integral role of transportation in this new style of commerce (Jackson 1985, 159).

In the United States, roads initiated related social trends that added to Americans' dependence on petroleum. Most important, between 1945 and 1954, nine million people moved to suburbs. The majority of the suburbs were connected to urban access by only the automobile. Between 1950 and 1976, central city population grew by ten million, whereas suburban growth was eighty-five million. Housing developments and the shopping/strip-mall culture that accompanied decentralization of the population made the automobile a virtual necessity. Shopping malls, suburbs, and fast-food restaurants became the American norm through the end of the twentieth century, making American reliance on petroleum complete (Kay 1997, 220–25).

Hitting the Road

The earliest suburbs did not necessitate infrastructure such as road construction. However, the rapid expansion of American interest in living outside of urban areas combined with the growing American passion for the automobile by the 1920s required the development of American roads. Although the motorcar was the quintessential private instrument, its owners had to operate it over public spaces. Who would pay for these public thoroughfares? After a period of acclimation, Americans viewed highway building as a form of social and economic development that was necessary for almost any community to succeed. They justified public financing for such projects on the theory that roadway improvements would pay for themselves by increasing property-tax revenues along the route. At this time, asphalt, macadam, and concrete were each used on different roadways

By the 1920s, the congested streets of urban areas pressed road building into other areas. Most urban regions soon proposed express streets without stop lights or intersections. These aesthetically conceived roadways, normally following the natural topography of the land, soon took the name parkways. Long Island and Westchester County, New York, used parkways with bridges and tunnels to separate them from local cross-traffic. The Bronx River Parkway (1906), for instance, follows a river park and forest; it also is the first roadway to be declared a national historic site. In addition to pleasure driving, such roads stimulated automobile commuting (Jackson 1985, 166).

The Federal Road Act of 1916 offered funds to states that organized highway departments, designating 200,000 miles of road as primary and thus eligible for federal funds. More importantly, ensuing legislation also created a Bureau of Public Roads to plan a highway network to connect all cities of 50,000 or more inhabitants. Some states adopted gasoline taxes to help finance the new roads. By 1925, the value of highway construction projects exceeded $1 billion. Expansion continued through the Great Depression, with road building becoming integral to city and town development.

Robert Moses of New York defined a new role as road builder and social planner. Through his work in the greater New York City area from 1928 to 1960, Moses created a model for a metropolis that included and even emphasized the automobile as opposed to mass transportation. This was a dramatic change in the motivation of urban design. Historian Clay McShane wrote, "In their headlong search for modernity through mobility, American urbanites made a decision to destroy the living environments of nineteenth-century neighborhoods by converting their gathering places into traffic jams, their playgrounds into motorways, and their shopping places into elongated parking lots" (McShane 1994, 38).

Consumption Meets the Road

During the twentieth century, planners and designers gave Americans what they wanted: a life and landscape married to the automobile. Most details of the new, planned landscape reflected this social dynamic. For instance, whereas drivers through the 1930s often slept in roadside yards, developers soon took advantage of this opportunity by devising the roadside camp or motel. Independently owned tourist camps graduated from tents to cabins, which were often called "motor courts." After World War II, the form became a motel in which all the rooms were tied together in one structure. Still independently owned, by 1956, there were 70,000 motels nationwide. Best Western and Holiday Inn soon used ideas of prefabrication to create chains of motels throughout the nation. Holiday Inn defined this new part of the auto landscape by emphasizing uniformity so that travelers felt as if they were in a familiar environment regardless of where they traveled (Belasco 1979, 44–56).

The auto landscape, of course, needed to effectively incorporate its essential raw material: petroleum. The gas station, which originally existed as little more than a roadside shack, mirrors the evolution of the automobile-related architecture in general. By the 1920s, filling stations had integrated garages and service facilities. These facilities were privately owned and uniquely constructed. By the mid-1930s, oil giants such as Shell and Texaco developed a range of prototype gas stations that would recreate the site as a showroom for tires, motor oil, and other services. George Urich introduced the nation's first self-service gas station in California in 1947. By the end of the 1900s, gas stations had been further streamlined to include convenience stores and the opportunity to pay at the pump. The gas station experience would steadily become less personalized.

As cars became more familiar in everyday Americans' lives, planners and developers formalized refueling stations for the human drivers, as well. Food stands informally provided refreshment during these early days, but soon restaurants were developed that used marketing strategies from the motel and petroleum industries. Diners and family restaurants sought prime locations along frequently traveled roads; however, these forms did not alter dining patterns significantly. In 1921, White Castle hamburgers combined the food stand with the restaurant to create an eating place that could be put almost anywhere. Drive-in restaurants would evolve around the idea of quick service, often allowing drivers to remain in their automobile. Fast food as a concept, of course, derives specifically from Ray Kroc and the McDonald's concept that he marketed out of California beginning in 1952. Clearly, the idea of providing service to automobile drivers had created an entire offshoot of the restaurant industry (Kay 1997, 54–59).

While most roadside building types evolved gradually, the drive-in theater was deliberately invented. Richard M. Hollingshead, Jr. of New Jersey believed that entertainment needed to incorporate the automobile. Hollingshead patented the first drive-in theater in 1933, but the invention would not proliferate until the 1950s. Viewing films outdoors in one's car became a symbol of the culture of consumption that overtook the American middle class during the post-war era. Of course, it also established the automobile as a portable, private oasis where youth could express their sexuality as well as experiment with drugs and alcohol.

The Interstate Highway System

The Federal Road Act of 1916 began a century of road building that some historians have called the "largest construction feat of human history," and the American road system unfolded throughout the early twentieth century. Highway building was intensified in the 1950s when President Dwight D. Eisenhower included a national system of roads in his preparedness plans for nuclear attack. Road development for this purpose became a vital portion of Cold War economic growth in the interrelated economy that Ike dubbed the "military-industrial complex." This connection cleared the way for the federal funds to create the world's greatest highway system: the Interstate Highway Act spanned the nation with large-scale roads that could aid with mass evacuations in the event of a nuclear war. In the meantime, however, the roads became the corridor for trade and industry that allowed trucking to supplement and eventually replace railroading (Lewis 1997, 6–12).

Why did such inspiration come to Ike? In 1920, Eisenhower led troops across the American road system in a military call for new roads. Then he witnessed the spectacle of Hitler's Autobahn first hand. When he became president, he worked with automobile manufacturers and others to devise a 1956 plan to connect America's future to the automobile. The interstate highway system was the most expensive public works project in human history. The public rationale for this hefty project revolved around fear of nuclear war: such roadways would assist in exiting urban centers in the event of such a calamity. The emphasis, however, was clearly economic expansion. At the cost of many older urban neighborhoods, often occupied by minority groups, the huge wave of concrete was unrolled that linked all the major cities of the nation.

Conclusion: The Lessons of the View in the Rearview Mirror

Linking our everyday existence to petroleum helped to define the politics, warfare, and foreign diplomacy of the late twentieth century. Although a separate essay will investigate how these decisions play out in the early twenty-first century, petroleum dominated American foreign diplomacy after the 1973 oil embargo.

Sources and Further Reading: Belasco, *Americans on the Road*; Jacobs, *The Death and Life of Great American Cities*; Jackson, *Crabgrass Frontier: The Suburbanization of the United States*; Kay, *Asphalt Nation*; Kunstler, *Geography of Nowhere: The Rise and Decline of America's Man-Made Landscape*; Lewis, *Divided Highways*.

RACHEL CARSON AND CHANGING AMERICANS' VIEW OF CHEMICALS

Time Period: 1950–1970
In This Corner: Rachel Carson, environmentalists, portions of the scientific community
In the Other Corner: Chemical industry, portions of the scientific community
Other Interested Parties: American public, political officials
General Environmental Issue(s): Chemicals, environment and popular culture, modern environmentalism

The American confidence in new technologies was seen clearly in the rapid adoption of many untested chemicals into everyday twentieth century life. Many of these chemicals significantly improved aspects of America's standard of living. Some of these, however, were eventually determined to carry with them significant health and environmental impacts.

Given the positive aspects of chemical use, convincing American consumers of the dangers of some of them was profoundly difficult. Scientists such as Rachel Carson led this intellectual awakening and may, in the process, have saved many lives.

DDT and the Chemical Future

In 1939, scientists were stunned when a few grains of white powder miraculously wiped out colonies of mosquito larvae. The powder had been developed years earlier but never tied to a specific purpose. With neither testing nor consideration of potential dangers, dichlorodiphenyltrichloroethane (DDT) left the laboratory for use as an insecticide.

The chemical seemed to be one of the great examples of the ability of humans to use technology to eliminate limitations of the natural world around them. During World War II, for instance, B-25 bombers sprayed DDT before invasions in the Pacific to kill off insects and help reduce the possibility of disease for the American soldiers who would follow. After the war, application of DDT almost singlehandedly wiped out malaria in the developed world and drastically reduced it elsewhere. Paul Müller, the chemist who first turned it on unsuspecting flies, won a Nobel Prize in 1948 for his work. In 1970, the National Academy of Sciences estimated that, during the previous decades, DDT had likely saved more than 500 million lives from malaria (Opie 1998, 413–15).

DDT became a major part of the modernist view that nature could be made "pestless." By the late 1950s, DDT production was nearly five times the production of the World War II era. Town and municipal authorities liberally sprayed DDT on American suburbs to eradicate tent caterpillars, gypsy moths, and the beetles that carried Dutch elm disease. It was an important tool in creating the suburban aesthetic of green nature composed of only what we wanted. Often, residents would see the plume of spray passing through their neighborhoods and would shield neither themselves nor their children and pets from possible effects from overspray. In fact, the event of the spraying rig on a truck or airplane was viewed only with relief at what the chemical would do to bothersome insects.

Slowly, however, observers began to note that DDT killed indiscriminately. It became clear that the insecticide could not be controlled or limited once it entered an ecosystem. In addition, the new lessons of ecology demonstrated that some species identified as pests actually played an important role in larger food chains and ecological webs that connected them with species that humans cared about, or even with ourselves. The connection that it had on human life, however, required that the nation be given a lesson in systems ecology.

Carson Alters the Chemical Paradigm

Pollution composed the most frequent environmental complaint before 1950, but its nuisance derived primarily from physical discomfort more than from a scientific correlation with human illness. Scientific inquiry was required to definitively connect health problems to pollution and chemicals, but most scientists were housed in the industries that were responsible for creating such toxicity. Middle-class Americans became concerned with such information by the late 1950s. The suburban ideal had helped to create an expectation of safety in Americans. One of the first writers to take advantage of this increased interest among middle-class Americans was a government biologist named Rachel Carson. She began writing about nature for general readers in the late 1950s. Then, in 1962, Rachel Carson's *Silent Spring* erupted onto the public scene to become a bestseller after first being serialized in *The New Yorker*.

Often, literary critics compare the impact of *Silent Spring* on the American scene with that of *Uncle Tom's Cabin*, which contributed to the start of the Civil War after its publication in the 1850s. "Without this book," wrote former Vice President Al Gore, "the environmental movement might have been long delayed or never have developed at all" (Carson 2002, xix). From the day it hit bookstores, *Silent Spring* fueled a vigorous public debate about the use of chemicals, and the role of technology in our nation's future, that continues today. Carson's scientific findings brought into question basic assumptions that Americans had about their own safety and many of the chemicals that they used to create their comfortable standard of living.

At the time of the writing of *Silent Spring*, Carson had no intention of starting a large-scale public movement. Growing up in western Pennsylvania, Carson studied nature and read scientific books. She focused on marine biology and, although she began at the Pennsylvania College for Women in the mid-1920s as an English major, she graduated with a degree in biology. She earned a master's degree in marine zoology from Johns Hopkins University and went on to take a job as a junior aquatic biologist for the U.S. Bureau of Fisheries. For her first stab at writing for the general public, Carson turned to her passion for the sea. Published in 1941, *Under the Sea-Wind* sold fewer than 2,000 copies. During the next decade, however, Carson became recognized as a scientist who could write for a broader reading audience. Colleagues began alerting her to scientific issues that might make good topics for her writing, topics that would appeal to nonscientific audiences. Her second book, *The Sea Around Us*, topped the best-seller list and received the prestigious National Book Award. She had achieved literary success with writing based on natural science.

She wanted to use her recognition, however, to help counter specific problems in the natural environment. Carson began to consider a book that would deal with the scientific community's nagging questions about the effect of pesticides on the land and its residents. While working on this project, Carson received the diagnosis that she had breast cancer in 1960. The project became even more important to her; however, now she was forced to contend with a mastectomy and subsequent radiation treatments that left her nauseated and bedridden. Although she did not know what had caused her own cancer, her personal health problems made her more committed than ever to her book on pesticide use (Lear 1997, 403–19).

In addition to these personal difficulties, Carson was forced to deal with harsh criticism from those that would be most effected by her findings: the chemical industry. These companies began an unparalleled assault on Carson's credibility during the early 1960s. When the *New Yorker* began serializing Carson's book, Velsicol Corporation, which manufactured the pesticide chlordane, threatened to sue the magazine for libel. Lawsuits were also threatened against Carson.

Overall, however, the cultural attitude toward chemical progress was beginning to change. In July 1962, news broke that a drug given to thousands of pregnant women in Europe for morning sickness had been proven to cause widespread birth defects. Public opinion swayed as newspapers and magazines ran photographs of malformed babies. Inadvertently, this story likely helped establish the groundswell of support for Carson's *Silent Spring*. In a single summer, chemical science had fallen from its pedestal of unchallenged confidence and progress. Here is a portion of what Carson wrote:

> The "control of nature" is a phrase conceived in arrogance, born of the Neanderthal age of biology and philosophy, when it was supposed that nature exists for the convenience of man. The concepts and practices of applied entomology for the most part date from that Stone Age of science. It is our alarming misfortune that so primitive a science has armed itself with the most modern and terrible weapons, and that in turning them against the insects it has also turned them against the earth. (Carson 2002, 12–14)

Creating a Public Reaction

Silent Spring went on sale September 27, 1962, and quickly became a national bestseller. Its popularity remained strong throughout the fall. Although critics accused Carson of overstating the dangers of chemicals, Supreme Court Justice William O. Douglas, echoed by others, called *Silent Spring* "the most important chronicle of this century for the human race." Carson's work had struck a national nerve.

Although she was weakened by her own illness, Carson appeared on television and before Congress to answer her critics. As a result, the broad-based grassroots support for her findings grew. When she appeared on CBS television in April 1963, an estimated fifteen million Americans tuned in. They heard Carson explain, "We still talk in terms of conquest. I think we're challenged, as mankind has never been challenged before, to prove our maturity and our mastery, not of nature but of ourselves." Many alarmed Americans wrote to the USDA, the PHS, and the Food and Drug Administration (FDA) insisting that action be taken to ensure their personal health safety (Lear 1997, 450–56).

When President Kennedy's Science Advisory Committee released its report on the subject in 1963, its findings backed Carson's thesis and severely criticized the government and the chemical industry. Overall, the report called for "orderly reductions of persistent pesticides." For the first time, the connection had been forged between public concern and federal policy. Americans looked to the federal government to ensure their personal health and safety.

Carson died in 1964 at age 56. Her efforts brought the chemical industry under new federal oversight. More important, however, Carson's line of questioning aroused an overall awareness in the American public. "Rachel Carson's legacy has less to do with pesticides than with awakening of environmental consciousness," said biographer Linda Lears. "She changed the way we look at nature. We now know we are a part of nature, and we can't damage it without it coming back to bite us" (Carson 2002).

Generating a Popular Environmental Movement

Carson's descriptions were based in the science of ecology that had emerged in the 1930s. Combined with her skills as a writer and storyteller, however, the ecological principles were

welcomed by a general public more interested than ever in understanding its place in the natural environment. Her scientific findings added credibility and meaning to students involved
in "back to nature" movements. As students grew willing to question nearly every aspect of
their existence, the human role in nature came under increasing scrutiny. Was our lifestyle
sustainable? Did we take more than we gave to the environment? Were we good stewards of
the world around us? Many Americans started local recycling projects or became active in
local politics. Others checked out completely. Such disenfranchised people would become
known as hippies, but the intent behind their actions was often based around forming what
became known as a "counterculture."

Of course, a counterculture movement is organized around ideas of a cultural norm to
which discontented individuals react. Many youth in the 1960s rejected government, the organization of society, and the war in Vietnam. Similar to the ethic of Carson's writing, the
counterculture that they created rejected the assumption of new technologies as an automatic
good. Young people in particular rejected the stable patterns of middle-class life that their
parents had created in the decades after World War II. Some disenchanted youth plunged
into radical political activity; many more embraced new standards of dress and sexual
behavior.

Although not a cohesive cultural movement with manifestos and leaders, hippies
expressed their desire for change with communal or nomadic lifestyles and by renouncing
the symbols of middle-class American life such as labor-saving household technologies and
mass-produced, packaged food products. These types of products became known as symbols
of the "establishment." Instead of the preferences of mass American culture, hippies strove
for simplicity and self-sufficiency. They sought the opportunity to entirely disengage from
the "establishment." The most successful hippies were likely those who took up residence on
one of many communes that sprang up around the nation. Modeled after the transcendental
farms and camps of the early 1800s, communes grew their own food organically and allowed
residents to live a life totally outside of American culture.

Conclusion

Carson's attack was very specific about its criticism of pesticides. As a symbol of modern progress though, the indictment of these chemicals became a clear statement about the priorities
of American life. Combined with the 1960s counterculture, this critique became an important portion of the modern environmental movement.

Sources and Further Reading: Carson, *Silent Spring*; Gottlieb, *Forcing the Spring: The Transformation
of the American Environmental Movement*; Lear, *Rachel Carson: Witness for History*; Opie, *Nature's
Nation*.

EPILOGUE

In the shadow of Hurricane Katrina in 2005 and growing attention to the implications of global climate change, fire came as never before to residential California in the fall of 2007. Only in such an arid region could small fires grow to threaten nearly two million acres, and, in the region of California near San Diego and Los Angeles where suburbs reach into the driest, most volatile areas of scrub brush, million-dollar homes go up as quickly as tall pines. The need for housing in this area has forced developers to press the envelope of the ecology of southern California.

After a very dry summer and with heavier than usual Santa Ana winds, the outcome could be horrible. Residents always knew this, but the upsides of living in dreamy California far outweighed such possibilities.

October 2007 was the first time that the worst-case scenario played out: one-half million acres burned, approximately 2,000 homes destroyed, and seven deaths. Financial losses have been estimated in excess of $1 billion. If the fires had burned in the same locations in 1980, the *New York Times* estimated that approximately 61,000 homes would have been within a mile of the fire, by 2000, the figure would be 106,000 homes, and, in 2007, more than 125,000 homes were threatened.

Although for generations reformers have called for curtailing residential patterns in such fragile ecosystems, the fires of 2007 will surely make this problem one of the next great debates in a book like this one. Already housing codes are being reconsidered and emergency plans revised. However, the fact remains that millions of people must find a way to call this delicate region home.

So many of the debates recounted in these pages pitted reformers against profit-centered business interests. Particularly during the early twentieth century, reformers, unions, journalists, and politicians sought to demand safer, healthier, and more efficient methods and products from manufacturers. In many instances, reformers have called for the involvement of a governmental authority, whether on the local, state, national, or international levels. In most of these stories, such involvement has greatly helped the plight of many consumers. In fact, in recent years, many large corporations have clearly come to feel that it is important to present an environmentally friendly image to the public.

In the energy sector, which is dominated by some of the nation's most powerful corporations, industry leaders have realized that consumer demands for change are also in the best interest of their long-term profits. Biofuels, particularly the use of ethanol, as well as increased fuel efficiency in autos and the availability of hybrid cars and a variety of sustainable products demonstrate that many companies believe green business can be good business.

Does this mean American manufacturers have responded to a century of increasing calls for reform and regulation by admitting that consumers do have some say in how they do business? Or is it another episode of "greenwashing" that only shows a temporary change? Only time will tell.

And only time will tell how well government officials work with planners to implement the lessons of the 2007 fires in California and the floodwaters of 2005 in New Orleans. It is critical, however, that each of these events and the debates and issues from which they grow be viewed in historical context. This volume seeks to contribute to this context.

BIBLIOGRAPHY

Readers are encouraged to consult any of the environmental history essays at http://nationalhumanitiescenter.org/tserve/nattrans/nattrans.htm.

Abruzzi, W. S. "The Social and Ecological Consequences of Early Cattle Ranching in the Little Colorado River Basin," *Human Ecology* 23 (1995): 75–98.

Adams, D. A. *Renewable Resource Policy: The Legal-Institutional Foundation*. Washington, DC: Island Press, 1993.

Adams, J. A. *The American Amusement Park Industry*. Boston, MA: Twayne Publishers, 1991.

Adams, S. P. *The US Coal Industry in the Nineteenth Century*, http://eh.net/encyclopedia/article/adams.industry.coal.us.

Albion, R. G. *Forests and Sea Power*. Cambridge, MA: Harvard University Press, 1926.

Albright, H. M., Cahn, R. *The Birth of the National Park Service: The Founding Years, 1913–33*. Salt Lake City, UT: Howe Brothers, 1985.

Albright, H. M., Schenck, M. A. *Creating the National Park Service: The Missing Years*. Norman, OK: University of Oklahoma Press, 1999.

Alverson, W. S., Waller, D., and Kuhlmann, W. *Wild Forests: Conservation Biology and Public Policy*. Washington, DC: Island Press, 1994.

Ambler, M. *Breaking the Bonds: Indian Control of Energy Development*. Lawrence, KS: University Press of Kansas, 1990.

Ambrose, S. E. *Undaunted Courage: Meriwether Lewis, Thomas Jefferson, and the Opening of the American West*. New York: Simon & Schuster, 1996.

American Frontiers: A Public Lands Journey. "Energy from Public Lands," http://americanfrontiers.net/energy/Energy2.php.

American State Papers. *Gallatin's Report on Roads and Canals*, http://www.union.edu/PUBLIC/ECODEPT/kleind/eco024/documents/internal/internal_callendar.htm.

Anderson, Jr., O. E. *The Health of a Nation: Harvey W. Wiley and the Fight for Pure Food*. Chicago, IL: University of Chicago Press, 1958.

Anderson, T. H. *The Movement and the Sixties*. New York, NY: Oxford University Press, 1995.

Andreas, P. *Border Games: Policing the U.S.–Mexico Divide*. Cornell, NY: Cornell University Press, 2000.

Andrews, R. N. L. *Managing the Environment, Managing Ourselves*. New Haven, CT: Yale University Press, 1999.

Annis, S. "Evolving Connectedness among Environmental Groups and Grassroots Organizations in Protected Areas of Central America," *World Development* 20, no. 4 (1992): 587–95.

Antarctic Conservation Act of 1978. 16 U.S.C. §§2401–2413, October 28, as amended 1996. Antarctic Treaty. Scientific Committee on Antarctic Research, www.scar.org/treaty/.

Anti-Defamation League. 2007. http://www.adl.org/.

APVA Preservation Virginia. *Old Cape Henry Lighthouse*, http://www.apva.org/capehenry/origin.php.

Ashworth, W. *The Late, Great Lakes: An Environmental History*. New York, NY: Knopf, 1986.

Associated Press. "PETA Claims Victory as Fashion House Drops Fur," 2006.

Athansiou, T. *Divided Planet: The Ecology of Rich and Poor*. Athens, GA: University of Georgia Press, 1998.

Atomic Archive. *Report on the Trinity Test by General Groves*, http://www.atomicarchive.com/Docs/Trinity/Groves.shtml.

Aurand, H. W. *Coalcracker Culture: Work and Values in Pennsylvania Anthracite, 1835–1935*. Harrisburg, PA: Susquehanna University Press, 2003.

Australian Center for the Moving Image. *Edison: The Invention of the Phonograph and the Electric Light Bulb*, http://www.acmi.net.au/AIC/EDISON_INVENT.html.

Bailey, A. J. *The Chessboard of War: Sherman and Hood in the Autumn Campaigns of 1864*. Lincoln, NE: University of Nebraska Press, 2000.

Bailey, A. J. *War and Ruin: William T. Sherman and the Savannah Campaign*. Wilmington, DE: Scholarly Resources, 2003.

Ballard, J. N., ed. *The History of the U.S. Army Corps of Engineers*. New York, NY: Diane Publishers, 1998.

Banerjee, N. "Tanking on a Coal Mining Practice as a Matter of Faith," *The New York Times*, October 28, 2006.

Barbalace, R. C. *Environmental Justice and the NIMBY Principle*, http://environmentalchemistry.com/yogi/hazmat/articles/nimby.html.

Barney, W. L. *The Passage of the Republic: An Interdisciplinary History of Nineteenth-Century America*. Lexington, MA: D. C. Heath, 1987.

Barry, J. *Rising Tide*. New York: Simon & Schuster, 1998.

Barry, T., and Brown, H. *The Challenge of Cross-border Environmentalism*. Albuquerque, NM: The Resource Center, 1994.

Bartram, W. *Travels*. New York, NY: Dover, 1983.

BBC News. "Iceland Violates Ban on Whaling," 2006, http://news.bbc.co.uk/2/hi/europe/6074230.stm.

Beale, E. F. The report of Edward Fitzgerald Beale to the Secretary of War concerning the wagon road from Fort Defiance to the Colorado River; April 26, 1858. 35th Congress, 1st Session, House of Representatives, Executive Document, no. 124, 137–281, in *Uncle Sam's Camels: The Journal of May Humphreys Stacey Supplemented by the Report of Edward F. Beale (1857–1858)*. Glorieta, NM: Rio Grande Press, 1970.

Beilharz, E. A., and Lopez, C. U., eds. *We Were 49ers! Chilean Accounts of the California Gold Rush*. Pasadena, CA: Ward Ritchie Press, 1976.

Belasco, J. *Americans on the Road*. Cambridge, MA: MIT Press, 1979.

Benedict, R. E. *Ozone Diplomacy*. Cambridge, MA: Harvard University Press, 1991.

Benfield, F. K., Terris J., and Vorsanger, N. *Solving Sprawl: Models of Smart Growth in Communities Across America*. National Resource Defense Council. Washington, DC: Island Press, 2001.

Bennett, Jr. L. *Before the Mayflower*. New York, NY: Penguin, 1993.

Benton, T. H. *Thrilling Sketch of the Life of Col. J. C. Fremont*. London, UK: J. Field, 1850.

Bergon, F., ed. *The Journals of Lewis and Clark*. New York, NY: Penguin Books, 1989.

Berman, M. *All That Is Solid Melts into Air*. New York, NY: Penguin, 1988.

Berry, W. *The Unsettling of America: Culture & Agriculture*. Washington, DC: Sierra Club Books, 2004.

Betts, R. B. *In Search of York: The Slave Who Went to the Pacific With Lewis and Clark*. Boulder, CO: University of Colorado Press, 2000.

Billington, R. A., and Hedges, J. B. *Westward Expansion: A History of the American Frontier*. New York, NY: The Macmillan Company, 1949.

Black, B. "Organic Planning: Ecology and Design in the Landscape of TVA," in *Environmentalism in Landscape Architecture*, ed. Michel Conan, Washington DC: Dumbarton Oaks, 2000a.

Black, B. *Petrolia: The Landscape of America's First Oil Boom.* Baltimore, MD: Johns Hopkins University Press, 2000b.

Black, B. *Contesting Gettysburg: Preserving an American Shrine.* Chicago, IL: Center for American Places, University of Chicago Press. Forthcoming.

Black, B. C. "Addressing the Nature of Gettysburg: Addition and Detraction in Preserving an American Shrine." RECONSTRUCTION, online international journal of contemporary culture, Winter 2006. Available at: http://reconstruction.eserver.org/072/black.shtml.

Black, R. "Did Greens Help Kill the Whale?", 2007, http://news.bbc.co.uk/2/hi/science/nature/6659401.stm.

Blaut, J. M. *Colonizer's Model of the World: Geographic Diffusionism and Eurocentric History.* New York, NY: The Guilford Press, 1993.

Boli, J., and Thomas, G. *Constructing World Culture: International Nongovernmental Organizations since 1875.* Stanford, CA: Stanford University Press, 1999.

Bormann, F. H., Balmori, D., Geballe, G. T., and Vernegaard, L. *Redesigning the American Lawn.* New Haven, CT: Yale University Press, 1995.

Boyer, P. *By The Bomb's Early Light.* Chapel Hill, NC: University of North Carolina Press, 1994.

Brack, D. *International Trade and the Montreal Protocol.* London, UK: Earthscan/James and James Press, 1996.

Bradsher, K. *High and Mighty: SUVs: The World's Most Dangerous Vehicles and How They Got That Way.* New York, NY: Public Affairs, 2002.

Brady, L. "The Wilderness War: Nature and Strategy in the American Civil War." *Environmental History,* vol. 10, no. 3 (July 2005).

Bragg, W. H. *Griswoldville.* Macon, GA: Mercer University Press, 2000.

Brands, H. W. *The Age of Gold.* New York, NY: Doubleday, 2002.

Brehm, V. M. "Environment, Advocacy, and Community Participation: MOPAWI in Honduras," *Development in Practice* 10, no. 1 (2000): 94–98.

Brennan, T. J., Palmer, K. L., Kopp, R. J., and Krupnick, A. J. *A Shock to the System—Restructuring America's Electricity Industry.* Washington, DC: Resources for the Future, 1996.

Broder, J. M. "Rule to Expand Mountaintop Coal Mining," *The New York Times,* August 23, 2007, Late edition—final, sec. A, p. 1.

Brown, D. *Bury My Heart at Wounded Knee: An Indian History of the American West,* NY: Owl Books, 2001.

Brown, L. *Seeds of Change.* New York, NY: Praeger Publishers, 1970.

Bruegmann, R. *Sprawl: A Compact History.* Chicago, IL: University of Chicago Press, 2005.

Brinkley, D. *Wheels for the World: Henry Ford, His Company and a Century of Progress.* New York, NY: Viking, 2003.

Broad, W. *The Universe Below.* New York, NY: Simon & Schuster, 1997.

Brooks, H. A. *The Prairie School.* New York, NY: W. W. Norton, 1996.

Brower, M. *Cool Energy: Renewable Solutions to Environmental Problems,* revised ed. Cambridge, MA: MIT Press, 1992.

Bryant, B. I. *Environmental Justice: Issues, Policies and Solutions.* Washington, DC: Island Press, 1995.

Bryant, Jr., K. L., ed. *Railroads in the Age of Regulation, 1900–1980.* New York: Facts on File, 1988.

Bryson, J. M. *Strategic Planning for Public and Nonprofit Organizations.* San Francisco, CA: Jossey-Bass Publishers, 1995.

Buckley, G. L. *Extracting Appalachia: Images of the Consolidation Coal Company, 1910–1945.* Akron, OH: Ohio University Press, 2004.

Budowski, G. *Peace through Parks,* United Nations Environment Program, 2007, http://www.unep.org/OurPlanet/imgversn/144/budowski.html.

Bullard, R., Lewis, J., and Chavis, B. *Unequal Protection: Environmental Justice and Communities of Color.* San Francisco, CA: Sierra Club Books, 1994.

Burian, S. J., Nix, S. J., Pitt, R. E., and Durrans, S. R. "Urban Wastewater Management in the United States: Past, Present, and Future," *Journal of Urban Technology* 7, no. 3 (2000): 33–62.

Burke, J. F. *Mestizo Democracy: The Politics of Crossing Borders*. College Station, TX: Texas A & M University Press, 2004.

Burton, L. *American Indian Water Rights and the Limits of Law*. Lawrence, KS: University Press of Kansas, 1991.

Bush, G. W. Letter from June 13, 2001, http://www.usemb.se/Environment/letter.html.

Cagan, J., et al. *Field of Schemes: How the Great Stadium Swindle Turns Public Money into Private Profit*. New York, NY: Common Courage Press, 1998.

California Department of Water Resources. "All-American Canal Lining Agreement Signed." News Release, Sacramento, CA, January 29, 2002.

California Department of Water Resources. "Coachella Canal and All-American Canal Lining Projects," 2007, http://wwwdpla.water.ca.gov/sd/environment/canal_linings.html.

Callcott, G. H. *Maryland & America: 1940 to 1980*. Baltimore and London: The Johns Hopkins University Press, 1985.

Calloway, C. G. *First Peoples*. Boston, MA: Bedford, 1999.

Calthorpe, P. *The Next American Metropolis*. New York, NY: Princeton Architectural Press, 1993.

Cantelon, P., and Williams R. C. *Crisis Contained: Department of Energy at Three Mile Island*. Carbondale, IL: Southern Illinois University Press, 1982.

Carleson, R. R. *Paradise Paved: The Challenge of Growth in the New West*. Salt Lake City, UT: University of Utah Press, 1996.

Carlsen, L. "After NAFTA—CAFTA and AFTA." Americas Program. Center for International Policy, September 19, 2005, http://americas.irc-online.org/am/655.

Carlton, W. "New England Masts and the King's Navy," *The New England Quarterly* 12, no. 1 (1939), 4–18.

Caro, R. A. *The Power Broker: Robert Moses and the Fall of New York*. New York, NY: Alfred A. Knopf, 1970.

Carr, E. *Wilderness by Design*. Lincoln, NE: University of Nebraska Press, 1988.

Carson, R. *Silent Spring*. New York, NY: Mariner Books, 2002.

Carstensen, V. *The Public Lands: Studies in the History of the Public Domain*. Madison, WI: University of Wisconsin Press, 1963.

Catlin, L. G. *Letters and Notes*, Letter No. 2: Mouth of Yellow Stone, Upper Missouri. Can be found at http://catlinclassroom.si.edu/interviews/al-batis.html.

Catton, T. *Wonderland: An Administrative History of Mount Ranier National Park*, 1996, http://www.nps.gov/history/history/online_books/mora/adhi/adhi.htm.

Center for the State of the Parks, Park Assessments. *Waterton-Glacier International Peace Park*, 2002. http://www.npca.org/stateoftheparks/glacier/.

Cerritos College, http://www.cerritos.edu/soliver/American%20Identities/Trail%20of%20Tears/quotes.htm.

Chadwick, D. *Yellowstone to Yukon*. Washington, DC: National Geographic Society, 2007.

Chamberlain, K. P. *Under Sacred Ground: A History of Navajo Oil, 1922–1982*. Albuquerque, NM: University of New Mexico Press, 2000.

Chase, A. *Playing God in Yellowstone: The Destruction of America's First National Park*. Orlando, FL: Harcourt Brace & Company, 1986.

Chernow, R. *Titan: The Life of John D. Rockefeller, Sr.* New York, NY: Random House, 1998.

Chester, C. C. "Landscape Vision and the Yellowstone to Yukon Conservation Initiative," in *Conservation Across Borders: Biodiversity in an Interdependent World*, chap. 4. Washington, DC: Island Press, 2006, 134–216.

Child Lead Poisoning and the Lead Industry, http://www.sueleadindustry.homestead.com/.

Christianson, G. *Greenhouse: The 200-Year Story of Global Warming*. New York, NY: Penguin, 2000.

Cincinnati Children's Hospital Medical Center. *History of Lead Advertising*, http://www.cincinnatichildrens.org/research/project/enviro/hazard/lead/leadadvertising/industry-role.htm.

Clapham, P. J., and Baker, C. S. "Modern Whaling," in *Encyclopedia of Marine Mammals*, eds. W. F. Perrin, B. Wursig, and J. G. M. Thewissen. New York, NY: Academic Press, 2002, 1328–32.

Clark, A. M., Friedman E. J., and Hochstetler K. "The Sovereign Limits of Global Civil Society: A Comparison of NGO Participation in UN World Conferences on the Environment, Human Rights, and Women," *World Politics*, no. 51 (1998): 1–35.

Clark, Jr., C. E. *The American Family Home.* Chapel Hill, NC: University of North Carolina Press, 1986.

Clarke, C. G. *The Men of the Lewis and Clark Expedition: A Biographical Roster of the Fifty-one Members and a Composite Diary of Their Activities from all the Known Sources.* Glendale, CA: A. H. Clark, 1970.

Clary, D. A. *Timber and the Forest Service.* Lawrence, KS: University of Kansas Press, 1986.

Clawson, M. "The Bureau of Land Management, 1947–1953," *The Cruiser* [newsletter of the Forest History Society], 10, no. 3 (1987): 3–6.

Clearwater. *Hudson River PCB Pollution Timeline*, http://www.clearwater.org/news/timeline.html.

Clough-Riquelme, J., and Bringas, N. L., eds. *Equity and Sustainable Development: Reflections from the US–Mexico Border.* La Jolla, CA: Center for U.S.–Mexican Studies, 2006.

CNN. "Bush OKs 7000-mile Border Fence," *Politics*, October 26, 2006, http://www.cnn.com/2006/POLITICS/10/26/border.fence/index.html.

CNN. "One-on-one with Angelina Jolie," *Anderson Cooper Blog 360°*, June 19, 2006, http://www.cnn.com/CNN/Programs/anderson.cooper.360/blog/2006/06/one-on-one-with-angelina-jolie.html.

CNN. "Senate Immigration Bill Suffers Crushing Defeat," June 28, 2007, http://www.cnn.com/2007/POLITICS/06/28/immigration.congress/index.html.

Coates, P. *Trans-Alaskan Pipeline Controversy: Technology, Conservation, and the Frontier.* Anchorage: University of Alaska Press, 1993.

Cody, B. A. *CRS Report for Congress*, "Major Federal Land Management Agencies: Management of Our Nation's Lands and Resources," 1995, http://www.ncseonline.org/NLE/CRSreports/Natural/nrgen-3.cfm?&CFID=8734533&CFTOKEN=91528013.

Colignon, R. A. *Power Plays.* Albany, NY: State University of New York Press, 1997.

Collin, R. H. *Theodore Roosevelt's Caribbean: The Panama Canal, The Monroe Doctrine, and the Latin American Context.* Baton Rouge, LA: Louisiana State University Press, 1990.

Colorado Plateau Land Use History of North America. *Native Americans and the Environment: A Survey of Twentieth Century Issues with Particular Reference to Peoples of the Colorado Plateau and Southwest*, http://cpluhna.nau.edu/Research/native_americans4.htm.

Colorado State Forest Service. "About Wildfire: Introduction and History," 2007, http://csfs.colostate.edu/wildfire.htm.

Colten, C. *Transforming New Orleans and Its Environs.* Pittsburgh, PA: University of Pittsburgh Press, 2001.

Colton, H. S. "Some Notes on the Original Condition of the Little Colorado River: A Side Light on the Problem of Erosion," *Museum Notes of the Museum of Northern Arizona* 10 (1937): 17–20.

Columbian Exposition. *The World's Columbian Exposition: Idea, Experience, Aftermath*, http://xroads.virginia.edu/~MA96/WCE/title.html.

Columbus, C. *The Four Voyages.* New York, NY: Penguin, 1969.

Commission for Environmental Cooperation. *Who We Are*, 2007, http://www.cec.org/who_we_are/index.cfm?varlan=english.

Commission for Labor Cooperation. *Objectives, Obligations, and Principles*, 2007, http://www.naalc.org/english/objective.shtml.

Connolly, J. A. *Three Years in the Army of the Cumberland: The Letters and Diary of Major James A. Connolly*, ed. P. M. Angle. 1928. Reprint, Bloomington, IN: Indiana University Press, 1996.

Conservation Law Foundation. *Early History of CLF's Fight to Clean up Boston Harbor 1983–1986*, http://www.clf.org/programs/cases.asp?id=188.

Contra Costa Times, 2003. "Rodeo Residents' Beef with PETA Must Wait." http://www.beefusa.org/newsrodeoresidentsbeefwithpetamustwait13689.aspx.

Conway, G. *The Doubly Green Revolution.* Ithaca, NY: Cornell University Press, 1998.

Conzen, M., ed. *The Making of the American Landscape.* Boston, MA: Unwin Hyman Publishers, 1990.

Cooke, M. *Report of the Great Plains Drought Area Committee*, http://newdeal.feri.org/hopkins/hop27.htm.

Cooper, G. *Air-Conditioning America.* Baltimore, MD: Johns Hopkins University Press, 2002.

Cornelius, W. *Death at the Border: The Efficacy and the "Unintended" Consequences of U.S. Immigration Control Policy 1993–2000*. San Diego, CA: The Center for Comparative Immigration Studies, University of California, 2001.

Cowdrey, A. *This Land, This South*. Lexington, KY: University Press of Kentucky, 1983.

Cowles, H. C. *Ecology and the American Environment*, http://memory.loc.gov/ammem/award97/icuhtml/aepsp4.html.

Creese, W. L. *TVA's Public Planning*. Knoxville, TN: University of Tennessee Press, 1990.

Creighton, M. S. *Rites and Passages: The Experience of American Whaling, 1830–1870*. Cambridge, 1995.

Creighton, M. S. *The Colors of Courage: Gettysburg's Forgotten History: Immigrants, Women, And African Americans in the Civil War's Defining Battle*. NY: Basic Books, 2006.

Cronon, W. *Changes in the Land*. New York, NY: W. W. Norton, 1991a.

Cronon, W. *Nature's Metropolis*. New York, NY: W. W. Norton, 1991b.

Cronon, W., ed. *Uncommon Ground: Rethinking the Human Place in Nature*. New York, NY: Norton, 1996.

Crosby, A. *Ecological Imperialism*. New York, NY: Cambridge University Press, 1986.

Crosby, A. *America's Forgotten Pandemic: The Influenza of 1918*. New York, NY: Cambridge University Press, 1990.

Culp, P. W. *Restoring the Colorado Delta with the Limits of the Law of the River*. Tucson, AZ: Udall Center for Studies in Public Policy, The University of Arizona, 2000.

Cunfer, G. *On the Great Plains: Agriculture and the Environment*. College Station, TX: Texas A & M University, 2005.

Custer, G. A. *My Life on the Plains: Or Personal Experiences with the Indians*. Norman: University of Oklahoma Press, 1977.

Cutright, P. *Theodore Roosevelt: The Making of a Conservationist*. Urbana, IL: University of Illinois Press, 1985.

Dana, C. W. *The Great West, or the Garden of the World*. Boston, MA: Wentworth and Co., 1857.

Dana, S. T., and Fairfax, S. K. *Forest and Range Policy: Its Development in the United States*. 2nd ed. New York, NY: McGraw-Hill, 1980.

Dangerfield, G. *The Awakening of American Nationalism, 1815–1828*. New York, NY: Harper and Row, 1965.

Darrah, W. C. *Pithole, the Vanished City*. Gettysburg, PA: 1964.

Darst, R. G. *Smokestack Diplomacy: Cooperation and Conflict in East-West Environmental Politics*. Cambridge, MA: MIT Press, 2001.

Darwin, C. *Origin of Species*. New York, NY: Signet Classics, 2003.

Dary, D. *The Buffalo Book: The Full Saga of the American Animal*. Athens, OH: Swallow Press/Ohio University Press, 1989.

Davis, D. *When Smoke Ran Like Water*. New York, NY: Basic Books, 2002.

Davis, J., ed. *The Earth First! Reader: Ten Years of Radical Environmentalism*. Salt Lake City, UT: Peregrine Smith Books, 1991.

Davis, S. G. *Spectacular Nature*. Berkeley, CA: University of California Press, 1997.

Davison, C. *White Pines for the Royal Navy*, http://www.nhssar.org/essays/Whtpines.html.

Day, D. *The Whale War*. San Francisco, CA: Sierra Club Books, 1987.

De Tocqueville, A. *Democracy in America*, http://xroads.virginia.edu/~HYPER/DETOC/home.html.

Demarest, Jr. D. P. *"The River Ran Red": Homestead, 1892*. Pittsburgh, PA: University of Pittsburgh Press, 1992.

DeVoto, B., ed. *The Journals of Lewis and Clark*. Boston, MA: Houghton Mifflin Co., 1997.

Diamond, J. *Guns, Germs and Steel: The Fates of Human Societies*. New York, NY: W. W. Norton, 1997.

Dietsch, T. V. "Assessing the Conservation Value of Shade-Grown Coffee: A Biological Perspective Using Neotropical Birds," *Endangered Species Update* 17 (2000): 122–30, University of Michigan, School of Natural Resources.

Dolin, E. J. *Political Waters: The Long, Dirty, Contentious, Incredibly Expensive but Eventually Triumphant History of Boston Harbor—A Unique Environmental Success Story*. Cambridge, MA: University of Massachusetts Press, 2004.

Domer, D., ed. *Lawrence on the Kaw: A Historical and Cultural Anthology.* Lawrence, KS: University Press of Kansas, 2000.

Donahue, D. L. *The Western Range Revisited: Removing Livestock from Public Lands to Conserve Native Biodiversity.* Legal History of North America Series, vol. 5. Norman, OK: University of Oklahoma Press, 1999. http://ipl.unm.edu/cwl/fedbook/taylorgr.html.

Dorsey, K. *The Dawn of Conservation Diplomacy.* Seattle: University of Washington Press, 1994.

Douglass, F. *My Bondage and My Freedom.* Urbana, IL: University of Illinois Press, 1987.

Dowie, M. *American Foundations: An Investigative History.* Cambridge, MA: MIT Press, 2001.

Downing, A. J. *A Treatise on the Theory and Practice of Landscape Gardening,* 9th ed. New York: Orange Judd, 1875. Reprint Little Compton, RI: Theophrastus Publishers, 1977.

Doyle, J. *Taken for a Ride: Detroit's Big Three and the Politics of Air Pollution.* New York, NY: Four Walls Eight Windows, 2000.

Dreyer, E. L. *Early Ming China: A Political History 1355–1435.* Stanford, CA: Stanford University Press, 1982.

Duany, A., and Plater-Zyberk, E. *Suburban Nation: The Rise of Sprawl and the Decline of the American Dream.* New York, NY: North Point Press, 2000.

Dunlop, B., and Scully, V. *Building a Dream: The Art of Disney Architecture.* New York, NY: Harry N. Abrams, 1996.

Dysart, B. C., and Clawson, M. *Managing Public Lands in the Public Interest.* New York, NY: Praeger Publishers, 1988.

Earth First! 2007. http://www.earthfirst.org/.

Eaton, D. J., and Anderson, J. M. *The State of the Rio Grande/Rio Bravo: A Study of Water Resource Issues along the Texas/Mexico Border.* Tucson, AZ: University of Arizona Press, 1986.

Edwards, M., and Hulme, D. *Beyond the Magic Bullet: NGO Performance and Accountability in the Post-Cold War World.* West Hartford, CT: Kumarian Press, 1996.

Egan, T. "New Fight in Old West: Farmers vs. Condo City." *The New York Times,* October 3, 1989.

Eichstaedt, P. H. *If You Poison Us: Uranium and Native Americans.* Sante Fe, NM: Crane Books, 1994.

El Nasser, H., and Overberg, P. "A Comprehensive Look at Sprawl in America," *USA Today,* February 22, 2001. http://www.usatoday.com/news/sprawl/main.htm.

Elkington, J., and Fennell, S. "Partners for Sustainability: Business-NGO Relations and Sustainable Development," *Greener Management International,* no. 24 (1998): 48–61.

Emerson, R. W. *Nature.* http://oregonstate.edu/instruct/phl302/texts/emerson/nature-contents.html.

Environmental Literacy Council. *Superfund,* http://www.enviroliteracy.org/article.php/329.html.

Erikson, K. *A New Species of Trouble—The Human Experience of Modern Disasters.* New York, NY: W. W. Norton, 1994.

Etheridge, E. W. *Sentinel for Health.* Berkeley, CA: University of California Press, 1992.

Eureka Country, Nevada Nuclear Waste Office. 2007. http://www.yuccamountain.org/new.htm.

Everhart, W. C. *The National Park Service.* Boulder, CO: Westview Press, 1983.

Ewers, J. C. *The Blackfeet: Raiders on the Northwestern Plains.* Norman, OK: University of Oklahoma Press, 1958.

Ewing, F. E. *America's Forgotten Statesman: Albert Gallatin.* New York, NY: Vantage Press, 1959.

Fairmont Water Works Interpretive Center. http://www.fairmountwaterworks.com.

Fall, J. *Drawing the Line: Nature, Hybridity and Politics in Transboundary Spaces.* Burlington, VT: Ashgate Publishing Company, 2005.

Fattig, P. "Forest Speaker Says He's No Terrorist," *Mail Tribune: Southern Oregon's News Source,* January 23, 2003, Local page. http://archive.mailtribune.com/archive/2003/0123/local/stories/07local.htm.

Federal Land and Policy Management Act. http://www.blm.gov/flpma/ and http://www.blm.gov/flpma/snapshot.htm.

Federal Wildlife Laws Handbook. http://ipl.unm.edu/cwl/fedbook/taylorgr.html.

Federation of American Scientists. *NSC-68,* http://www.fas.org/irp/offdocs/nsc-hst/nsc-68.htm.

Feller, D. *The Jacksonian Promise: America, 1815–1840.* Baltimore, MD: Johns Hopkins University Press, 1995.

Finkel, M. "From Yellowstone to Yukon," *Audubon* (1999): 44–53.

Fisher, H. *Wolf Wars: The Remarkable Inside Story of the Restoration of Wolves to Yellowstone.* Missoula, MT: Fischer Outdoor Discoveries, 2003.

Fishman, R. *Bourgeois Utopias: The Rise and Fall of Suburbia.* New York, NY: Basic Books, 1989.

Fixico, D. L. *The Invasion of Indian Country in the Twentieth Century: American Capitalism & Tribal Natural Resources.* Niwot, CO: University Press of Colorado, 1998.

Flink, J. J. *The Automobile Age.* Cambridge, MA: MIT Press, 1990.

Flores, D. "Bison Ecology and Bison Diplomacy: The Southern Plains from 1800–1850," *Journal of American History* 78 (1991): 2.

Flores, D. "The West That Was, and the West That Can Be," *High Country News.* August 18, 1997. http://www.hcn.org/servlets/hcn.Article?article_id=3560.

Floudas, D. A., and Rojas, L. F. "Some Thoughts on NAFTA and Trade Integration in the American Continent," *International Problems* LII (2000): 371–89.

Food and Agriculture Organization of the United Nations. "Women and the Green Revolution," 2007. http://www.fao.org/FOCUS/E/Women/green-e.htm.

Foreman, D. *Confessions of an Eco-Warrior.* NY: Three River Press, 1993.

Foreman, D. *Ecodefense: A Field Guide to Monkeywrenching.* 3rd ed. Chico, CA: Abbzugg Press, 1994.

Foresta, R. A. *America's National Parks and Their Keepers.* Washington, DC: Resources for the Future, 1985.

Formwalt, L. W. "Benjamin Henry Latrobe and the Revival of the Gallatin Plan of 1808," *Pennsylvania History* 48, no. 1 (19:1) 99–128.

Fox, S. *The American Conservation Movement.* Madison, WI: University of Wisconsin Press, 1981.

Franz, D. "The Yellowstone Wolves Win One," *E Magazine* 11 (2000): 14.

Freedmen's Act. http://www.multied.com/documents/Freedman.html.

Freese, B. *Coal: A Human History.* New York: Penguin Books, 2004.

Freydkin, D. "Celebrity Activists Put Star Power to Good Use," *USA Today.* June 22, 2006. http://usatoday.com/life/people/2006-06-22-celebcharities-main-x.htm.

Fri, R. "The Corporation as Nongovernment Organization." *Columbia Journal of World Business* 27, no. 3/4 (1992): 90–96.

Fridell, G. *Fair Trade Coffee: The Prospects and Pitfalls of Market-Driven Social Justice.* Studies in Comparative Political Economy and Public Policy. Toronto, Canada: University of Toronto Press, 2007.

Frisvold, G. B., and Caswell, M. F. "Trans-boundary Water Management: Game-Theoretic Lessons for Projects on the U.S.–Mexico Border," *Agricultural Economics* 24 (2000): 101–11.

Furze, B., de Lacy, T., and Birckhead, J. *Culture, Conservation and Biodiversity: The Social Dimension of Linking Local Level Development and Conservation through Protected Areas.* Hoboken, NJ: John Wiley & Sons, 1996.

Gadd, B. "The Yellowstone to Yukon Landscape," in *A Sense of Place: Issues, Attitudes and Resources in the Yellowstone to Yukon Ecoregion,* A. Harvey, ed. Canmore, Canada: Yellowstone to Yukon Conservation Initiative, 1998, 9–18.

Gadsden Purchase Treaty. http://www.yale.edu/lawweb/avalon/diplomacy/mexico/mx1853.htm.

Gallery of Lead Pollution Promotions. http://www.uwsp.edu/geo/courses/geog100/lead-ads.htm.

Gardner, J. S., and Sainato, P. "Mountaintop Mining and Sustainable Development in Appalachia," *Mining Engineering* 48 (2007): 48–55.

Garner, J. S., ed. *The Midwest in American Architecture.* Chicago, IL: University of Illinois Press, 1991.

Garwin, R. L., and Charpak, G. *Megawatts and Megatons: A Turning Point in the Nuclear Age.* New York, NY: Knopf, 2001.

Gatell, O., and Goodman, P. *Democracy and Union: The United States, 1815–1877.* New York, NY: Holt, Rinehart, and Winston, 1972.

Gates, P. W. *History of Public Land Law Development.* Washington, DC: U.S.G.P.O. for the Public Land Law Review Commission, 1968.

Gates, Jr., H. L., ed. *The Classic Slave Narratives.* New York, NY: Penguin, 1987.

Gelbspan, R. *The Heat Is On: The Climate Crisis, the Cover-Up, the Prescription.* NY: Perseus Books, 1998.

Gelletly, L. *Mexican Immigration: The Changing Face of North America.* Broomall, PA: Mason Crest Publishers, 2004.

Geraint, J., and Sheard, R. *Stadia: A Design and Development Guide*. Boston, MA: Architectural Press, 1997.

Gerard, D. *1872 Mining Law: Digging A Little Deeper*, PERC Policy Series, PS-11. Bozeman, MT: Political Economy Research Center, 1997.

German MAB National Committee, ed. *Full of Life: UNESCO Biosphere Reserves—Model Regions for Sustainable Development. The German Contribution to the UNESCO Programme Man and the Biosphere (MAB)*. New York, NY: Springer, 2005.

Giddens, P. *Early Days of Oil*. Gloucester, MA: Peter Smith, 1964.

Gifford, J. *The Exceptional Interstate Highway System*, http://onlinepubs.trb.org/onlinepubs/trnews/trnews244newvision.pdf.

Gilman, C. *The Poetry of Traveling in the United States*. New York: S. Colman, 1838.

Ginsburg, S. *Nuclear Waste Disposal: Gambling on Yucca Mountain*. Walnut Creek, CA: Aegean Park Press, 1994.

Glatthaar, J. T. *The March to the Sea and Beyond: Sherman's Troops in the Savannah and Carolinas Campaign*. New York, NY: New York University Press, 1985.

Glave, D. "A Garden So Brilliant with Colors, So Original in Its Design," *Environmental History* 8, no. 3 (2003): 395–411.

Goldish, M. *Gray Wolves: Return to Yellowstone*. New York, NY: Bearport Publishing Company, 2008.

Goodale, U. M., Lanfer A. G., Stern, M. J., Margoluis, C., and Fladeland, M., eds. *Transboundary Protected Areas: The Viability of Regional Conservation Strategies*. Binghamton, NY: The Haworth Press, 2003.

Gordon, R. B., and Malone, P. M. *The Texture of Industry*. New York, NY: Oxford, 1994.

Gordon, R., and VanDorn, P. *Two Cheers for The 1872 Mining Law*. Washington, DC: CATO Institute, 1998.

Gorman, H. *Redefining Efficiency: Pollution Concerns, Regulatory Mechanisms, and Technological Change in the U.S. Petroleum Industry*. Akron, OH: University of Akron Press, 2001.

Gottleib, R. *Forcing the Spring: The Transformation of the American Environmental Movement*. Washington, DC: Island Press, 1993.

Gowda, M., Rajeev, V., and Easterling, D. "Nuclear Waste and Native America: The MRS Sitting Exercise," *Risk: Health, Safety & Environment* 229 (1998): 229–58.

Graebner, N. B., ed. *Manifest Destiny*. New York, NY: Bobbs Merrill, 1968.

Graham, F. *The Adirondack Park: A Political History*. New York, NY: Random House, 1978.

Grinde, D. A., and Johansen, B. E. *Ecocide of Native America: Environmental Destruction of Indian Lands and Peoples*. Sante Fe, NM: Clear Light, 1995.

Griswold, D. T. *NAFTA at 10: An Economic and Foreign Policy Success*, Cato Institute, 2002. http://www.freetrade.org/node/87.

Gruen, V., and Smith, L. *Shopping Towns USA: The Planning of Shopping Centers*. New York, NY: Van Nostrand Reinhold Company, 1960.

Gumprecht, B. "Transforming the Prairie: Early Tree Planting in an Oklahoma Town," *Historical Geography* 29 (2001): 116–34.

Gura, P. F., and Myerson, J., eds. *Critical Essays on American Transcendentalism*. London: G. K. Hall, 1982.

Gutfreund, O. D. *20th Century Sprawl: Highways and the Reshaping of the American Landscape*. New York, NY: Oxford University Press, 2005.

Hage, W. *Storm Over Rangelands: Private Rights in Federal Lands*. Bellevue, WA: Free Enterprise Press, 1989.

Hague, J. A., ed. *American Character and Culture*. New York: Eveett Edwards Press, 1964.

Haines, F. *The Buffalo: The Story of American Bison and Their Hunters from Prehistoric Times to the Present*. New York, NY: Crowell, 1970. Reprint Norman, OK: University of Oklahoma Press, 1995.

Hales, P. B. *William Henry Jackson and the Transformation of the American Landscape*. Philadelphia, PA: Temple University Press, 1988.

Hampton, W. *Meltdown: A Race against Nuclear Disaster at Three Mile Island: A Reporter's Story*. Cambridge, MA: Candlewick Press, 2001.

Hanford Downwinders Litigation Information Resource. http://www.downwinders.com/index.html.

Hardin, G. "The Tragedy of the Commons," *Science* 162 (1968): 1243–48.

Hargrove, E. C. *Prisoners of Myth.* Princeton, NJ: Princeton University Press, 1994.

Hargrove, E. C., and Conkin, P. K., eds. *TVA: Fifty Years of Grass-Roots Bureaucracy.* Knoxville, TN: University of Tennessee Press, 1984.

Harris, D. "Celebs and Charity: Trendiness or Benevolence? The Hottest Trend in Hollywood is Taking Up a Cause," *ABC News,* January 14, 2007. http://abcnews.go.com/Entertainment/WNT/story?id=2794458&page=1.

Harris, T. "Texas Patrols the Mexican Border—Virtually," *ABC News,* June 6, 2006. http://abcnews.go.com/US/story?id=2044968&page=1.

Hart, J. F., ed., *Our Changing Cities.* Baltimore, MD: Johns Hopkins University Press, 1991.

Hartzog, Jr., G. B. *Battling for the National Parks.* Mt. Kisco, NY: Moyer Bell, 1988.

Harvey, M. *Wilderness Forever: Howard Zahniser and the Path to the Wilderness Act.* Seattle, WA: University of Washington Press, 2005.

Haycox, Jr., E. "Building The Transcontinental Railroad, 1864–1869," *Montana: The Magazine of Western History,* 45 (2001).

Hayes, D. L. "The All-American Canal Lining Project: A Catalyst for Rational and Comprehensive Groundwater Management on the United States–Mexico Border," *Natural Resources Journal* 31 (1991): 806–15.

Hays, S. P. *Beauty, Health, and Permanence: Environmental Politics in the United States, 1955–85.* New York, NY: Cambridge University Press, 1993.

Hays, S. P. *Conservation and the Gospel of Efficiency.* Pittsburgh, PA: University of Pittsburgh Press, 1999.

Heacox, K. "Antarctica. The Last Continent," *National Geographic Destinations.* April 1, 1999.

Helms, D. "Conserving the Plains: The Soil Conservation Service in the Great Plains," Reprinted from *Agricultural History* 64 (1990): 58–73. http://www.nrcs.usda.gov/about/history/articles/ConservingThePlains.html.

Henderson, H. L., and Woolner, D. B., eds. *FDR and the Environment.* New York, NY: Palgrave, 2004.

Herzog, L. A., ed. *Shared Space: Rethinking the U.S.–Mexico Border Environment.* La Jolla, CA: Center for U.S.–Mexican Studies, University of California, San Diego, 2000.

Hickman and Eldredge. http://www.forester.net/msw_9909_brief_history.html.

Hicks, D. A. *The Limits of Celebrity Activism.* 2007. http://www.religion-online.org/showarticle.asp?title=3331.

Hietala, T. R. *Anxiety and Aggrandizement: The Origins of American Expansion in the 1840s.* Ann Arbor, MI: University Microfilms International, 1981.

Higgs, E. *Nature by Design: People, Natural Process, and Ecological Restoration.* Cambridge, MA: Massachusetts Institute of Technology Press, 2003.

Higham, J. *Critical Issues in Ecotourism: Understanding a Complex Tourism Phenomenon.* Burlington, MA: Elsevier, 2007.

Hine, R. V., and Faragher, J. M. *The American West: A New Interpretive History.* New Haven, CT: Yale University Press, 2000.

Hines, T. S. "The Imperial Mall: The City Beautiful Movement and the Washington Plan of 1901–1902," in *The Mall in Washington, 1791–1991.* Washington, DC: National Gallery of Art, 1991.

Hirschhorn, J. S. *Sprawl Kills: How Blandburbs Steal Your Time, Health, and Money.* New York: Sterling & Ross, 2005.

Hirt, P. W. *A Conspiracy of Optimism: Management of the National Forests since World War Two.* Lincoln, NE: University of Nebraska Press, 1994.

Hohman, E. *The American Whaleman.* Clifton, NJ: Augustus M. Kelley, 1928.

Holliday, J. S. *The World Rushed In.* New York, NY: Simon & Schuster, 1981.

Honey, M. *Ecotourism and Sustainable Development: Who Owns Paradise?* Washington, DC: Island Press, 1999.

Hopkins, J. "'Slow Food' Movement Gathers Momentum," *USA Today.* November 25, 2003.

Horgan, P. *Great River: The Rio Grande in North American History: Volume 1, Indians and Spain. Volume 2, Mexico and the United States.* Hanover, NH: Wesleyan University Press, 1984.

Hornaday's report. http://etext.lib.virginia.edu/railton/roughingit/map/figures3/bufhornaday.html.

Horowitz, D. *Jimmy Carter and the Energy Crisis of the 1970s.* New York, NY: St. Martin's Press, 2005.

Horsman, R. *Race and Manifest Destiny.* Cambridge: Harvard University Press, 1981.

Hughes, T. *Networks of Power: Electrification in Western Society, 1880–1930.* Baltimore, MD: Johns Hopkins University Press, 1983.

Hughes, T. *American Genesis.* New York, NY: Penguin, 1989.

Hunter, L. C., and Bryant, L. *A History of Industrial Power in the United States, 1780–1930,* vol 3, The Transmission of Power. Cambridge, MA: MIT Press, 1991, 207–8.

Hurley, A. *Environmental Inequalities: Class, Race, and Industrial Pollution in Gary, Indiana, 1945–1980.* Chapel Hill, NC: University of North Carolina Press, 1995.

Immigrant Solidarity Network. *2006 Report on Migrant Deaths at the US–Mexico Border.* November 18, 2006. http://www.immigrantsolidarity.org/cgi-bin/datacgi/database.cgi?file=Issues&report=SingleArticle&ArticleID=0646.

Imperial Irrigation District. *Water Department.* 2007. http://www.iid.com/Water_Index.php.

International Boundary and Water Commission. *International Boundary & Water Commission: United States and Mexico, United States Section,* 2007. http://www.ibwc.state.gov/home.html.

International Conference on Transboundary Protected Areas as a Vehicle for International Co-operation. *Parks for Peace.* Conference Proceedings. September 16–18, 1997. Somerset West, South Africa.

Inventory of Conflict and Environment. *The Buffalo Harvest,* http://www.american.edu/TED/ice/buffalo.htm.

Irwin, W. *The New Niagara.* University Park, PA: Pennsylvania State University Press, 1996.

Isaacs, J. C. "The Limited Potential of Ecotourism to Contribute to Wildlife Conservation," *The Ecologist* 28 (2000): 61–69.

Ise, J. *Our National Park Policy: A Critical History.* Baltimore, MD: Johns Hopkins University Press, 1961.

Isenberg, A. *The Destruction of the Bison.* New York, NY: Cambridge University Press, 2001.

Jackson, D. C. *Building the Ultimate Dam.* Lawrence, KS: University of Kansas Press, 1995.

Jackson, K. T. *Crabgrass Frontier: The Suburbanization of the United States.* New York, NY: Oxford University Press, 1985.

Jackson, W. H. *Time Exposure: The Autobiography of William Henry Jackson.* Tucson, AZ: The Patrice Press, 1994.

Jackson, W. H., and Fielder, J. *Colorado, 1870–2000.* New York, NY: Westcliffe Publishing, 1999.

Jacobs, J. *The Death and Life of Great American Cities.* New York, NY: Vintage Books, 1961.

Jefferson, T. "Observations on the Whale-Fishery," in *Public and Private Papers.* New York, NY: Vintage Books, 1990.

Jenkins, V. S. *The Lawn.* Washington, DC: Smithsonian Institution Press, 1994.

Johnson, D. "Yellowstone Will Shelter Wolves Again," *The New York Times.* June 17, 1994. http://query.nytimes.com/gst/fullpage.html?res=9401E7D7173DF934A25755C0A962958260&n=Top%2fReference%2fTimes%20Topics%2fOrganizations%2fS%2fSierra%20Club.

Johnson, P. M., and Beaulieu, A. *The Environment and NAFTA: Understanding and Implementing the New Continental Law.* Washington, DC: Island Press, 1996.

Jones, H. R. *John Muir and the Sierra Club: The Battle for Yosemite.* San Francisco, CA: Sierra Club, 1965.

Jones, L. C. "Assessing Transboundary Environmental Impacts on the U.S.–Mexican and U.S.–Canadian Borders," *Journal of Borderlands Studies* 12 (1997): 81.

Jones, L. Y., ed. *The Essential Lewis and Clark.* New York, NY: Ecco Press, 2000.

Jordan, T., and Kaups, M. *The American Backwoods Frontier.* Baltimore, MD: Johns Hopkins University Press, 1986.

Jorgensen, J. G. *Native Americans and Energy Development II.* Boston, MA: Anthropology Resource Center & Seventh Generation Fund, 1984.

Josephson, P. R. *Red Atom: Russia's Nuclear Power Program from Stalin to Today.* New York, NY: W. H. Freeman, 2000.

Kamauro, O. *Ecotourism: Suicide or Development? Voices from Africa #6: Sustainable Development, UN Non-Governmental Liaison Service.* United Nations News Service, 1996.

Karliner, J. *A Brief History of Greenwash.* CorpWatch, http://www.greenwashing.net/.

Kasson, J. *Amusing the Million: Coney Island at the Turn of the Century.* New York, NY: Hill and Wang, 1978.

Kay, J. H. *Asphalt Nation.* Berkeley, CA: University of California Press, 1997.

Keller, C. *Philanthropy Betrayed.* www.apspub.com/proceedings/1441/Keller.pdf.

Kellogg, P. U. *The Pittsburgh Survey.* http://www.clpgh.org/exhibit/stell30.html.

Kelly, G. "Grazing Act Still at Work to Protect Grasslands," *Denver Rocky Mountain News.* November 2, 1999.

Kennedy, R. F. *Crimes Against Nature: How George W. Bush and His Corporate Pals Are Plundering the Country and Hijacking Our Democracy.* New York, NY: Harper Collins, 2004.

Kennett, L. B. *Marching through Georgia: The Story of Soldiers and Civilians during Sherman's Campaign.* New York, NY: Harper Collins, 1995.

Kern, S. *The Culture of Time and Space, 1880–1918.* Cambridge, MA: Harvard University Press, 1983.

Kinder, C. *Preventing Lead Poisoning in Children,* http://www.yale.edu/ynhti/curriculum/units/1993/5/93.05.06.x.html#b.

Kirby, A. *Extinction Nears for Whales and Dolphins.* 2003. http://news.bbc.co.uk/2/hi/science/nature/3024785.stm.

Kirby, J. T. *The American Civil War: An Environmental View.* http://www.nhc.rtp.nc.us:8080/tserve/nattrans/ntuseland/essays/amcwar.htm.

Klinkenborg, V. "The New Range Wars," *Mother Jones.* 2003. http://www.motherjones.com/news/featurex/2003/09/nrw.pdf.

Klobuchar, A. *Uncovering the Dome.* New York, NY: Waveland Press, 1986.

Knight, R., and Landres, P. *Stewardship Across Boundaries.* Washington, DC: Island Press, 1998.

Kraut, A. M. *Silent Travelers: Germs, Genes, and the Immigrant Menace.* New York, NY: Basic Books, 1994.

Krech III, S. *Buffalo Tales: The Near Extermination of the American Bison,* http://www.nhc.rtp.nc.us/tserve/nattrans/ntecoindian/essays/buffalo.htm.

Krech III, S. *The Ecological Indian: Myth and History.* New York, NY: W. W. Norton, 1999.

Kremen, C., Merenlender, A. M., and Murphy, D. D. "Ecological Monitoring: A Vital Need for Integrated Conservation and Development Programs in the Tropics," *Conservation Biology* 8, no. 2 (1994): 388–97.

Kunstler, J. H. *The Geography of Nowhere: The Rise and Decline of America's Man-Made Landscape.* New York, NY: Touchstone Books, 1993.

Kuppenheimer, L. B. *Albert Gallatin's Vision of Democratic Stability: An Interpretive Profile.* Westport, CT: Praeger Publishers, 1996.

Kushida, H. "Searching for Federal Aid: The Petitioning Activities of the Chesapeake and Delaware Canal Company," *Japanese Journal of American Studies* 14 (2003): 87–103. http://www.soc.nii.ac.jp/jaas/periodicals/JJAS/PDF/2003/No.14-087.pdf.

Labaree, B. *America and the Sea.* Mystic, CT: Mystic Seaport, 1999.

Lahusen, C. *The Rhetoric of Moral Protest: Public Campaigns, Celebrity Endorsement, and Political Mobilization.* New York, NY: Walter de Gruyter and Company, 1996.

Lamar, H., ed. *The New Encyclopedia of the American West.* New Haven, CT: Yale University Press, 1998.

Lavathes, L. *When China Ruled the Seas: The Treasure Fleet of the Dragon Throne 1405–1433.* Oxford, UK: Oxford University Press, 1994.

Leahy, P. Senate Speech 2004. http://leahy.senate.gov/press/200404/042604a.html

Lear, L. *Rachel Carson: Witness for History.* New York, NY: Owl Books, 1997.

Lederman, D., Maloney, W., and Serven, L. *Lessons from NAFTA for Latin America and the Caribbean.* Palo Alto, CA: Stanford University Press, 2005.

Lee, M. *Earth First!: Environmental Apocalypse.* Syracuse, NY: Syracuse University Press, 1995.

Leffler, W. L. *Deepwater Petroleum Exploration and Production.* New York, NY: PennWell, 2003.

Legler, D., and Korab, C. *Prairie Style: Houses and Gardens by Frank Lloyd Wright and the Prairie School.* New York, NY: Stewart, Tabori and Chang, 1999.

Legrain, P. *Immigrants: Your Country Needs Them.* Princeton, NJ: Princeton University Press, 2007.

Lemann, N. *The Promised Land: The Great Black Migration and How It Changed America.* New York, NY: Vintage Books, 1992.

Lemlich, C. *Life in the Shop,* http://www.ilr.cornell.edu/trianglefire/texts/stein_ootss/ootss_cl.html?location=Sweatshops+and+Strikes.

Leopold, A. *A Sand County Almanac, and Sketches Here and There.* [1948] New York, NY: Oxford University Press, 1987.

Leshy, J. D. *The Mining Law: A Study in Perpetual Motion.* Washington, DC: Resources For The Future, 1987.

Letts, C. W., Ryan, W. P., and Grossman, A. *High Performance Nonprofit Organizations: Managing Upstream for Greater Impact.* New York, NY: John Wiley and Sons, 1999.

Levesque, S. M. "The Yellowstone to Yukon Conservation Initiative: Reconstructing Boundaries, Biodiversity, and Beliefs," in *Reflections on Water: New Approaches to Transboundary Conflicts and Cooperation,* eds. J. Blatter and H. Ingram. Cambridge, MA: MIT Press, 2001, 123–62.

Lewis, D. R. *Neither Wolf nor Dog: American Indians, Environment, and Agrarian Change.* New York, NY: Oxford University Press, 1994.

Lewis, J. *Lead Poisoning: A Historical Perspective,* http://www.epa.gov/history/topics/perspect/lead.htm.

Lewis, M. J. "The First Design for Fairmount Park," *The Pennsylvania Magazine of History and Biography* 130.3 (2006): 33 pars. http://www.historycooperative.org/journals/pmh/130.3/lewis.html.

Lewis, T. *Divided Highways.* New York, NY: Penguin Books, 1997.

Library of Congress. *The Extermination of the American Bison,* http://memory.loc.gov/learn/features/timeline/riseind/west/bison.html.

Library of Congress. *Treaty for Russion for the Purchase of Alaska,* http://www.loc.gov/rr/program/bib/ourdocs/Alaska.html.

Liebs, C. H. *Main Street to Miracle Mile.* Baltimore, MD: Johns Hopkins University Press, 1995.

Liftin, K. T. *Ozone Discourses.* New York, NY: Columbia University Press, 1984.

Limerick, P. N. *Legacy of Conquest.* New York, NY: W. W. Norton, 1987.

Little, C. *Jefferson and Nature.* Baltimore: Johns Hopkins, 1988.

Livernash, R. "The Growing Influence of NGOs in the Developing World," *Environmental Conservation* 34, no. 5 (1992): 12–43.

Loeb, P. *Moving Mountains: How One Woman and Her Community Won Justice from Big Coal.* Lexington, KY: The University Press of Kentucky, 2007.

Los Angeles Department of Water and Power. *The Story of the Los Angeles Aqueduct,* http://www.ladwp.com/ladwp/cms/ladwp001006.jsp.

Lorey, D. E. *The U.S.–Mexican Border in the Twentieth Century.* Washington, DC: SR Books, 1999.

Lovins, A. *Soft Energy Paths.* New York, NY: Harper Collins, 1979.

Low, N., and Gleeson, B. *Justice, Society and Nature: An Exploration of Political Ecology.* New York, NY: Routledge, 1998.

Lowenthal, D. *George Perkins Marsh: Prophet of Conservation.* Seattle, WA: University of Washington Press, 2000.

Lower East Side Tenement Museum. *Health and Disease,* chap. 4. http://www.tenement.org/encyclopedia/diseases_cholera.htm.

Lowitt, R. *The New Deal and the West.* Norman, OK: University of Oklahoma Press, 1984.

MacArthur, J. R. *The Selling of "Free Trade": NAFTA, Washington, and the Subversion of American Democracy.* Berkeley, CA: University of California Press, 2001.

Macfarlane, A. M., and Ewing, R. C., eds. *Uncertainty Underground: Yucca Mountain and the Nation's High Level Nuclear Waste.* Cambridge, MA: MIT Press, 2006.

Machlis, G. E., Kaplan, A. B., Tuler, S. P., Bagby, K. A., and McKendry, J. E. *Burning Questions: A Social Science Research Plan for Federal Wildland Fire Management.* Report to the National Wildfire Coordinating Group. Contribution Number 943 of the Idaho Forest, Wildlife and Range Experiment Station. College of Natural Resources, University of Idaho, Moscow, 2002.

Mackintosh, B. *The National Parks: Shaping the System.* Washington, DC: National Park Service, 1991.

Magoc, C. *Yellowstone*. Santa Fe, NM: University of New Mexico Press, 1999.

Maher, N. *Nature's New Deal*. New York: Oxford University Press, 2007.

Maher, N. "Neil Maher on Shooting the Moon," *Environmental History* 9.3 (2004): 12 pars.

Makah: http://www.historylink.org/essays/output.fm?file_id=5301.

Malin, *The Grassland of North America*, http://www.kancoll.org/books/malin/mgchap02.htm.

Marcello, P. *Ralph Nader: A Biography*. Westport, CT: Greenwood.

Markell, D. L., and Knox, J. H., eds. *Greening NAFTA: The North American Commission for Environmental Cooperation*. Stanford Law and Politics. Stanford, CA: Stanford University Press, 2003.

Markowitz, G., Rosner, D. *Deceit and Denial: The Deadly Politics of Industrial Pollution*. Berkeley, CA: University of California Press, 2002.

Marks, R. B. *Origins of the Modern World*. Oxford, UK: Rowan and Littlefield, 2002.

Marsh, G. P. *Lectures on the English Language*. New York, NY: Charles Scribner's Sons, 1859, 1884, 1887.

Marsh, G. P. *The Camel—His Organization, Habits and Uses*. Boston, MA: Gould and Lincoln, 1856 (for material on importing the camel to America, see Chapters XVII, XVIII, and Appendix D).

Marsh, G. P. *Man and Nature*. Cambridge, MA: The Harvard University Press, 1965 (notated reprint of the original 1864 edition).

Marsh, G. P. *The Earth as Modified by Human Action: Man and Nature*. New York: Scribner, Armstrong, and Co., 1976 (straight reprint of the 1874 edition).

Marston, E. "The Old West Is Going Under," *High Country News*. April 27, 1998. http://www.hcn.org/servlets/hcn.Article?article_id=4105.

Martin, A. *Railroads Triumphant: The Growth, Rejection and Rebirth of a Vital American Force*. New York, NY: Oxford University Press, 1992.

Martin, R. "'New Urban Islands Dot the West," *Monday Business Roundup*. September 24, 2007. New West-Boulder. http://www.newwest.net/city/article/new_urban_islands_dot_the_west/C94/L94/.

Marx, L. *The Machine in the Garden*. New York, NY: Oxford University Press, 1964.

Massachusetts Turnpike Authority. *The Big Dig*, http://www.masspike.com/bigdig/background/index.html.

Massey, D. *Smoke and Mirrors: U.S. Immigration Policy in the Age of Globalization*. New York, NY: Russell Sage Foundation Publications, 2001.

Matthiessen, P. *Arctic National Wildlife Refuge: Seasons of Life and Land*. Seattle, WA: The Mountaineers Books, 2005.

May, E. R. *American Cold War Strategy*. Boston, MA: Bedford Books, 1993.

May, E. T. *Homeward Bound*. New York, NY: Basic Books, 1988.

Mayer, F. W. *Interpreting NAFTA: The Science and Art of Political Analysis*. New York, NY: Columbia University Press, 1998.

Mayr, E. *The Growth of Biological Thought*. Cambridge, MA: Harvard University Press, 1982.

McCullough, D. *Path Between the Seas*. New York, NY: Touchstone, 1977.

McCullough, D. *The Johnstown Flood*. New York, NY: Simon & Schuster, 1968.

McCullough, D. *Great Bridge*. New York, NY: Simon & Schuster, 1983.

McGreevy, P. V. *Imagining Niagara*. Amherst, MA: University of Massachusetts Press, 1994.

McGrory, K. C. *Who Controls Public Lands?: Mining, Forestry, and Grazing Policies, 1870–1990*. Chapel Hill, NC: University of North Carolina Press, 1996.

McGuire, T., Lord, W. B., and Wallace, M. G., eds. *Indian Water in the New West*. Tucson, AZ: University of Arizona Press, 1993.

McHarg, I. *Design with Nature*. New York, NY: John Wiley and Sons, 1992.

McHugh, T. *The Time of the Buffalo*. New York, NY: Knopf, 1972.

McKee, B. "As Suburbs Grow, So Do Waistlines," *The New York Times*. September 4, 2003.

McKinsey, E. *Niagara Falls: Icon of the American Sublime*. Cambridge, MA: Cambridge University Press, 1985.

McLaren, D. *Rethinking Tourism and Ecotravel: The Paving of Paradise and What You Can Do to Stop It*. West Hartford, CT: Kamarian Press, 1998.

McMahon, P. "'Cause Coffees' Produce a Cup with an Agenda," *USA Today*. July 25, 2001. http://www.usatoday.com/money/general/2001-07-26-coffee-usat.htm.

McNamee, T. *The Return of the Wolf to Yellowstone*. New York, NY: Henry Holt and Company, 1997.

McNeil, J. R. *Something New Under the Sun: An Environmental History of the Twentieth-Century World*. New York, NY: W. W. Norton, 2001.

McPhee, J. *Assembling California*. New York, NY: Farrar, Straus and Giroux, 1993.

McShane, C. *Down the Asphalt Path*. New York, NY: Columbia University Press, 1994.

Medicine Crow, J. *From the Heart of the Crow Country: The Crow Indians' Own Stories*. New York, NY: Orion Books, 1992.

Meikle, J. L. *American Plastic: A Cultural History*. New Brunswick, NJ: Rutgers University Press, 1997.

Meilander, P. C. *Towards a Theory of Immigration*. New York, NY: Palgrave Macmillian, 2001.

Melosi, M. *Coping with Abundance*. New York, NY: Knopf, 1985.

Melosi, M. *Sanitary City*. Baltimore, MD: Johns Hopkins University Press, 1999.

Melville, H. *Moby Dick* (originally published as *The Whale*). New York, NY: Harper and Brothers, 1851.

Merchant, C. *Green versus Gold*. New York, NY: Island Press, 1998.

Merchant, C. *Major Problems in American Environmental History*. New York, NY: Heath, 2005.

Merrill, K. *Public Lands and Political Meaning: Ranchers, the Government, and the Property between Them*. Los Angeles, CA: University of California Press, 2002.

Merrill, K. *The Oil Crisis of 1973–1974: A Brief History with Documents*. New York, NY: Bedford Books, 2007.

Metlar, G. W. *Northern California, Scott and Klamath Rivers, Their Inhabitants and Characteristics, Historical Features, Arrival of Scott and his Friends, Mining Interests*. Yreka, CA: Yreka Union Office, 1856.

Millennium Whole Earth Catalog. San Francisco, CA: Harper, 1998.

Miller, B. *Coal Energy Systems*. Burlington, MA: Elsevier Academic Press, 2005.

Miller, C. *Gifford Pinchot and the Making of Modern Environmentalism*. New York, NY: Shearwater Books, 2004.

Miller, C. E. *Jefferson and Nature*. Baltimore, MD: Johns Hopkins University Press, 1988.

Miller, J. *My Life amongst the Indians*. Chicago, IL: Morril, Higgins, and Co., 1892.

Miller, J. *Germs: Biological Weapons and America's Secret War*. New York, NY: Simon & Schuster, 2001.

Miller, J. B. *An Evolving Dialogue: Theological and Scientific Perspectives on Evolution*. London: Trinity Press International, 2001.

Miller, P., ed. *The Transcendentalists: An Anthology*. Cambridge: Harvard University Press, 1950.

Miller, J. *An Evolving Dialogue: Theological and Scientific Perspectives on Evolution*. London: Trinity Press International, 2001.

Milligan, S. "US Senate Passes Bill to Build Mexican Border Fence," *The Boston Globe*, Nation, September 30, 2006. http://www.boston.com/news/nation/articles/2006/09/30/us_senate_passes_bill_to_build_mexican_border_fence/.

Mitchell, J. G. "When Mountains Move," *National Geographic*. March 2006. http://www7.nationalgeographic.com/ngm/0603/feature5/index.html.

Miyanishi, K., and Johnson, E. A. "A Re-examination of the Effects of Fire Suppression in the Boreal Forest," *Conservation Biology* 16 (2001): 1177–78.

Montrie, C. *To Save the Land and People: A History of Opposition to Surface Coal Mining in Appalachia*. Chapel Hill, NC: University of North Carolina Press, 2003.

Moorhouse, J. C., ed. *Electric Power: Deregulation and the Public Interest*. San Francisco, CA: Pacific Research Institute for Public Policy, 1986.

Morris, E. *Theodore Rex*. New York, NY: Random House, 2001.

Morrison, E. *J. Horace McFarland*. Harrisburg, PA: Pennsylvania Historical and Museum Commission, 1995.

Moss, S., and deLeiris, L. *Natural History of the Antarctic Peninsula*. New York, NY: Columbia University Press, 1988.

Motavalli, J. *Forward Drive: The Race to Build "Clean" Cars for the Future*. San Francisco, CA: Sierra Club Books, 2001.

"Mr. Coal's Story." http://www.sip.ie/sip019B/Mr_%20Coal's%20Story_files/Mr_%20Coal's%20Story. htm, Ohio State University.

Muhn, J., and Hanson R. S., eds. *Opportunity and Challenge: The Story of BLM*. Washington, DC: U.S. Printing Office, 1988.

Muir, J. *Travels in Alaska*, http://www.sierraclub.org/john_muir_exhibit/writings/travels_in_alaska/.

Mulvaney, K. *The Whaling Season: An Inside Account of the Struggle to Stop Commercial Whaling*. Washington DC: Island Press, 2003.

Mumford, L. *Technics and Civilization*. New York, NY: Harcourt, 1963.

Nabhan, G. P. *Coming Home to Eat: The Pleasures and Politics of Local Foods*. New York, NY: W. W. Norton, 2002.

Naess, A. "The Shallow and the Deep, Long-Range Ecology Movement," *Inquiry* 16 (1973): 95–100.

Nash, R. *Wilderness and the American Mind*. New Haven, CT: Yale University Press, 1982.

Nash, Roderick, ed. *American Environmentalism*. New York, NY: Cambridge University Press, 2001.

National Conservation Training Center. *Origins of the U.S. Fish and Wildlife Service*, http://training. fws.gov/history/origins.html.

National Council for Science and the Environment. *Chippewa Treaty Rights: History and Management in Minnesota and Wisconsin*, http://ncseonline.org/nae/docs/chippewa.html and http://ncseonline. org/NAE/fishing.html.

National Humanities Center. http://www.nhc.rtp.nc.us/pds/amerbegin/contact/contact.htm.

National Interagency Fire Center. www.nifc.gov.

National Park Service. *Flood Witnesses*, http://www.nps.gov/archive/jofl/witness.htm.

National Park Service. *Fort Raleigh*, http://www.nps.gov/history/history/online_books/hh/16/hh16d2.htm.

National Park Service. *Mount Rainier*, http://www.nps.gov/archive/mora/adhi/adhit.htm.

National Park Service. "Reading 1: Flour Milling," in *Wheat Farms, Flour Mills, and Railroads: A Web of Interdependence*. http://www.cr.nps.gov/nr/twhp/wwwlps/lessons/106wheat/106facts1.htm.

National Park Service. "Reading 1: The Work at Hopewell Furnace," in *Hopewell Furnace: A Pennsylvania Iron-Making Plantation*. http://www.cr.nps.gov/nr/twhp/wwwlps/lessons/97hopewell/ 97facts1.htm.

National Park Service. "Reading 2: The Bonanza Farms of North Dakota," in *Wheat Farms, Flour Mills, and Railroads: A Web of Interdependence*. http://www.cr.nps.gov/nr/twhp/wwwlps/lessons/ 106wheat/106facts2.htm.

National Park Service. *Yellowstone National Park Wolf Restoration*. 2007. http://www.nps.gov/yell/ naturscience/wolfrest.htm.

National Research Council. *Hardrock Mining on Federal Lands, Committee on Hardrock Mining on Federal Lands*. Washington, DC: National Academy Press, 1999.

Natural Resources Defense Commission. *Historic Hudson River Cleanup to Begin After Years of Delay, But Will General Electric Finish the Job?*, http://www.nrdc.org/water/pollution/hhudson.asp

New Jersey Lighthouse Society. *An Act for the Establishment and Support of Lighthouse, Beacons, Buoys, and Public Piers*, http://www.njlhs.org/historicdocs/act.htm.

Newkirk, I. "The ALF: Who, Why, and What?" in *Terrorists or Freedom Fighters? Reflections on the Liberation of Animals*, S. Best and A. J. Nocella, eds. Seattle, WA: Lantern Press, 2004, 341.

Niven, J. *The Coming of the Civil War, 1837–1861*. Arlington Heights, IL: Harlan Davidson, 1990.

Norman, J., MacLean, H. L., and Kennedy, C. A. "Comparing High and Low Residential Density: Life Cycle Analysis of Energy Use and Greenhouse Gas Emissions," *Journal of Urban Planning and Development* 132 (2006): 10–21.

Norrell, B. "Yucca Mountain Lawsuit Filed," *Indian Country Today*, March 11, 2005. http://www. indiancountry.com/content.cfm?id=1096410530.

Norris, F. *The Octopus*. New York, NY: Penguin, 1986.

North American Development Bank. 2007. http://www.nadbank.org/.

Novak, B. *Nature and Culture*. New York, NY: Oxford University Press, 1980.

NPR. *Teaching Evolution: A State-by-State Debate*, http://www.npr.org/templates/story/story. php?storyId=4630737.

Nugent, W. "Western History, New and Not So New," Reprinted from the *OAH Magazine of History* 9 (1994).

Numbers, R. L. *The Creationists: The Evolution of Scientific Creationism.* Berkeley, CA: University of California Press, 1991.

Nye, D. *Technological Sublime.* Boston, MA: MIT Press, 1996.

Nye, D. *Electrifying America: Social Meanings of a New Technology.* Boston, MA: MIT Press, 1999.

Oblinger, U. W. *Letter from Uriah W. Oblinger to Mattie V. Oblinger and Ella Oblinger,* April 6, 1873. Courtesy of the Nebraska State Historical Society, Oblinger Family Collection.

Ohlemacher, S. "Number of Immigrants Hits Record 37.5 Million," Associated Press. September 12, 2007. http://news.yahoo.com/s/ap/20070912/ap_on_go_ot/census_demographics.

Oliens, R. M., and Davids, D. *Oil and Ideology: The American Oil Industry, 1859–1945.* Chapel Hill, NC: University of North Carolina Press, 1999.

Olmsted, F. L. "Draft of Preliminary Report upon the Yosemite and Big Tree Grove" and "Letter on the Great American Park of the Yosemite." Typed transcriptions. Frederick Law Olmsted Papers. Manuscript Division, Library of Congress, Washington, DC.

Olmsted, F. L. Letter, *New York Evening Post,* June 18, 1868.

Olmsted, F. L. "Yosemite and the Mariposa Grove: A Preliminary Report, 1865," *Landscape Architecture* 43, no. 1 (1952).

Olmsted, F. L. *The Papers of Frederick Law Olmsted, Volume Five: The California Years, 1863–1865,* ed. V. P. Ranney. Baltimore, MD: Johns Hopkins University Press, 1990.

Olmsted, F. L. *Yosemite and the Mariposa Grove: A Preliminary Report, 1865.* Yosemite, CA: Yosemite Association, 1995. http://www.yosemite.ca.us/history/olmsted/report.html.

Opie, J. *Nature's Nation.* New York, NY: Harcourt Brace, 1998.

Oster, S. M. *Strategic Management for Nonprofit Organizations.* New York, NY: Oxford University Press, 1995.

Oxfam. "Dumping without Borders: How U.S. Agricultural Policies are Destroying the Livelihoods of Mexican Corn Farmers." 2003. http://www.oxfam.org/en/files/pp030827_corn_dumping.pdf.

Painter, N. I. *Exodusters.* New York, NY: W. W. Norton, 1992.

Parrington, V. L. *The Romantic Revolution in America, 1800–1860.* Norman, OK: University of Oklahoma Press, 1987.

Patton, contained in Soderlund. *William Penn, and the Founding of Pennsylvania.*

PBS. *Timeline: Life and Death of the Electric Car,* http://www.historynet.com/exploration/great_migrations/3036611.html.

Peace Parks Foundation. "What Is the International Status of Peace Parks?" 2007. http://www.peace-parks.org/faq.php?mid=451&pid=302.

Peffer, W. A. *The Farmer's Side.* New York, 1891.

Pendergrast, M. *Uncommon Grounds: The History of Coffee and How It Transformed Our World.* New York, NY: Basic Books, 1999.

People for the Ethical Treatment of Animals. 2007. http://www.peta.org/.

Perkins, J. H. *Geopolitics and the Green Revolution: Wheat, Genes, and the Cold War.* New York, NY: Oxford University Press, 1997.

Peschard-Sverdrup, A. *U.S.–Mexico Transboundary Water Management: The Case of the Rio Grande/Rio Bravo.* CSIS Monograph. Washington, DC: Center for Strategic & International Studies, 2003.

Peterson, C. S. "A Portrait of Lot Smith—Mormon Frontiersman," *The Western Historical Quarterly* 1 (1970): 393–414.

Peterson, C. S. *Take Up Your Mission: Mormon Colonizing along the Little Colorado River 1870–1900.* Tucson, AZ: University of Arizona Press, 1973.

Petrikin, J. S. *Environmental Justice,* New York, NY: Greenhaven, 1995.

Petrini, C., and Padovani, G. *Slow Food Revolution: A New Culture for Eating and Living.* New York, NY: Rizzoli, 2006.

Petroski, H. *The Evolution of Useful Things.* New York, NY: Knopf, 1992, 100–101.

Pfanz, H. *Gettysburg.* Chapel Hill, NC: University of North Carolina Press, 1993.

Pinchot, G. *Breaking New Ground*. New York, NY: Island Press, 1998.

Pinchot, G. *The Use of the National Forests*. New York: Intaglio Press, 1907.

Pinchot, G. *A Primary of Forestry*, http://www.forestry.auburn.edu/sfnmc/class/pinchot.html.

Pitt, J. "Two Nations, One River: Managing Ecosystem Conservation in the Colorado River Delta," *Natural Resources Journal* 40 (2000): 855–57.

Pollan, *Omnivore's Dilemma*. New York: Penguin Books, 2007.

Pollan, M. *Second Nature*. New York, NY: Delta, 1992.

Poole, Jr., R. W., ed. *Unnatural Monopolies: The Case for Deregulating Public Utilities*. Lexington, MA: Lexington Books, 1985.

Posewitz, J. "Yellowstone to the Yukon (Y2Y): Enhancing Prospects for a Conservation Initiative," *International Journal of Wilderness* 4 (1998): 25–27.

Pratt, J. *Offshore Pioneers: Brown & Root and the History of Offshore Oil and Gas*. Houston, TX: Gulf Professional Publishing, 1997.

Preston, J. "U.S. Farmers Go Where Workers Are: Mexico," *International Herald Tribune*. September 4, 2007. http://www.iht.com/articles/2007/09/04/america/export.php.

Price, J. *Flight Maps*. New York, NY: Basic Books, 2000.

Price, M. "Ecopolitics and Environmental Nongovernmental Organizations in Latin America," *Geographic Review* 84, no. 1 (1994): 42–59.

Price, M. F. *Mountain Research in Europe: Overview of MAB Research from the Pyrenees to Siberia*. United Nations Educational, Scientific and Cultural Organization. New York, NY: The Parthenon Publishing Group, 1995.

Priest, T. *The Offshore Imperative: Shell Oil's Search for Petroleum in Postwar America*. Houston, TX: Texas A & M University Press, 2007.

Princen, T., and Finger, M. *Environmental NGOs in World Politics*. London, UK: Routledge, 1994.

Pyne, S. *Fire in America: A Cultural History of Wildland and Rural Fire*. Princeton, NJ: Princeton University Press, 1982.

Pyne, S. J. *Vestal Fire: An Environmental History, Told through Fire, of Europe and Europe's Encounter with the World*, 1997.

Pyne, S. J. *Tending Fire: Coping With America's Wildland Fires*. Washington, DC: Island Press, 2004.

Pyne. S. J. *History with Fire in Its Eye: An Introduction to Fire in America*, http://www.nhc.rtp.nc.us/tserve/nattrans/ntuseland/essays/fire.htm.

Quest, R. "On the Trail of the Celebrity Activist," CNN. September 1, 2005. http://edition.cnn.com/2005/WORLD/europe/08/11/quest/.

Rabe, B. *George Beyond Nimby: Hazardous Waste Siting in Canada and the United States*, Brookings Institute, November 1994.

Raber, P., ed. *The Archaic Period in Pennsylvania*. Harrisburg, PA: Pennsylvania Historical and Museum Commission, 1998.

Rainforest Alliance. "Leading Conservation Groups Release Guidelines for Growing Earth-Friendly Coffee," Press Release. May 30, 2001. www.ra.org/news/2001/coffee-principles.html.

Raustalia, K. "States, NGOs, and International Environmental Institutions," *International Studies Quarterly* 41 (1997): 719–40.

Rediker, M. *Between the Devil and the Deep Blue Sea*. New York, NY: Cambridge University Press, 1987.

Redstone Arsenal. *Women and War*, http://www.redstone.army.mil/history/women/welcome.html.

Reece, E. "Moving Mountains: The Battle for Justice Comes to the Coal Fields of Appalachia," *Orion Magazine*. January/February 2006. http://www.orionmagazine.org/index.php/articles/article/166/.

Reece, E. *Lost Mountain: A Year in the Vanishing Wilderness: Radical Strip Mining and the Devastation of Appalachia*. New York, NY: Penguin Group, 2006.

Reid, J. *Rio Grande*. Austin, TX: University of Texas Press, 2004.

Reiger, J. *American Sportsmen and the Origins of Conservation*. Norman, OK: University of Oklahoma Press, 1988.

Reisner, M. *Cadillac Desert*. New York, NY: Penguin, 1993.

Relph, E. *The Modern Urban Landscape*. Baltimore, MD: Johns Hopkins University Press, 1987.

Rettie, D. F. *Our National Park System: Caring for America's Greatest Natural and Historic Treasures.* Urbana, IL: University of Illinois Press, 1995.

Reuss, M. *Water Resources Administration in the United States: Policy, Practice, and Emerging Issues.* Ann Arbor, MI: Michigan State University Press, 1993.

Rhodes, R. *Audubon.* New York, NY: Knopf, 2004.

Rice, O. K. *The Allegheny Frontier.* Lexington, KY: University of Kentucky Press, 1970.

Ridenour, J. M. *The National Parks Compromised: Pork Barrel Politics and America's Treasures.* Merrillville, IN: ICS Books, 1994.

Riegel, R. E. *Young America, 1830–1840.* Norman, OK: University of Oklahoma Press, 1949.

Ries, L. A., and Stewart, J. S. *This Venerable Document,* http://www.phmc.state.pa.us/bah/dam/charter/charter.html.

Rifkin, J. *The Hydrogen Economy.* New York, NY: Penguin, 2003.

Riis, J. *How the Other Half Lives,* http://www.cis.yale.edu/amstud/inforev/riis/title.html.

Ringholz, R. C. *Uranium Frenzy, Boom and Bust on the Colorado Plateau,* Albuquerque, NM: University of New Mexico Press, 1989.

Risser, P. G., and Cornelison, K. D. *Man and the Biosphere.* Norman, OK: University of Oklahoma Press, 1979.

Rivera, J. A. *Acequia Culture: Water, Land, and Community in the Southwest.* Albuquerque, NM: University of New Mexico Press, 1998.

Robb, J., and Riebsame, W. E., eds. *Atlas of the New West: Portrait of a Changing Region.* New York, NY: W. W. Norton, 1997.

Robbins, J. "Resurgent Wolves Now Considered Pests by Some," *Wyoming Journal.* March 7, 2006.

Robbins, R. M. *Our Landed Heritage: The Public Domain, 1776–1970.* Lincoln, NE: University of Nebraska Press, 1976.

Robbins, W. G. *Colony and Empire: The Capitalist Transformation of the American West.* Lawrence, KS: University Press of Kansas, 1995.

Roberts, P. *Anthracite Coal Communities.* 1904. Reprint, Greenwood Publishers, 1970.

Rohrbough, M. J. *The Land Office Business: The Settlement and Administration of America's Public Lands, 1789–1837.* New York, NY: Oxford University Press, 1968.

Rohrbough, M. J. *Days of Gold: The California Gold Rush and the American Nation.* Berkeley, CA: University of California Press, 1997.

Rohter, L. "Canal Project Sets Off U.S.–Mexico Clash Over Water for Border Regions," *The New York Times,* page 2, October 2, 1989.

Rome, A. *The Bulldozer in the Countryside: Suburban Sprawl and the Rise of American Environmentalism.* New York: Cambridge University Press, 2001.

Roosevelt, T. *The Roosevelt Corollary to the Monroe Doctrine.* 1904. http://www.theodore-roosevelt.com/trmdcorollary.html.

Roper, L. W. *FLO: A Biography of Frederick Olmsted.* Baltimore, MD: John Hopkins University Press, 1973.

Rosenberg, C. *The Cholera Years: The United States in 1832, 1849, and 1866.* Chicago, IL: University of Chicago Press, 1962.

Rosensweig, R., and Blackmar, E. *The Park and the People: A History of Central Park.* Ithaca, NY: Cornell University Press, 1998.

Rosner, D. "The Living City: Engineering Social and Urban Change in New York City, 1865 to 1920," *Bulletin of the History of Medicine* 73 (1999): 124–29.

Rosner, D., ed. "Introduction: Hives of Sickness," in *Hives of Sickness: Public Health and Epidemics in New York City.* Piscataway, NJ: Rutgers University Press, 1995.

Rosove, M. H. *Let Heroes Speak: Antarctic Explorers, 1772–1922.* Annapolis, MD: Naval Institute Press, 2000.

Roth, L. M. *A Concise History of American Architecture.* New York, NY: Harper and Row, 1970.

Rothman, H. K. *Preserving Different Pasts: The American National Monuments.* Urbana, IL: University of Illinois Press, 1989.

Rothman, H. K. *The Greening of a Nation.* New York, NY: Harcourt, 1998.

Rothman, H. K. *Saving the Planet: The American Response to the Environment in the 20th Century.* Chicago, IL: Ivan R. Dee, 2000.

Rottenberg, D. *In the Kingdom of Coal: An American Family and the Rock That Changed the World.* New York, NY: Routledge, 2003.

Rowley, W. D. *U.S. Forest Service Grazing and Rangelands: A History.* College Station, TX: Texas A & M University Press, 1985.

Roy, A. *The Coal Mines.* New York, NY: Robison, Savage & Co., 1876.

Royster, J. "Water Quality and the Winters Doctrine." http://www.ucowr.siu.edu/updates/pdf/V107_A9.pdf.

Runte, A. *National Parks: The American Experience.* 3rd ed. Lincoln, NE: University of Nebraska Press, 1997.

Rupp, L. *Mobilizing Women for War.* Princeton: Princeton University Press, 1978.

Rush, E. *Annexing Mexico: Solving the Border Problem through Annexation and Assimilation.* Jamul, CA: Level 4 Press, 2007.

Russell, E. *War and Nature: Fighting Humans and Insects with Chemicals from World War I to Silent Spring.* New York, NY: Cambridge University Press, 2001.

Russell, E., and Tucker, R. P., eds. *Natural Enemy, Natural Ally: Toward an Environmental History of War.* Corvallis, OR: Oregon State University Press, 2005.

Rybczynski, W. "Suburban Despair," *Slate.* November 7, 2005. www.slate.com/id/2129636/?nav=tap3.

Sabin, P. *Crude Politics: The California Oil Market, 1900–1940.* Berkeley, CA: University of California Press, 2005.

Salm, J. "Coping with Globalization: A Profile of the Northern NGO Sector." *Nonprofit and Voluntary Sector Quarterly* 28, no. 4 (1999): 87–103.

Sanchea Munguia, V., ed. *El Revestimiento del Canal Todo Americano: Competencia o cooperacion por el agua en la frontera Mexico–Estados Unidos?* Tijuana, Baja California, Mexico: El Colegio de la Frontera Norte, 2004.

Savage, W. S. *Blacks in the West.* Westport, CT: Greenwood Press, 1976.

Scarce, R. *Eco-Warriors: Understanding the Radical Environmental Movement.* Chicago, IL: Nobel Press, 2006.

Scharff, V. *Taking the Wheel: Women and the Coming of the Motor Age.* New York: Free Press; Toronto: Collier Macmillan Canada, 1991.

Schiffer, M. B., Butts, T. C., and Grimm, K. K. *Taking Charge: The Electric Automobile in America.* Washington, DC: Smithsonian Institution Press, 1994.

Schlebecker, J. T. *Whereby We Thrive: A History of American Farming, 1607–1972.* Ames, IA: Iowa State University Press, 1975.

Schlosser, E. *Fast Food Nation: The Dark Side of the All-American Meal.* New York, NY: Houghton Mifflin Company, 2001.

Schuyler, D. *Apostle of Taste: Andrew Jackson Downing, 1815–1852.* Baltimore, MD: Johns Hopkins University Press, 1996.

Schweder, T. "Protecting Whales by Distorting Uncertainty: Non-precautionary Mismanagement?" *Fisheries Research* 52 (2001): 217–25.

Science Education Resource Center. *Impacts of Resource Development on Native American Lands,* http://serc.carleton.edu/research_education/nativelands/index.html.

Sears, S. *Sacred Places.* New York, NY: Oxford University Press, 1989.

Sellars, R. W. *Preserving Nature in the National Parks: A History.* New Haven, CT: Yale University Press, 1997.

Sellers, C. C. *Mr. Peale's Museum.* New York, NY: Regina Ryan Publishing Enterprises, 1979.

Semonin, P. *American Monster: How the Nation's First Prehistoric Creature Became a Symbol of National Identity.* New York, NY: New York University Press, 2000.

Shankland, R. *Steve Mather of the National Parks.* 3rd ed. New York, NY: Knopf, 1976.

Shaw, R. E. *Canals for a Nation: The Canal Era in the United States, 1790–1860.* Lexington, KY: University of Kentucky Press, 1990.

Shear, L. "Celebrity Activists," *WireTap Magazine.* May 2006. http://www.wiretapmag.org/stories/36850/.

Sheehan, B. *Seeds of Extinction; Jeffersonian Philanthropy and the American Indian.* New York, NY: Norton, 1973.

Sheppard, M. *Cloud by Day: The Story of Coal and Coke and People.* Pittsburgh, PA: University of Pittsburgh Press, 2001.

Sheriff, C. *The Artificial River: The Erie Canal and the Paradox of Progress, 1817–1862.* New York: Hill and Wang, 1997.

Shiva, V. *The Violence of the Green Revolution: Ecological Degradation and Political Conflict in Punjab.* New Delhi, India: Zed Press, 1992.

Sierra Nevada Earth First! "History of Earth First!" 2007. http://www.sierranevadaearthfirst.org/.

Singer, P. *Animal Liberation.* New York, NY: Harper Collins, 1975.

Singer, P. *In Defense of Animals: The Second Wave.* Malden, MA: Blackwell Publishing, 2006.

Sixeas, V. M. "Saving the Rio Grande," *Environment* 42 (2000): 7.

Sloane, D. C. *The Last Great Necessity.* Baltimore, MD: Johns Hopkins University Press, 1995.

Slotkin, R. *The Fatal Environment: The Myth of the Frontier in the Age of Industrialization, 1800–1890.* Norman: University of Oklahoma Press, 1998.

Slow Food International. 2007. http://www.slowfood.com/.

Slow Food U.S.A. 2007. http://www.slowfoodusa.org/.

Smil, V. *Energy in China's Modernization: Advances and Limitations.* Armonk, NY: M. E. Sharpe, 1988.

Smil, V. *Energy in World History.* Boulder, CO: Westview Press, 1994.

Smith, D. *Mining America: The Industry and the Environment, 1800–1980.* Lawrence, KS: Kansas University Press, 1987.

Smith, D. W., and Ferguson, G. *Decade of the Wolf: Returning the Wild to Yellowstone.* Guilford, CT: The Lyons Press, 2005.

Smith, D. W., Stahler, D. R., Guernsey, D. S., Metz, M., Nelson, A., Albers, E., and McIntyre, R. Smithsonian Migratory Bird Center, First Sustainable Coffee Congress overview paper. *Yellowstone Wolf Project, Annual Report 2006.* National Park Service, U.S. Department of the Interior. Yellowstone Center for Resources. Yellowstone National Park, Wyoming, 2007.

Smith, H. N. *Virgin Land: The American West as Symbol and Myth.* Cambridge, MA: Harvard University Press, 1978.

Smith, T. *Making the Modern.* Chicago, IL: University of Chicago Press, 1993. http://www.smokey-bear.com/vault/wartime_prevention.asp.

Smith, T. G. *Green Republican: John Saylor and Preservation of America's Wilderness.* Pittsburgh, PA: University of Pittsburgh Press, 2006.

Sobel, R. *Conquest and Conscience: The 1840's.* New York, NY: Crowell, 1971.

Society of Plactics Industry. *History of Plastics,* http://www.plasticsindustry.org/industry/history.htm.

Soderlund, J. R. ed. *William Penn and the Founding of Pennsylvania, 1680–1684: A Documentary History.* Philadelphia, PA: University of Pennsylvania Press, 1983.

Solnit, R. *Savage Dreams: A Journey into the Hidden Wars of the American West.* San Francisco, CA: Sierra Club Books, 1994.

Soule, M. E., and Terborgh, J. *Continental Conservation: Scientific Foundations of Regional Reserve Networks.* Washington, DC: Island Press, 1999.

Spence, C. "The Golden Age of Dredging," *Western Historical Quarterly* (1980): 403–14.

Spence, M. D. *Dispossessing the Wilderness: Indian Removal and the Making of the National Parks.* New York, NY: Oxford University Press, 2000.

Stanford University. *The Influenza Pandemic of 1918.* http://www.stanford.edu/group/virus/uda/.

State Library of North Carolina. *First English Settlement in the New World,* http://statelibrary.dcr.state.nc.us/nc/ncsites/english1.htm.

Steen, H. K. *The U.S. Forest Service: A History.* Seattle, WA: University of Washington Press, 1976.

Stegner, W. *The American West as Living Space.* Ann Arbor, MI: University of Michigan Press, 1987.

Stegner, W. *Mormon Country.* Lincoln, NE: Bison Books, 2003.

Steinberg, T. *Nature Incorporated: Industrialization and the Water of New England.* New York, NY: Cambridge University Press, 1991.

Steinberg, T. *Down to Earth.* New York: Oxford University Press, 2002.

Stephanson, A. *Manifest Destiny: American Expansionism and the Empire of Right.* New York: Hill and Wang, 1995.

Stern, W. E., and Long, D. W. "U.S. Supreme Court Upholds 1995 Department of the Interior Grazing Regulations." 2000. http://www.modrall.com/articles/article_68.html#.

Stevens, D. L. *A Homeland and a Hinterland.* 1991. http://www.cr.nps.gov/history/online_books/ozar/hrs4.htm.

Stevens, J. E. *Hoover Dam.* Norman, OK: University of Oklahoma Press, 1988.

Stevenson, E. *Park Maker: A Life of Frederick Law Olmsted.* New York, NY: Macmillan, 1999.

Stewart, M. *"What Nature Suffers to Groe": Life, Labor and Landscape on the Georgia Coast, 1680–1920.* Athens, GA: University of Georgia Press, 1996.

Stiles, J. *Brave New West: Morphing Moab at the Speed of Greed.* Tucson, AZ: University of Arizona Press, 2007.

Stilgoe, J. *Common Landscapes of America.* New Haven, CT: Yale University Press, 1982.

Stilgoe, J. *Metropolitan Corridor: Railroads and the American Scene.* New Haven, CT: Yale University Press, 1983.

Stilgoe, J. R. *Borderland.* New Haven, CT: Yale University Press, 1988.

Stilgoe, J. *Alongshore.* New Haven: Yale University Press, 1998.

Stine, J. K. *Mixing the Waters: Environment, Politics, and the Building of the Tennessee-Tombigbee Waterway.* Akron, OH: University of Akron Press, 1993.

Stokes, K. M. *Man and the Biosphere: Toward a Coevolutionary Political Economy.* Armonk, NY: M. E. Shape, 1992.

Stonehouse, B. *The Last Continent: Discovering Antarctica.* New York, NY: W. W. Norton, 2000.

Stradling, D. *Smokestacks and Progressives: Environmentalists, Engineers, and Air Quality in America, 1881–1951.* Baltimore, MD: Johns Hopkins University Press, 1999.

Stratton, D. *Tempest over Teapot Dome: The Story of Albert B. Fall.* Norman, OK: University of Oklahoma Press, 1998.

Strohmeyer, J. *Extreme Conditions: Big Oil and the Transformation of Alaska.* Anchorage, AK: University of Alaska Press, 1997.

Sutter, P. S. *Driven Wild: How the Fight Against Automobiles Launched the Modern Wilderness Movement.* Seattle, WA: University of Washington Press, 2002.

Swain, D. C. *Wilderness Defender: Horace M. Albright and Conservation.* Chicago, IL: University of Chicago Press, 1970.

Tansley. *Ecology and the American Environment,* http://memory.loc.gov/ammem/award97/icuhtml/aepsp6.html.

Tarbell, I. *All in the Day's Work: An Autobiography.* Champaign, IL: University of Illinois Press, 2003.

Tarr, J. *The Search for the Ultimate Sink.* Akron, OH: University of Akron Press, 1996.

Tarr, J., ed. *Devastation and Renewal.* Pittsburgh, PA: University of Pittsburgh Press, 2003.

Taylor Grazing Act 43 U.S.C. §§315-316o, June 28, 1934, as amended 1936, 1938, 1939, 1942, 1947, 1948, 1954, and 1976. http://www4.law.cornell.edu/uscode/html/uscode43/usc_sup_01_43_10_8A_20_I.html.

Taylor, B., Chait, R., and Holland, T. "The New Work of the Nonprofit Board," *Harvard Business Review* (1996): 36–46.

Taylor, J. M. *Bloody Valverde: A Civil War Battle on the Rio Grande, February 21, 1862.* Albuquerque, NM: 1999.

Teacher Oz's Kingdom of History. *Women and the Home Front during World War II,* http://www.teacheroz.com/WWIIHomefront.htm.

Tempest-Williams, T. *Refuge.* New York: Vintage, 1992. http://www.ratical.org/radiation/inetSeries/TTW_C1-BW.html.

Terrie, P. *Forever Wild: A Cultural History of Wilderness in the Adirondacks.* Syracuse, NY: Syracuse University Press, 1994.

Teysott, G., ed. *The American Lawn*. Princeton, NJ: Princeton Architectural Press, 1999.

Thoreau, H. D. *Walden*. 1854. http://eserver.org/thoreau/walden00.html.

Thoreau, H. D. *The Maine Woods*. 1864. http://eserver.org/thoreau/mewoods.html.

Tomes, N. *The Gospel of Germs: Men, Women, and the Microbe in American Life*. Cambridge, MA: Harvard University Press, 1998.

Trachtenberg, A. *Brooklyn Bridge*. Chicago, IL: University of Chicago Press, 1979.

Trachtenberg, A. *Incorporation of America*. New York, NY: Hill and Wang, 1982.

Transboundary Parks. http://www.eoeartth.org/article/Transboundary_protected_areas.

Travis, W. R. *New Geographies of the American West: Land Use and the Changing Patterns of Place*. Washington, DC: Island Press, 2007.

Tremble, M. *The Little Colorado River*, in *Riparian Management: Common Threads and Shared Interests*, eds. B. Tellman, H. J. Cortner, M. G. Wallace, L. F. DeBano, R. H. Hamre. USDA Forest Service, Rocky Mountain Forest and Range Experiment Station, Fort Collins, CO, 1993, 283–89.

Trimble, S. *The People: Indians of the American Southwest*. Santa Fe, NM: School of American Research Press, 1993.

Turner, F. J. *The United States, 1830–1850*. New York, NY: Henry Holt, 1934. http://xroads.virginia.edu/~HYPER/TURNER/.

Twain, M. *Roughing It*. London, UK: George Routledge, 1871.

Tyler, A. F. *Freedom's Ferment: Phases of American Social History to 1860*. New York, NY: Harper and Row, 1962.

U.N. Educational Scientific and Cultural Organization. "The World Network of Biosphere Reserves," *UNESCO Courier*. May 1997.

U.N. Educational Scientific and Cultural Organization. *UNESCO's Man and the Biosphere Programme (MAB)*. 2007a. http://www.unesco.org/mab/mabProg.shtml.

U.N. Educational Scientific and Cultural Organization. *UNESCO Man and Biosphere Programme FAQs*. 2007b. http://www.gov.mb.ca/conservation/wno/status-report/fa-8.19.pdf.

UNEP, Ozone Secretariat. *The 1987 Montreal Protocol on Substances that Deplete the Ozone Layer (as agreed in 1987)*. 2004. http://ozone.unep.org/Ratification_status/montreal_protocol.shtml.

UNEP, Ozone Secretariat. *Evolution of the Montreal Protocol*. 2007. http://ozone.unep.org/Ratification_status/index.shtml.

United Farm Workers. *Address by Cesar Chavez*, http://www.ufw.org/_page.php?menu=research&inc=history/10.html.

U.S. Antarctic Program External Panel of the National Science Foundation. *Antarctica—Past and Present*, www.nsf.gov/pubs/1997/antpanel/antpan05.pdf.

U.S. Army Corps of Engineers. *Hydrogeomorphic Approach to Assessing Wetland Functions: Guidelines for Developing Regional Guidebooks*, http://el.erdc.usace.army.mil/wetlands/pdfs/trel02–3.pdf.

U.S. Army Corps of Engineers. *Navigational Improvements before the Civil War*, http://www.usace.army.mil/inet/usace-docs/eng-pamphlets/ep870-1-13/c-1.pdf.

U.S. Central Intelligence Agency. "Antarctica," *The World Factbook*. https://www.cia.gov/library/publications/the-world-factbook/geos/ay.html.

U.S. Department of Energy. "DOE Awards $3 Million Contract to Oak Ridge Associated Universities for Expert Review of Yucca Mountain Work." Press Release. March 31, 2006. www.energy.gov/news/3418.htm.

U.S. Department of Health and Human Services. *Health*. http://www.nlm.nih.gov/exhibition/phs_history/intro.html.

U.S. Department of State. *Purchase of Alaska, 1867*, http://www.state.gov/r/pa/ho/time/gp/17662.htm.

U.S. Department of the Interior. *Ecological Issues on Reintroducing Wolves into Yellowstone National Park*. National Park Service, SuDoc I 29.80:22, 1993.

U.S. Environmental Protection Agency. *Mountaintop Mining and Valley Fills in Appalachia: Final Programmatic Environmental Impact Statement*. 2006. www.epa.gov/region3/mtntop/index.htm.

U.S. Environmental Protection Agency. *Milestones in Garbage*. http://www.epa.gov/msw/timeline_alt.htm.

U.S. Environmental Protection Agency. *The Birth of Superfund*. http://www.epa.gov/superfund/20years/ch2pg2.htm.

U.S. Environmental Protection Agency. *Record of Decision*. http://www.epa.gov/hudson/d_rod.htm#record.

U.S. Fish and Wildlife Service. *A Guide to the Laws and Treaties of the United States for Protecting Migratory Birds*. http://www.fws.gov/migratorybirds/intrnltr/treatlaw.html.

U.S.–Canadian Wildlife Protection Treaties in the Progressive Era. Seattle: Univ. of Washington Press, 1998.

U.S. Geological Survey. *Wetlands of the United States*. http://www.npwrc.usgs.gov/resource/wetlands/uswetlan/century.htm.

U.S. Global Change Research Program. *U.S. National Assessment of the Potential Consequences of Climate Variability and Change*, http://www.usgcrp.gov/usgcrp/nacc/background/meetings/forum/greatplains_summary.html.

U.S. Government Accountability Office. *Illegal Immigration: Border-Crossing Deaths Have Doubled Since 1995*. GAO-06-770. Washington, DC: U.S. Government Accountability Office, 2006.

University at Buffalo Libraries. *Love Canal Collection*. http://ublib.buffalo.edu/libraries/projects/lovecanal/.

University of Virginia. *The Diseased City*, Benjamin Henry, Notes, http://xroads.virginia.edu/~MA96/forrest/WW/fever.html.

Unrau, H. *Gettysburg Administrative History*. Washington, DC: National Park Service, 1991.

Utley, R. *Cavalier in Buckskin: George Armstrong Custer and the Western Military Frontier*. Norman, OK: University of Oklahoma Press, 1988.

Valavenes, P. *Hysplex*. Berkeley, CA: University of California Press, 1999.

Vandenbosch, R., and Vandenbosch, S. E. *Nuclear Waste Stalemate: Political and Scientific Controversies*. Salt Lake City, UT: University of Utah Press, 2007.

Vergara, R. "NGOs: Help or Hindrance for Community Development in Latin America?" *Community Development Journal* 29, no. 4 (1994): 322–28.

Vieyra, D. I. *Fill 'Er Up: An Architectural History of America's Gas Stations*. New York, NY: Macmillan, 1979.

Virginia Places. *Canals of Virginia*, http://www.virginiaplaces.org/transportation/canals.html.

Vivanco, L. "Ecotourism, Paradise Lost—A Thai Case Study." *The Ecologist* 32 (2002): 28–30.

Voice of America. "Celebrity Activism: Publicity Stunt or Sincere Care?" Washington, February 21, 2007. http://www.voanews.com/english/archive/2007-02/2007-02-21-voa38.cfm?CFID=137892992&CFTOKEN=74360178.

Voight Jr. W. *Public Grazing Lands: Use and Misuse by Industry and Government*. New Brunswick, NJ: Rutgers University Press, 1976.

Wall, D. *Earth First! And the Anti-Roads Movement: Radical Environmentalism and Comparative Social Movements*. London: Routledge, 2001.

Wallace, *Report of the Great Plains Drought Area*, http://newdeal.feri.org/hopkins/hop27.htm.

Walters, R. *Albert Gallatin: Jeffersonian Financier and Diplomat*. New York, NY: Macmillan, 1957.

Ward, J. A. *Railroads and the Character of America*. Knoxville, TN: University of Tennessee Press, 1986.

Washington Post. "As Border Crackdown Intensifies, A Tribe Is Caught in the Crossfire." September 15, 2006.

Waste of the West. http://www.wasteofthewest.com/Chapter1.html.

Weaver, D. *Ecotourism*. Milton, Australia: John Wiley and Sons Australia, 2002.

Weber, S. J. "In Mexico, U.S. and Canada, Public Support for NAFTA Surprisingly Strong, Given Each Country Sees Grass as Greener on the Other Side." *World Public Opinion*. January 23, 2006. http://www.worldpublicopinion.org/pipa/articles/brlatinamericara/161.php?nid=&id=&pnt=161&lb=brla.

Weeks, J. *Gettysburg: Memory, Market, and an American Shrine*. Princeton: Princeton University Press, 2003.

Weiner, D. R. *Models of Nature: Ecology, Conservation, and Cultural Revolution in Soviet Russia*. Bloomington, IN: Indiana University Press, 1988.

Weiss, T. G., and Gordenker, L. *NGOs, the UN, and Global Governance*. Boulder, CO: Lynne Rienner Publishers, 1996.

Wellman, P. I. *The House Divides: The Age of Jackson and Lincoln*. Garden City, NY: Doubleday, 1966.

Wendell Cox Consultancy. "Demographia," *The Public Purpose*. 2002. www.demographia.com/dbx-intlair.htm.

Wennberg, R. N. *God, Humans, and Animals: An Invitation to Enlarge Our Moral Universe*. Grand Rapids, MI: Eerdmans Publishing Co., 2003.

Wertime, R. *Citadel on the Mountain*. New York, NY: Farrar, Straus, and Giroux, 2000.

West, E. *The Contested Plains: Indians, Goldseekers, and the Rush to Colorado*. Lawrence, KS: University of Kansas Press, 2000.

West, E. *A New Look at the Great Plains*, http://www.historynow.org/09_2006/historian2.html.

Wheeler, H. W. *Buffalo Days: Forty Years in the Old West: The Personal Narrative of a Cattleman Indian Fighter, and Army Officer Colonel Homer W. Wheeler*. New York, NY: The Bobbs-Merrill Company, 1925.

Wheelwright, J. *Degrees of Disaster: Prince William Sound, How Nature Reels and Rebounds*. New Haven, CT: Yale University Press, 1996.

White, R. *It's Your Misfortune and None of My Own*. Norman, OK: University of Oklahoma Press, 1991.

White, R. *Organic Machine*. New York, NY: Hill and Wang, 1996.

White House. "President Signs Yucca Mountain Bill." Press Release. July 23, 2002. http://www.whitehouse.gov/news/releases/2002/07/20020723-2.html.

Whitman, W. "Passage to India," in *Leaves of Grass*. 1855. http://www.bartleby.com/142/183.html.

Wild, A. *Coffee: A Dark History*. New York, NY: W. W. Norton, 2004.

Wilderness Society. *Yellowstone to Yukon*. 2007. http://www.wilderness.org/WhereWeWork/Montana/y2y.cfm?TopLevel=Y2Y.

Wiley, P., and Gottlieb, R. *Empires in the Sun: The Rise of the New American West*. Tucson, AZ: University of Arizona Press, 1985.

Williams, M. *Americans and Their Forests*. New York, NY: Cambridge University Press, 1992.

Wilson, A. *The Culture of Nature*. Cambridge, MA: Blackwell, 1992.

Wirth, C. L. *Parks, Politics, and the People*. Norman, OK: University of Oklahoma Press, 1980.

Wise Uranium Project. *Uranium Mining and Indigenous People*, http://www.wise-uranium.org/uip.html.

Wishart, D. *The Fur Trade of the American West*. Lincoln, NE: University of Nebraska Press, 1979.

Wong, P. "McCormick's Revolutionary Reaper," *Illinois History*. December 1992. http://www.lib.niu.edu/ipo/1992/ihy921205.html.

Wooster, R. *The Military and United States Indian Policy 1865–1903*. New Haven CT: Yale University Press, 1988.

Workman, D. P. *PETA Files: The Dark Side of the Animal Rights Movement*. Bellevue, WA: Merril Press, 2003.

Worster, D. *Dust Bowl: The Southern Plains in the 1930s*. New York, NY: Oxford University Press, 1979.

Worster, D. *Nature's Economy*. New York, NY: Cambridge University Press, 1994.

Worster, D. *A River Running West: Life of John Wesley Powell*, New York, NY: Oxford University Press, 2000.

Wright, A. *The Death of Ramon Gonzalez: The Modern Agricultural Dilemma*. Austin, TX: University of Texas Press, 1992.

Wright, G. *Building the Dream*. Cambridge, MA: MIT Press, 1992.

Wuerthner, G. *Wildfire: A Century of Failed Forest Policy. Foundation for Deep Ecology*. Washington, DC: Island Press, 2006.

Yellowstone National Park. Forty-Second Congress. Session II Ch. 21–24. 1872. March 1, 1872. Chap. XXIV.

Yergin, D. *The Prize: The Epic Quest for Oil, Money & Power*. New York, NY: Free Press, 1993.

Young, J. H. *The Medical Messiahs: A Social History of Health Quackery in Twentieth-Century America.* Princeton, NJ: Princeton University Press, 1967.

Young, J. H. *Pure Food: Securing the Federal Food and Drugs Act of 1906.* Princeton, N.J.: Princeton University Press, 1989.

Zbicz, D. C. *Transboundary Cooperation in Conservation: A Global Survey of Factors Influencing Cooperation between Internationally Adjoining Protected Areas.* PhD thesis, Duke University, 1999.

Zimmerman, M. E., Callicott, J. B., Sessions, G., Warren, K. J., and Clark, J., eds. *Environmental Philosophy: From Animal Rights to Radical Ecology.* Englewood Cliffs, NJ: Prentice-Hall, 1993.

Zuniga, V., and Hernandez-Leon, R., eds. *New Destinations: Mexican Immigration in the United States.* New York, NY: Russell Sage Foundation Publications, 2006.

OTHER RESOURCES

Dams, GeoGuide Online, National Geographic, http://www.nationalgeographic.com/resources/ngo/education/geoguide/dams/.http://environment.about.com/library/weekly/blrenew4.htm.

INDEX

About the Authors

BRIAN BLACK is Associate Professor in the departments of history and environmental studies at Penn State University, Altoona. He is the author of *Nature and the Environment in Nineteenth-Century American Life* and *Nature and the Environment in Twentieth-Century American Life*, each with Greenwood. In addition, Black is the author of *Petrolia: The Landscape of America's First Oil Boom* (2003) and the forthcoming *Contesting Gettysburg: Preserving a Cherished American Landscape*.

DONNA L. LYBECKER is an Assistant Professor of Political Science at Idaho State University. She received her M.A. from Tulane University and her Ph.D. from Colorado State University, where she focused on environmental politics and Latin America. Her research interests include decentralization of environmental policy, water policy in the West and along the U.S.–Mexico border, and cross-border environmental politics.